Implicit Curves and Surfaces:
Mathematics, Data Structures and Algorithms

Abel J.P. Gomes · Irina Voiculescu
Joaquim Jorge · Brian Wyvill · Callum Galbraith

Implicit Curves and Surfaces: Mathematics, Data Structures and Algorithms

Abel J.P. Gomes
Universidade da Beira Interior
Covilha
Portugal

Brian Wyvill
University of Victoria
Victoria BC
Canada

Irina Voiculescu
Oxford University Computing Laboratory (OUCL)
Oxford
United Kingdom

Callum Galbraith
University of Calgary
Calgary
Canada

Joaquim Jorge
Universidade Tecnica de Lisboa
Lisboa
Portugal

ISBN 978-1-4471-5877-6   ISBN 978-1-84882-406-5 (eBook)
DOI 10.1007/978-1-84882-406-5
Springer Dordrecht Heidelberg London New York

British Library Cataloguing in Publication Data
A catalogue record for this book is available from the British Library

© Springer-Verlag London Limited 2009
Softcover re-print of the Hardcover 1st edition 2009
Apart from any fair dealing for the purposes of research or private study, or criticism or review, as permitted under the Copyright, Designs and Patents Act 1988, this publication may only be reproduced, stored or transmitted, in any form or by any means, with the prior permission in writing of the publishers, or in the case of reprographic reproduction in accordance with the terms of licenses issued by the Copyright Licensing Agency. Enquiries concerning reproduction outside those terms should be sent to the publishers.
The use of registered names, trademarks, etc., in this publication does not imply, even in the absence of a specific statement, that such names are exempt from the relevant laws and regulations and therefore free for general use.
The publisher makes no representation, express or implied, with regard to the accuracy of the information contained in this book and cannot accept any legal responsibility or liability for any errors or omissions that may be made.

Cover design: KuenkelLopka GmbH

Printed on acid-free paper

Springer is part of Springer Science+Business Media (www.springer.com)

Preface

This book presents the mathematics, computational methods and data structures, as well as the algorithms needed to render implicit curves and surfaces. Implicit objects have gained an increasing importance in geometric modelling, visualisation, animation, and computer graphics due to their nice geometric properties which give them some advantages over traditional modelling methods. For example, the point membership classification is trivial using implicit representations of geometric objects—a very useful property for detecting collisions in virtual environments and computer game scenarios. The ease with which implicit techniques can be used to describe smooth, intricate, and articulatable shapes through blending and constructive solid geometry show us how powerful they are and why they are finding use in a growing number of graphics applications.

The book is mainly directed towards graduate students, researchers and developers in computer graphics, geometric modelling, virtual reality and computer games. Nevertheless, it can be useful as a core textbook for a graduate-level course on implicit geometric modelling or even for general computer graphics courses with a focus on modelling, visualisation and animation. Finally, and because of the scarce number of textbooks focusing on implicit geometric modelling, this book may also work as an important reference for those interested in modelling and rendering complex geometric objects.

Abel Gomes
Irina Voiculescu
Joaquim Jorge
Brian Wyvill
Callum Galbraith

March 2009

Acknowledgments

The authors are grateful to those who have kindly assisted with the editing of this book, in particular Helen Desmond and Beverley Ford (Springer-Verlag).

We are also indebted to Adriano Lopes (New University of Lisbon, Portugal), Afonso Paiva (University of São Paulo, Brazil), Bruno Araújo (Technical University of Lisbon, Portugal), Ron Balsys (Central Queensland University, Australia) and Kevin Suffern (University of Technology, Australia) who generously have contributed beautiful images generated by their algorithms; also to Tamy Boubekeur (Telecom ParisTech, France) for letting us to use the datasets of African woman and Moai statues (Figures 8.7 and 8.10).

Abel Gomes thanks the Computing Laboratory, University of Oxford, England, and CNR-IMATI, Genova, Italy, where he spent his sabbatical year writing part of this book. In particular, he would like to thank Bianca Falcidieno and Giuseppe Patanè for their support and fruitful discussions during his stage at IMATI. He is also grateful to Foundation for Science and Technology, Institute for Telecommunications and University of Beira Interior, Portugal.

Irina Voiculescu acknowledges the support of colleagues at the Universities of Oxford and Bath, UK, who originally enticed her to study this field and provided a stimulating discussion environment; also to Worcester College Oxford, which made an ideal thinking retreat.

Joaquim Jorge is grateful to the Foundation for Science and Technology, Portugal, and its generous support through project VIZIR.

Brian Wyvill is grateful to all past and present students who have contributed to the Implicit Modelling and BlobTree projects; also to the Natural Sciences and Engineering Research Council of Canada.

Callum Galbraith acknowledges the many researchers from the Graphics Jungle at the University of Calgary who helped shape his research. In particular, he would like to thank his PhD supervisor, Brian Wyvill, for his excellent experience in graduate school, and Przemyslaw Prusinkiewicz for his expert guidance in the domain of modelling plants and shells; also to the University of Calgary and the Natural Sciences and Engineering Research Council of Canada for their support.

Contents

Preface V

Acknowledgments VII

Part I Mathematics and Data Structures

1 Mathematical Fundamentals 7
  1.1 Introduction 7
  1.2 Functions and Mappings 8
  1.3 Differential of a Smooth Mapping........................ 9
  1.4 Invertibility and Smoothness 10
  1.5 Level Set, Image, and Graph of a Mapping 13
     1.5.1 Mapping as a Parametrisation of Its Image 13
     1.5.2 Level Set of a Mapping............................. 15
     1.5.3 Graph of a Mapping 20
  1.6 Rank-based Smoothness 24
     1.6.1 Rank-based Smoothness for Parametrisations 25
     1.6.2 Rank-based Smoothness for Implicitations 27
  1.7 Submanifolds 30
     1.7.1 Parametric Submanifolds 30
     1.7.2 Implicit Submanifolds and Varieties................. 35
  1.8 Final Remarks 40

2 Spatial Data Structures................................... 41
  2.1 Preliminary Notions 41
  2.2 Object Partitionings 43
     2.2.1 Stratifications..................................... 43
     2.2.2 Cell Decompositions 45
     2.2.3 Simplicial Decompositions 49
  2.3 Space Partitionings 51

		2.3.1	BSP Trees	52
		2.3.2	K-d Trees	55
		2.3.3	Quadtrees	58
		2.3.4	Octrees	60
	2.4	Final Remarks		62

Part II Sampling Methods

3	**Root Isolation Methods**			67
	3.1	Polynomial Forms		67
		3.1.1	The Power Form	68
		3.1.2	The Factored Form	68
		3.1.3	The Bernstein Form	69
	3.2	Root Isolation: Power Form Polynomials		72
		3.2.1	Descartes' Rule of Signs	73
		3.2.2	Sturm Sequences	74
	3.3	Root Isolation: Bernstein Form Polynomials		78
	3.4	Multivariate Root Isolation: Power Form Polynomials		81
		3.4.1	Multivariate Descartes' Rule of Signs	81
		3.4.2	Multivariate Sturm Sequences	82
	3.5	Multivariate Root Isolation: Bernstein Form Polynomials		82
		3.5.1	Multivariate Bernstein Basis Conversions	83
		3.5.2	Bivariate Case	83
		3.5.3	Trivariate Case	84
		3.5.4	Arbitrary Number of Dimensions	86
	3.6	Final Remarks		87

4	**Interval Arithmetic**			89
	4.1	Introduction		89
	4.2	Interval Arithmetic Operations		91
		4.2.1	The Interval Number	91
		4.2.2	The Interval Operations	91
	4.3	Interval Arithmetic-driven Space Partitionings		93
		4.3.1	The Correct Classification of Negative and Positive Boxes	94
		4.3.2	The Inaccurate Classification of Zero Boxes	96
	4.4	The Influence of the Polynomial Form on IA		98
		4.4.1	Power and Bernstein Form Polynomials	99
		4.4.2	Canonical Forms of Degrees One and Two Polynomials.	101
		4.4.3	Nonpolynomial Implicits	104
	4.5	Affine Arithmetic Operations		105
		4.5.1	The Affine Form Number	105
		4.5.2	Conversions between Affine Forms and Intervals	106
		4.5.3	The Affine Operations	107

		4.5.4 Affine Arithmetic Evaluation Algorithms 108
	4.6	Affine Arithmetic-driven Space Partitionings 109
	4.7	Floating Point Errors 111
	4.8	Final Remarks 114

5 Root-Finding Methods 117
    5.1 Errors of Numerical Approximations 118
        5.1.1 Truncation Errors 118
        5.1.2 Round-off Errors 119
    5.2 Iteration Formulas 119
    5.3 Newton-Raphson Method................................... 120
        5.3.1 The Univariate Case 121
        5.3.2 The Vector-valued Multivariate Case.................. 123
        5.3.3 The Multivariate Case 124
    5.4 Newton-like Methods 126
    5.5 The Secant Method 127
        5.5.1 Convergence 128
    5.6 Interpolation Numerical Methods 131
        5.6.1 Bisection Method 131
        5.6.2 False Position Method 133
        5.6.3 The Modified False Position Method 136
    5.7 Interval Numerical Methods 136
        5.7.1 Interval Newton Method 136
        5.7.2 The Multivariate Case 139
    5.8 Final Remarks 139

Part III Reconstruction and Polygonisation

6 Continuation Methods 145
    6.1 Introduction 145
    6.2 Piecewise Linear Continuation 146
        6.2.1 Preliminary Concepts 146
        6.2.2 Types of Triangulations 147
        6.2.3 Construction of Triangulations 148
    6.3 Integer-Labelling PL Algorithms 151
    6.4 Vector Labelling-based PL Algorithms 156
    6.5 PC Continuation 164
    6.6 PC Algorithm for Manifold Curves 164
    6.7 PC Algorithm for Nonmanifold Curves 167
        6.7.1 Angular False Position Method....................... 168
        6.7.2 Computing the Next Point 168
        6.7.3 Computing Singularities............................. 169
        6.7.4 Avoiding the Drifting/Cycling Phenomenon........... 171
    6.8 PC Algorithms for Manifold Surfaces 173

Contents

- 6.8.1 Rheinboldt's Algorithm ... 173
- 6.8.2 Henderson's Algorithm ... 174
- 6.8.3 Hartmann's Algorithm ... 175
- 6.8.4 Adaptive Hartmann's Algorithm ... 179
- 6.8.5 Marching Triangles Algorithm ... 180
- 6.8.6 Adaptive Marching Triangles Algorithms ... 182
- 6.9 Predictor–Corrector Algorithms for Nonmanifold Surfaces ... 183
- 6.10 Final Remarks ... 186

7 Spatial Partitioning Methods ... 187
- 7.1 Introduction ... 187
- 7.2 Spatial Exhaustive Enumeration ... 188
  - 7.2.1 Marching Squares Algorithm ... 189
  - 7.2.2 Marching Cubes Algorithm ... 194
  - 7.2.3 Dividing Cubes ... 200
  - 7.2.4 Marching Tetrahedra ... 201
- 7.3 Spatial Continuation ... 207
- 7.4 Spatial Subdivision ... 208
  - 7.4.1 Quadtree Subdivision ... 208
  - 7.4.2 Octree Subdivision ... 211
  - 7.4.3 Tetrahedral Subdivision ... 213
- 7.5 Nonmanifold Curves and Surfaces ... 219
  - 7.5.1 Ambiguities and Singularities ... 220
  - 7.5.2 Space Continuation ... 221
  - 7.5.3 Octree Subdivision ... 221
- 7.6 Final Remarks ... 224

8 Implicit Surface Fitting ... 227
- 8.1 Introduction ... 227
  - 8.1.1 Simplicial Surfaces ... 227
  - 8.1.2 Parametric Surfaces ... 228
  - 8.1.3 Implicit Surfaces ... 230
- 8.2 Blob Surfaces ... 232
- 8.3 LS Implicit Surfaces ... 234
  - 8.3.1 LS Approximation ... 234
  - 8.3.2 WLS Approximation ... 238
  - 8.3.3 MLS Approximation and Interpolation ... 239
- 8.4 RBF Implicit Surfaces ... 249
  - 8.4.1 RBF Interpolation ... 249
  - 8.4.2 Fast RBF Interpolation ... 252
  - 8.4.3 CS-RBF Interpolation ... 252
  - 8.4.4 The CS-RBF Interpolation Algorithm ... 253
- 8.5 MPU Implicit Surfaces ... 255
  - 8.5.1 MPU Approximation ... 258
  - 8.5.2 MPU Interpolation ... 261

8.6 Final Remarks 261

Part IV Designing Complex Implicit Surface Models

9 Skeletal Implicit Modelling Techniques 267
9.1 Distance Fields and Skeletal Primitives 267
9.2 The *BlobTree* 270
9.3 Functional Composition Using f_Z Functions 271
9.4 Combining Implicit Surfaces 272
9.5 Blending Operations 274
  9.5.1 Hierarchical Blending Graphs 275
  9.5.2 Constructive Solid Geometry 277
  9.5.3 Precise Contact Modelling 279
  9.5.4 Generalised Bounded Blending 281
9.6 Deformations 284
9.7 BlobTree Traversal 284
9.8 Final Remarks 285

10 Natural Phenomenae-I: Static Modelling 287
10.1 *Murex Cabritii* Shell 288
10.2 Shell Geometry 288
10.3 *Murex Cabritii* 289
10.4 Modelling *Murex Cabritii* 290
  10.4.1 Main Body Whorl 291
  10.4.2 Constructing Varices 294
  10.4.3 Constructing Bumps 295
  10.4.4 Constructing Axial Rows of Spines 297
  10.4.5 Construction of the Aperture 298
10.5 Texturing the Shell 300
10.6 Final Model of *Murex Cabritii* 301
10.7 Shell Results 301
10.8 Final Remarks 301

11 Natural Phenomenae-II: Animation 303
11.1 Animation: Growing *Populus Deltoides* 303
11.2 Visualisation of Tree Features 305
  11.2.1 Modelling Branches with the *BlobTree* 306
  11.2.2 Modelling the Branch Bark Ridge and Bud-scale Scars . 308
11.3 Global-to-Local Modelling of a Growing Tree 309
  11.3.1 Crown Shape 310
  11.3.2 Shoot Structure 312
  11.3.3 Other Functions 313
11.4 Results 315
11.5 Final Remarks 316

References 319

Index 345

Part I

Mathematics and Data Structures

Overview

Part I introduces the mathematics of curves and surfaces, as well as the data structures suited to storing their sampled or discrete approximations.

We all have an intuitive idea of what curves and surfaces are. Curves can be traced by sliding a pencil along a protractor to draw an angle, or by making a piece of wire bend. The boundary of a square-shaped region, or the graph of a function $y = f(x)$ are also curves. When we think of surfaces, we might think of the surface of a magnifying glass, a parabolic antenna, or the outside cover of a globe. In some sense, most surfaces can be viewed as a sheet bent smoothly in space (possibly with self-intersections, singularities and other eccentricities). While we may have an intuitive perception of what a curve or a surface is, we need a mathematically accurate definition from which we can proceed to discuss their different properties.

Parametric Equations

In mathematical terms, a curve can be defined as a function that maps $\mathbb{R}^1$ into $\mathbb{R}^n$. This often leads to the definition of *parametric* curves and surfaces. Parametric curves defined as maps that take a single variable (the parameter) and map it into a vector in $\mathbb{R}^2$ are called *parametrised* or *parametrically defined* curves. The function gives information about the rate of traversal of the curved path according to the parameter's value.

For example, $f : [0, 2\pi] \to \mathbb{R}^2$, $f(\theta) = (x(\theta), y(\theta))$

$$\begin{cases} x = \sin\theta \\ y = \cos\theta \end{cases} \quad \text{where } \theta \in [0, 2\pi]$$

is a parametric form of the curve commonly known as the *unit circle*, shown in Figure 0.1(a). The same unit circle can be obtained through the following parametric equations:

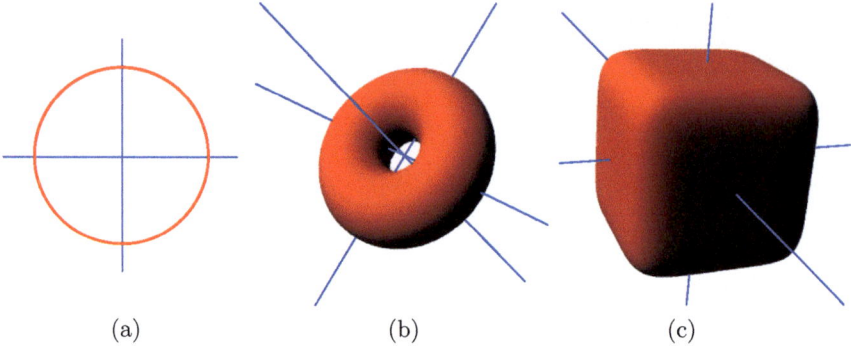

Fig. 0.1. (a) A unit circle; (b) a torus; (c) a blended cube.

$$\begin{cases} x = \dfrac{2t}{1+t^2} \\ y = \dfrac{1-t^2}{1+t^2} \end{cases} \text{where } t \in [0,1]$$

Parametric surfaces can be defined similarly, except their range is $\mathbb{R}^3$, and their expressions are defined in terms of *two* parameters instead of one. For example, the surface of the torus in Figure 0.1(b) is defined by the following parametric equations

$$f(u,v) = \begin{pmatrix} x(u,v) \\ y(u,v) \\ z(u,v) \end{pmatrix} = \begin{pmatrix} \cos u\,(R + r\cos v) \\ \sin u\,(R + r\cos v) \\ r\sin v \end{pmatrix}$$

i.e. it is defined in terms of two parameters u and v which, when varied over a region of the uv plane, yield a surface in $\mathbb{R}^3$; the real constants R and r are the major and minor radii of the torus.

Thus, parametric curves and surfaces can also be considered as images of subsets of $\mathbb{R}^1$ and $\mathbb{R}^2$ into $\mathbb{R}^2$ and $\mathbb{R}^3$, respectively, under a given function.

Explicit Equations

Another way of expressing curves and surfaces is through *explicit* equations. A function $f : \mathbb{R} \to \mathbb{R}$ of the form $y = f(x)$ has a unique value of y for each $x \in \mathbb{R}$. Therefore, such curves cannot self-intersect. Referring back to the unit circle example, it is only one of the (top or bottom) semicircles that can be expressed by an explicit equation at any one time:

$$y = \sqrt{1-x^2}$$

or

$$y = -\sqrt{1-x^2}$$

Similarly, some three-dimensional surfaces can be written by way of explicit functions that give $z = f(x, y)$.

Implicit Equations

A planar curve is said to be *defined implicitly* or in *Cartesian coordinates* when it is described as the set of solutions to an equation, usually in two variables $f(x, y) = 0$. For example, the equation $x^2 + y^2 - 1 = 0$ or $x^2 + y^2 = 1$ describes an implicit unit circle in $\mathbb{R}^2$. This is the same as the one in Figure 0.1(a).

An implicit curve in $\mathbb{R}^3$ is given by the set of solutions to a system of equations. For example, the system of two equations

$$\begin{cases} x^2 + y^2 + z^2 = 1 \\ z = 0 \end{cases}$$

creates a spatial circle which overlaps with the unit circle discussed above. Geometrically, this can be thought of as the intersection between the sphere $x^2 + y^2 + z^2 = 1$ and the plane $z = 0$. It can also be thought of as the intersection between the cylinder $x^2 + y^2 = 1$ and the plane $z = 0$ or, indeed, the intersection between the elliptic paraboloid $z = x^2 + y^2 - 1$ (see Figure 4.3) and the plane $z = 0$.

Analogously, an implicit surface in $\mathbb{R}^3$ is the set of solutions to an equation in at most three variables, $f(x, y, z) = 0$. For example, $x^6 + y^6 + z^6 = 1$ describes a blended cube in Figure 0.1(c); both equations $(R - \sqrt{x^2 + y^2})^2 + z^2 - r^2 = 0$ and $(x^2 + y^2 + z^2 + R^2 - r^2)^2 - 4R^2(x^2 + y^2) = 0$ are implicit equations of the same torus of minor radius r and major radius R in Figure 0.1(b).

The number of variables in an implicit equation does not determine the space in which the equation is to be considered, though it determines the least dimension of the space. For example, in $\mathbb{R}^2$ the equation $x^2 + y^2 = 1$ describes the unit circle; in $\mathbb{R}^3$ this is an equation in which z can take any value, and is therefore the equation of the unbounded cylinder of radius 1 centred around the z axis. Conversely, the same equation in $\mathbb{R}^2$ can be interpreted geometrically as a "slice" through its corresponding $\mathbb{R}^3$ shape, with a horizontal plane of arbitrarily fixed z.

Level Sets

Certain generalised forms of implicit surfaces are known in computer graphics as *isosurfaces* or *level sets* of a function $f : \mathbb{R}^3 \to \mathbb{R}$. An isosurface is the point set in $\mathbb{R}^3$ that satisfies the equation $f(x, y, z) = k$, where k is a constant. However, not all equations of this sort define a surface, or indeed define a point set at all. In some cases, this makes their parametrisation unachievable.

Final Remarks

In short, curves and surfaces are defined by functions. Their parametric representations are viewed as images of a function, whereas their implicit representations define them as level sets of a function. Whilst this book focuses specifically on implicit objects of the form $f(\mathbf{x}) = 0$ with $\mathbf{x} \in \mathbb{R}^n$, it is sometimes useful to regard them as explicit functions $f : \mathbb{R}^n \to \mathbb{R}$ with $t = f(\mathbf{x})$ and study the function's roots and range of values.

The following chapters will explain the ways in which the implicit objects provide a more general form than other representations. The parametric form is considered by many as being more useful because it explicitly produces the curve or surface by varying the parameters, i.e. it is a generative representation. According to the nature of each application, computer graphics researchers, developers and practitioners sometimes find it desirable to derive a parametric representation from an implicit one. Under very general conditions, the implicit function theorem (see Chapter 1 for details) shows that this is theoretically possible. Unfortunately, the theoretical existence of such a conversion does not always help in practice. Often it is not easy to determine a parametric representation from an implicit representation of a curve or surface, nor to perform the conversion the other way around. Surface modelling software tends to use one of the two representations systematically.

1
Mathematical Fundamentals

This chapter deals with mathematical fundamentals of curves and surfaces, and more generally manifolds and varieties.[1] For that, we will pay particular attention to their smoothness or, putting it differently, to their singularities (i.e. lack of smoothness). As will be seen later on, these shape particularities are important in the design and implementation of rendering algorithms for implicit curves and surfaces. Therefore, although the context is the differential topology and geometry, we are interested in their applications in geometric modelling and computer graphics.

1.1 Introduction

The rationale behind the writing of this chapter was to better understand the subtleties of the manifolds, in particular to exploit the smooth structure of manifolds (e.g. Euclidean spaces) through the study of the *intrinsic* properties of their subsets or subspaces, i.e. independently of any choice of local coordinates (e.g. spherical coordinates, Cartesian coordinates, etc.). As known, manifolds provide us with the proper category in which most efficiently one can develop a coordinate-free approach to the study of the intrinsic geometry of point sets. It is obvious that the explicit formulas for a subset may change when one goes from one set of coordinates to another. This means that any geometric equivalence problem can be viewed as the problem of determining whether two different local coordinate expressions define the same intrinsic subset of a manifold. Such coordinate expressions (or change of coordinates) are defined by mappings between manifolds.

Thus, by defining mappings between manifolds such as Euclidean spaces, we are able to uncover the local properties of their subspaces. In geometric

---

[1] A real, algebraic or analytic variety is a point set defined by a system of equations $f_1 = \cdots = f_k = 0$, where the functions f_i ($0 \leq i \leq k$) are real, algebraic or analytic, respectively.

A.J.P. Gomes et al. (eds.), *Implicit Curves and Surfaces: Mathematics, Data Structures and Algorithms*,
© Springer-Verlag London Limited 2009

modelling, we are particularly interested in properties such as, for example, local smoothness, i.e. to know whether the neighbourhood of a point in a submanifold is (visually) smooth, or the point is a singularity. In other words, we intend to study the relationship between smoothness of mappings and smoothness of manifolds. The idea is to show that a mathematical theory exists to describe manifolds and varieties (e.g. curves and surfaces), regardless of whether they are defined explicitly, implicitly, or parametrically.

1.2 Functions and Mappings

In simple terms, a function is a relationship between two variables, typically x and y, so it often denoted by $f(x) = y$. The variable x is the independent variable (also called primary variable, function argument, or function input), while the variable y is the dependent variable (secondary variable, value of the function, function output, or the image of x under f). Therefore, a function allows us to associate a unique output for each input of a given type (e.g. a real number).

In more formal terms, a function is a particular type of binary relation between two sets, say X and Y. The set X of input values is said to be the *domain* of f, while the set Y of output values is known as the *codomain* of f. The *range* of f is the set $\{f(x) : x \in X\}$, i.e. the subset of Y which contains all output values of f. The usual definition of a function satisfies the condition that for each $x \in X$, there is at most one $y \in Y$ such that x is related to y. This definition is valid for most elementary functions, as well as maps between algebraic structures, and more importantly between geometric objects, such as manifolds.

There are three major types of functions, namely, injections, surjections and bijections. An *injection* (or one-to-one function) has the property that if $f(a) = f(b)$, then a and b must be identical. A *surjection* (or onto function) has the property that for every y in the codomain there is an x in the domain such that $f(x) = y$. Finally, a *bijection* is both one-to-one and onto.

The notion of a function can be extended to several input variables. That is, a single output is obtained by combining two (or more) input values. In this case, the domain of a function is the Cartesian product of two or more sets. For example, $f(x, y, z) = x^2 + y^2 + z^2 = 0$ is a trivariate function (or a function of three variables) that outputs the single value 0; the domain of this function is the Cartesian product $\mathbb{R} \times \mathbb{R} \times \mathbb{R}$ or, simply, $\mathbb{R}^3$. In geometric terms, this function defines an implicit sphere in $\mathbb{R}^3$.

Functions can be even further extended in order to have several outputs. In this case, we have a component function for each output. Functions with several outputs or component functions are here called *mappings*. For example, the mapping $f : \mathbb{R}^3 \to \mathbb{R}^2$ defined by $f(x, y, z) = (x^2+y^2+z^2-1, 2x^2+2y^2-1)$ has two component functions $f_1(x, y, z) = x^2 + y^2 + z^2 - 1$ and $f_2(x, y, z) = 2x^2 + 2y^2 - 1$. These components represent a sphere and a cylinder in $\mathbb{R}^3$, respectively, so that, intuitively, we can say that f represents the point set that results from the intersection between the sphere and the cylinder.

Before proceeding any further, it is also useful to review how functions are classified in respect to the properties of their derivatives. Let $f : X \to Y$ be a mapping of X into Y, where X, Y are open subsets of $\mathbb{R}^m, \mathbb{R}^n$, respectively. If $n = 1$, we say that the function f is C^r (or C^r *differentiable* or *differentiable of class C^r*, or C^r *smooth* or *smooth of class C^r*) on X, for $r \in \mathbb{N}$, if the partial derivatives of f exist and are continuous on X, that is, at each point $\mathbf{x} \in X$. In particular, f is C^0 if f is continuous. If $n > 1$, the mapping f is C^r if each of the *component functions* f_i ($1 \leq i \leq n$) of f is C^r. We say that f is C^∞ (or just *differentiable* or *smooth*) if it is C^r for all $r \geq 0$. Moreover, f is called a C^r *diffeomorphism* if: (i) f is a homeomorphism[2] and (ii) both f and f^{-1} are C^r differentiable, $r \geq 1$ (when $r = \infty$ we simply say *diffeomorphism*). For further details about smooth mappings, the reader is referred to, for example, Helgason [182, p. 2].

1.3 Differential of a Smooth Mapping

Let U, V be open sets in $\mathbb{R}^m, \mathbb{R}^n$, respectively. Let $f : U \to V$ be a mapping with component functions $f_1, \ldots, f_n$. Note that f is defined on every point $\mathbf{p}$ of U in the coordinate system $x_1, \ldots x_m$. We call f smooth provided that all derivatives of the f_i of all orders exist and are continuous in U. Thus for f smooth, $\partial^2 f_i / \partial x_1 \partial x_2$, $\partial^3 f_i / \partial x_1^3$, etc., and $\partial^2 f_i / \partial x_1 \partial x_2 = \partial^2 f_i / \partial x_2 \partial x_1$, etc., all exist and are continuous. Therefore, a mapping $f : U \to V$ is *smooth* (or *differentiable*) if f has continuous partial derivatives of all orders. And we call f a *diffeomorphism* of U onto V when it is a bijection, and both f, f^{-1} are smooth.

Let $f : U \to V$ be a smooth (or differentiable or C^∞) and let $\mathbf{p} \in U$. The matrix

$$\mathrm{J}f(\mathbf{p}) = \begin{bmatrix} \partial f_1(\mathbf{p})/\partial x_1 & \partial f_1(\mathbf{p})/\partial x_2 & \cdots & \partial f_1(\mathbf{p})/\partial x_m \\ \vdots & \vdots & & \vdots \\ \partial f_n(\mathbf{p})/\partial x_1 & \partial f_n(\mathbf{p})/\partial x_2 & \cdots & \partial f_n(\mathbf{p})/\partial x_m \end{bmatrix}$$

where the partial derivatives are evaluated at $\mathbf{p}$, is called **Jacobian matrix** of f at $\mathbf{p}$ [68, p. 51]. The linear mapping $\mathrm{D}f(\mathbf{p}) : \mathbb{R}^m \to \mathbb{R}^n$ whose matrix is the Jacobian is called the **derivative** or **differential** of f at $\mathbf{p}$; the Jacobian $\mathrm{J}f(\mathbf{p})$ is also denoted by $[\mathrm{D}f(\mathbf{p})]$. It is known in mathematics and geometric design that every polynomial mapping f (i.e. mappings whose component functions

---

[2] In topology, two topological spaces are said to be *equivalent* if it is possible to transform one to the other by continuous deformation. Intuitively speaking, these topological spaces are seen as being made out of ideal rubber which can be deformed somehow. However, such a continuous deformation is constrained by the fact that the dimension is unchanged. This kind of transformation is mathematically called *homeomorphism*.

f_i are all polynomial functions) is smooth. If the components are rational functions, then the mapping is smooth provided none of the denominators vanish anywhere.

Besides, the composite of two smooth mappings, possibly restricted to a smaller domain, is smooth [68, p. 51]. It is worth noting that the chain rule holds not only for smooth mappings, but also for differentials. This fact provides us with a simple proof of the following theorem.

Theorem 1.1. *Let U, V be open sets in $\mathbb{R}^m, \mathbb{R}^n$, respectively. If $f : U \to V$ is a diffeomorphism, at each point $\mathbf{p} \in U$ the differential $\mathrm{D}f(\mathbf{p})$ is invertible, so that necessarily $m = n$.*

Proof. See Gibson [159, p. 9].

The justification for $m = n$ is that it is not possible to have a diffeomorphism between open subspaces of Euclidean spaces of different dimensions [58, p. 41]. In fact, a famous theorem of algebraic topology (Brouwer's invariance of dimension) asserts that even a homeomorphism between open subsets of $\mathbb{R}^m$ and $\mathbb{R}^n$, $m \neq n$, is impossible. This means that, for example, a point and a line cannot be homeomorphic (i.e. topologically equivalent) to each other because they have distinct dimensions.

Theorem 1.1 is very important not only to distinguish between two manifolds in the sense of differential geometry, but also to relate the invertibility of a diffeomorphism to the invertibility of the associated differential. More subtle is the hidden relationship between singularities and noninvertibility of the Jacobian. We should emphasise here that the direct inverse of Theorem 1.1 does not hold. However, there is a partial or local inverse, called the inverse mapping theorem, possibly one of the most important theorems in calculus. It is introduced in the next section, where we discuss the relationship between invertibility of mappings and smoothness of manifolds.

1.4 Invertibility and Smoothness

The smoothness of a submanifold that is the image of a mapping depends not only on smoothness but also the invertibility of its associated mapping. This section generalises such a relationship between smoothness and *invertibility* to mappings of several variables. This generalisation is known in mathematics as the *inverse mapping theorem*. This leads to a general mathematical theory for geometric continuity in geometric modelling, which encompasses not only parametric objects but also implicit ones. Therefore, this generalisation is *representation-independent*, i.e. no matter whether a submanifold is parametrically or implicitly represented.

Before proceeding, let us then briefly review the invertibility of mappings in the linear case.

1.4 Invertibility and Smoothness

Definition 1.2. *Let X, Y be Euclidean spaces, and $f : X \to Y$ a continuous linear mapping. One says that f is **invertible** if there exists a continuous linear mapping $g : Y \to X$ such that $g \circ f = \mathrm{id}_X$ and $f \circ g = \mathrm{id}_Y$ where id_X and id_Y denote the identity mappings of X and Y, respectively. Thus, by definition, we have:*

$$g(f(x)) = x \quad \text{and} \quad f(g(y)) = y$$

for every $x \in X$ and $y \in Y$. We write f^{-1} for the inverse of f.

But, unless we have an algorithm to evaluate whether or not a mapping is invertible, smoothness analysis of a point set is useless from the geometric modelling point of view. Fortunately, linear algebra can help us at this point. Consider the particular case $f : \mathbb{R}^n \to \mathbb{R}^n$. The linear mapping f is represented by a matrix $A = [a_{ij}]$. It is known that f is invertible iff A is invertible (as a matrix), and the inverse of A, if it exists, is given by

$$A^{-1} = \frac{1}{\det A} \mathrm{adj}\, A$$

where $\mathrm{adj}\, A$ is a matrix whose components are polynomial functions of the components of A. In fact, the components of $\mathrm{adj}\, A$ are subdeterminants of A. Thus, A is invertible iff its determinant $\det A$ is not zero.

Now, we are in position to define invertibility for differential mappings.

Definition 1.3. *Let U be an open subset of X and $f : U \to Y$ be a C^1 mapping, where X, Y are Euclidean spaces. We say that f is C^1-**invertible** on U if the image of f is an open set V in Y, and if there is a C^1 mapping $g : V \to U$ such that f and g are inverse to each other, i.e.*

$$g(f(x)) = x \quad \text{and} \quad f(g(y)) = y$$

for all $x \in U$ and $y \in V$.

It is clear that f is C^0-invertible if the inverse mapping exists and is continuous. One says that f is C^r-invertible if f is itself C^r and its inverse mapping g is also C^r. In the linear case, we are interested in linear invertibility, which basically is the strongest requirement that we can make. From the theorem that states that a C^r mapping that is a C^1 diffeomorphism is also a C^r diffeomorphism (see Hirsch [190]), it turns out that if f is a C^1-invertible, and if f happens to be C^r, then its inverse mapping is also C^r. This is the reason why we emphasise C^1 at this point. However, a C^1 mapping with a continuous inverse is not necessarily C^1-invertible, as illustrated in the following example:

Example 1.4. Let $f : \mathbb{R} \to \mathbb{R}$ be the mapping $f(x) = x^3$. It is clear that f is infinitely differentiable. Besides, f is strictly increasing, and hence has an inverse mapping $g : \mathbb{R} \to \mathbb{R}$ given by $g(y) = y^{1/3}$. The inverse mapping g is continuous, but not differentiable, at 0.

Let us now see the behaviour of invertibility under composition. Let $f : U \to V$ and $g : V \to W$ be invertible C^r mappings, where V is the image of f and W is the image of g. It follows that $g \circ f$ and $(g \circ f)^{-1} = f^{-1} \circ g^{-1}$ are C^r-invertible, because we know that a composite of C^r mappings is also C^r.

Definition 1.5. *Let $f : X \to Y$ be a C^r mapping, and let $\mathbf{p} \in X$. One says that f is **locally C^r-invertible** at $\mathbf{p}$ if there exists an open subset U of X containing $\mathbf{p}$ such that f is C^r-invertible on U.*

This means that there is an open set V of Y and a C^r mapping $g : V \to U$ such that $f \circ g$ and $g \circ f$ are the corresponding identity mappings of V and U, respectively. Clearly, a composite of locally invertible mappings is locally invertible. Putting this differently, if $f : X \to Y$ and $g : Y \to Z$ are C^r mappings, with $f(\mathbf{p}) = \mathbf{q}$ for $\mathbf{p} \in U$, and f, g are locally C^r-invertible at $\mathbf{p}$, $\mathbf{q}$, respectively, then $g \circ f$ is locally C^r-invertible at $\mathbf{p}$.

In Example 1.4, we used the derivative as a test for invertibility of a real-valued function of one variable. That is, if the derivative does not vanish at a given point, then the inverse function exists, and we have a formula for its derivative. The inverse mapping theorem generalises this result to mappings, not just functions.

Theorem 1.6. (Inverse Mapping Theorem) *Let U be an open subset of $\mathbb{R}^m$, let $\mathbf{p} \in U$, and let $f : U \to \mathbb{R}^n$ be a C^1 mapping. If the derivative $\mathrm{D}f$ is invertible, f is locally C^1-invertible at $\mathbf{p}$. If f^{-1} is its local inverse, and $\mathbf{y} = f(\mathbf{x})$, then $\mathrm{J}f^{-1}(\mathbf{y}) = [\mathrm{J}f(\mathbf{x})]^{-1}$.*

Proof. See Boothby [58, p. 43].

This is equivalent to saying that there exists open neighbourhoods U, V of $\mathbf{p}, f(\mathbf{p})$, respectively, such that f maps U diffeomorphically onto V. Note that, by Theorem 1.1, $\mathbb{R}^m$ has the same dimension as the Euclidean space $\mathbb{R}^n$, that is, $m = n$.

Example 1.7. Let U be an open subset of $\mathbb{R}^2$ consisting of all pairs (r, θ), with $r > 0$ and arbitrary θ. Let $f : U \to V \subset \mathbb{R}^2$ be defined by $f(r, \theta) = (r \cos\theta, r \sin\theta)$, i.e. V represents a circle of radius r in $\mathbb{R}^2$. Then

$$\mathrm{J}f(r, \theta) = \begin{bmatrix} \cos\theta & -r\sin\theta \\ \sin\theta & r\cos\theta \end{bmatrix}$$

and

$$\det \mathrm{J}f(r, \theta) = r\cos^2\theta + r\sin^2\theta = r.$$

Thus, $\mathrm{J}f$ is invertible at every point, so that f is locally invertible at every point. The local coordinates f_1, f_2 are usually denoted by x, y so that we usually write

$$x = r\cos\theta \quad \text{and} \quad y = r\sin\theta.$$

The local inverse can be defined for certain regions of Y. In fact, let V be the set of all pairs (x, y) such that $x > 0$ and $y > 0$. Then the inverse on V is given by
$$r = \sqrt{x^2 + y^2} \quad \text{and} \quad \theta = \arcsin \frac{y}{\sqrt{x^2 + y^2}}.$$

As an immediate consequence of the inverse mapping theorem, we have:

Corollary 1.8. *Let U be an open subset of $\mathbb{R}^n$ and $f : U \to \mathbb{R}^n$. A necessary and sufficient condition for the C^r mapping f to be a C^r diffeomorphism from U to $f(U)$ is that it be one-to-one and Jf be nonsingular at every point of U.*

Proof. Boothby [58, p. 46]. □

Thus, diffeomorphisms have nonsingular Jacobians. This parallel between differential geometry and linear algebra makes us to think of an algorithm to check whether or not a C^r mapping is a C^r diffeomorphism. So, using computational differentiation techniques and matrix calculus, we are able to establish smoothness conditions on a submanifold of $\mathbb{R}^n$.

Note that the domain and codomain of the mappings used in Theorem 1.1, Theorem 1.6 and its Corollary 1.8 have the same dimension. This may suggest that only smooth mappings between spaces of the same dimension are C^r invertible. This is not the case. Otherwise, this would be useless, at least for geometric modelling. For example, a parametrised k-manifold in $\mathbb{R}^n$ is defined by the *image* of a parametrisation $f : \mathbb{R}^k \to \mathbb{R}^n$, with $k < n$. On the other hand, an implicit k-manifold is defined by the *level set* of a function $f : \mathbb{R}^k \to \mathbb{R}$, i.e. by an equation $f(\mathbf{x}) = c$, where c is a real constant.

1.5 Level Set, Image, and Graph of a Mapping

Let us then review the essential point sets associated with a mapping. This will help us to understand how a manifold or even a variety is defined, either implicitly, explicitly, or parametrically. Basically, we have three types of sets associated with any mapping $f : U \subset \mathbb{R}^m \to \mathbb{R}^n$ which play an important role in the study of manifolds and varieties: *level sets*, *images*, and *graphs*.

1.5.1 Mapping as a Parametrisation of Its Image

Definition 1.9. (Baxandall and Liebeck [35, p. 26]) *Let U be open in $\mathbb{R}^m$. The **image** of a mapping $f : U \subset \mathbb{R}^m \to \mathbb{R}^n$ is the subset of $\mathbb{R}^n$ given by*
$$\text{Image } f = \{\mathbf{y} \in \mathbb{R}^n \mid \mathbf{y} = f(\mathbf{x}), \forall \mathbf{x} \in U\},$$
*being f a **parametrisation** of its image with parameters $(x_1, \ldots, x_m)$.*

This definition suggests that *practically any mapping is a "parametrisation" of something* [197, p. 263].

Example 1.10. The mapping $f : \mathbb{R} \to \mathbb{R}^2$ defined by $f(t) = (\cos t, \sin t)$, $t \in \mathbb{R}$, has an image that is the unit circle $x^2 + y^2 = 1$ in $\mathbb{R}^2$ (Figure 1.1(a)). A distinct function with the same image as f is the mapping $g(t) = (\cos 2t, \sin 2t)$.

Example 1.10 suggests that two or more distinct mappings can have the same image. In fact, it can be proven that there is an infinity of different parametrisations of any nonempty subset of $\mathbb{R}^n$ [35, p. 29]. Free-form curves and surfaces used in geometric design are just images in $\mathbb{R}^3$ of some parametrisation $\mathbb{R}^1 \to \mathbb{R}^3$ or $\mathbb{R}^2 \to \mathbb{R}^3$, respectively. The fact that an image can be parametrised by several mappings poses some problems to meet smoothness conditions when we patch together distinct parametrised curves or surfaces, simply because it is not easy to find a global reparametrisation for a compound curve or surface. Besides, the *smoothness of the component functions that describe the image of a mapping does not guarantee smoothness for its image.*

Example 1.11. A typical example is the cuspidal cubic curve that is the image of a smooth mapping $f : \mathbb{R}^1 \to \mathbb{R}^2$ defined by $t \mapsto (t^3, t^2)$ which presents a cusp at $t = 0$, Figure 1.2(a). Thus, the cuspidal cubic is not a smooth curve.

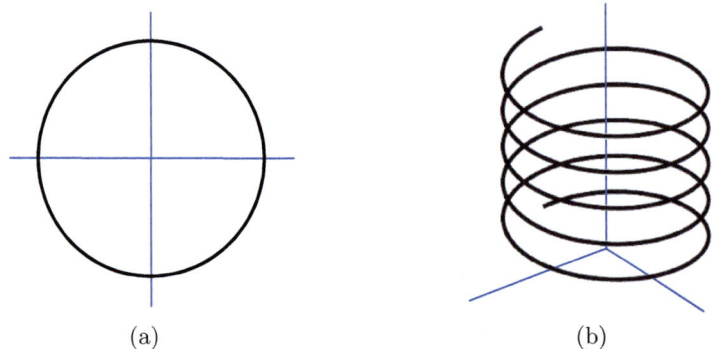

Fig. 1.1. (a) Image and (b) graph of $f(t) = (\cos t, \sin t)$.

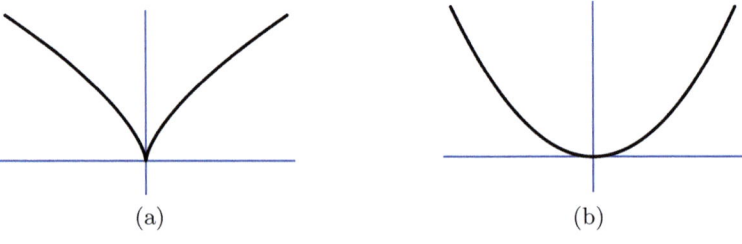

Fig. 1.2. (a) Cuspidal cubic $x^3 = y^2$ and (b) parabola $y = x^2$ as *images* of different parametrisations.

Conversely, the *smoothness of the image of a mapping does not imply that such a mapping is smooth*. The following example illustrates this situation.

Example 1.12. Let f, g and h be continuous mappings from $\mathbb{R}$ into $\mathbb{R}^2$ defined by the following rules:

$$f(t) = (t, t^2), \qquad g(t) = (t^3, t^6), \qquad \text{and} \qquad h(t) = \begin{cases} f(t), & t \geq 0, \\ g(t), & t < 0. \end{cases}$$

All three mappings have the same image, the parabola $y = x^2$ in $\mathbb{R}^2$, Figure 1.2(b). Their Jacobians are however distinct,

$$Jf(t) = [1 \quad 2t], \qquad Jg(t) = [3t^2 \quad 6t^5], \qquad \text{and} \qquad Jh(t) = \begin{cases} Jf(t), & t \geq 0, \\ Jg(t), & t < 0. \end{cases}$$

As polynomials, f, g are differentiable or smooth everywhere. Furthermore, because of $Jf(t) \neq [0 \quad 0]$ for any $t \in \mathbb{R}$, f is C^1-invertible everywhere. Consequently, its image is surely smooth. The function g is also smooth, but its Jacobian is null at $t = 0$, i.e. $Jg(0) = [0 \quad 0]$. This means that g is not C^1-invertible, or, equivalently, g has a singularity at $t = 0$, even though its image is smooth. Thus, a singularity of a mapping does not necessarily determine a singularity on its image. Even more striking is the fact that h is not differentiable at $t = 0$ (the left and right derivatives have different values at $t = 0$). This is so despite the smoothness of the image of h. This kind of situation where a smooth curve is formed by piecing together smooth curve patches is common in geometric design of free-form curves and surfaces used in industry.

The discussion above shows that every parametric smooth curve (in general, a manifold) can be described by several mappings, and that at least one of them is surely smooth and invertible, i.e. a diffeomorphism (see Corollary 1.8).

1.5.2 Level Set of a Mapping

Level sets of a mapping are varieties in some Euclidean space. That is, they are defined by equalities. Obviously, they are not necessarily smooth.

Definition 1.13. (Dineen [112, p. 6]) *Let U be open in $\mathbb{R}^m$. Let $f : U \subset \mathbb{R}^m \to \mathbb{R}^n$ and $\mathbf{c} = (c_1, \ldots, c_n)$ a point in $\mathbb{R}^n$. A **level set** of f, denoted by $f^{-1}(\mathbf{c})$, is defined by the formula*

$$f^{-1}(\mathbf{c}) = \{\mathbf{x} \in U \mid f(\mathbf{x}) = \mathbf{c}\}$$

In terms of coordinate functions $f_1, \ldots, f_n$ of f, we write

$$f(\mathbf{x}) = \mathbf{c} \iff f_i(\mathbf{x}) = c_i \quad \text{for } i = 1, \ldots, n$$

and thus

$$f^{-1}(\mathbf{c}) = \bigcap_{i=1}^{n}\{\mathbf{x} \in U \mid f_i(\mathbf{x}) = c_i\} = \bigcap_{i=1}^{n} f_i^{-1}(c_i).$$

The smoothness criterion for a variety defined as a level set of a vector-valued function is given by the following theorem.

Theorem 1.14. (Implicit Function Theorem, Baxandall [35, p. 145]) *A set $X \subseteq \mathbb{R}^m$ is a smooth variety if it is a level set of a C^1 function $f : \mathbb{R}^m \to \mathbb{R}$ such that $Jf(\mathbf{x}) \neq 0$ for all $\mathbf{x} \in X$.*

This theorem is a particular case of the implicit mapping theorem (IMT) for mappings which are functions. The IMT will be discussed later.

Example 1.15. The circle $x^2 + y^2 = 4$ is a variety in $\mathbb{R}^2$ that is a level set corresponding to the value 4 (i.e. point 4 in $\mathbb{R}$) of a function $f : \mathbb{R}^2 \to \mathbb{R}$ given by $f(x,y) = x^2 + y^2$. Its Jacobian is given by $Jf(x,y) = [2x \quad 2y]$ which is null at $(0,0)$. However, the point $(0,0)$ is not on the circle $x^2 + y^2 = 4$; hence the circle is a smooth curve.

Example 1.16. The sphere $x^2 + y^2 + z^2 = 9$ is a smooth surface in $\mathbb{R}^3$. It is the level set for the value 9 of a C^1 function $f : \mathbb{R}^3 \to \mathbb{R}$ defined by $f(x,y,z) = x^2 + y^2 + z^2$, and $Jf(x,y,z) \neq [0 \quad 0 \quad 0]$ at points on the sphere.

Example 1.17. Let $f : \mathbb{R}^3 \to \mathbb{R}$ be a function given by $f(x,y,z) = x^2 + y^2 - z^2$. Its level set corresponding to 0 is the right circular cone $z = \pm\sqrt{x^2 + y^2}$, whose apex is the point $(0,0,0)$ as illustrated in Figure 1.3(a). The Jacobian $Jf(x,y,z) = [2x \quad 2y \quad -2z]$ is null at the apex. Hence, the cone is not smooth at the apex, and the apex is said to be a singularity. Nevertheless, the level sets of the same function for which $x^2 + y^2 - z^2 = c \neq 0$ are smooth surfaces everywhere because the point $(0,0,0)$ is not on them. We have a hyperboloid of one sheet for $c > 0$ and a hyperboloid of two sheets for $c < 0$, as illustrated in Figure 1.3(b) and (c), respectively.

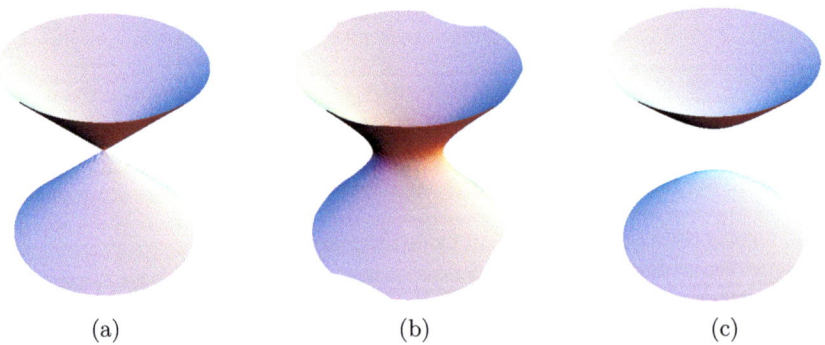

Fig. 1.3. (a) Cone $x^2 + y^2 - z^2 = 0$; (b) hyperboloid of one sheet $x^2 + y^2 - z^2 = a^2$; (c) hyperboloid of two sheets $x^2 + y^2 - z^2 = -a^2$.

Example 1.18. The Whitney umbrella with-handle $x^2 - zy^2 = 0$ in $\mathbb{R}^3$ (Figure 1.4) is not smooth. It is defined as the zero set of the function $f(x,y,z) = x^2 - zy^2$ whose Jacobian is $Jf(x,y,z) = [2x \quad -2yz \quad -y^2]$. It is easy to see that the Whitney umbrella is not smooth along the z-axis, i.e. the singular point set $\{(0,0,z)\}$ where the Jacobian is zero. This singular point set is given by the intersection $\{2x = 0\} \cap \{-2yz = 0\} \cap \{-y^2 = 0\}$, which basically is the intersection of two planes, $\{x = 0\}$ and $\{y = 0\}$, i.e. the z-axis.

The smoothness criterion based on the Jacobian is valid for functions and can be generalised to mappings. In this case, we have to use the implicit mapping theorem given further on. Even so, let us see an example of a level set for a general mapping, not a function.

Example 1.19. Let $f(x,y,z) = (x^2 + y^2 + z^2 - 1, 2x^2 + 2y^2 - 1)$ a mapping $f : \mathbb{R}^3 \to \mathbb{R}^2$ with component functions $f_1(x,y,z) = x^2 + y^2 + z^2 - 1$ and $f_2(x,y,z) = 2x^2 + 2y^2 - 1$. The set $f_1^{-1}(0)$ is a sphere of radius 1 in $\mathbb{R}^3$ while $f_2^{-1}(0)$ is a cylinder parallel to the z-axis in $\mathbb{R}^3$ (Figure 1.5). If $\mathbf{0} = (0,0)$ is

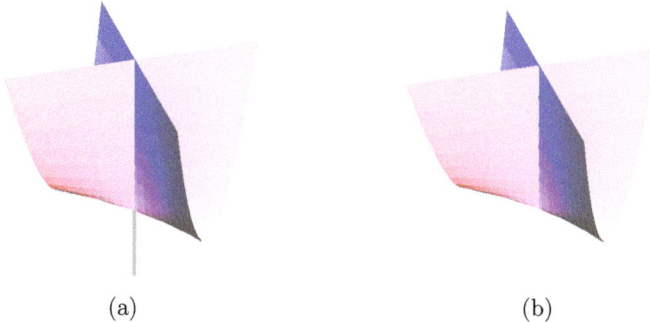

(a)   (b)

Fig. 1.4. (a) Whitney umbrella with-handle $x^2 - zy^2 = 0$; (b) Whitney umbrella without-handle $\{x^2 - zy^2 = 0\} - \{z < 0\}$.

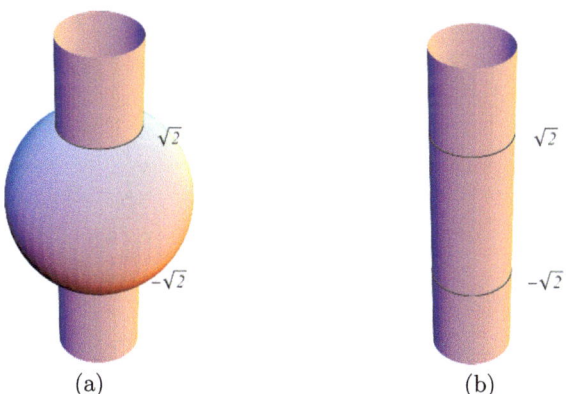

(a)   (b)

Fig. 1.5. Two circles as the intersection of a cylinder and sphere in $\mathbb{R}^3$.

the origin in $\mathbb{R}^2$, the level set

$$f^{-1}(\mathbf{0}) = f^{-1}(0,0) = f_1^{-1}(0) \cap f_2^{-1}(0)$$

is the intersection of a sphere and a cylinder in $\mathbb{R}^3$. This intersection consists of two circles that can be obtained by solving the equations $f_1(x,y,z) = f_2(x,y,z) = 0$. Such circles are in the planes $z = \sqrt{2}$ and $z = -\sqrt{2}$.

Let us see now the role of the differentiability in the local structure of level sets defined by general mappings as in Example 1.19. As noted in [112, p. 11], by taking into account the linear approximation of differentiable functions and standard results on solving systems of linear equations, we start to recognise and accept that level sets are locally graphs.

Let $f : U \subset \mathbb{R}^m \to \mathbb{R}^n$, U an open subset of $\mathbb{R}^m$, $f = (f_1, \ldots, f_n)$, $\mathbf{c} = (c_1, \ldots, c_n)$. We assume that f is differentiable. Let us consider the level set $f^{-1}(\mathbf{c}) = \bigcap_{i=1}^{n} f_i^{-1}(c_i)$, i.e. the set whose points $(x_1, \ldots, x_m) \in U$ satisfy the equations

$$f_1(x_1, \ldots, x_m) = c_1$$
$$\vdots \qquad\qquad (1.1)$$
$$f_n(x_1, \ldots, x_m) = c_n.$$

We have m unknowns $(x_1, \ldots, x_m)$ and n equations. If each component function f_i is linear, we have a system of linear equations and the rank of the matrix gives us the number of linearly independent solutions, and information enough to identify a complete set of independent variables. The Implicit Mapping Theorem states that all this information can be *locally* obtained for differentiable mappings. This is due to the fact that differentiable mappings, by definition, enjoy a good local linear approximation.

If $\mathbf{p} \in f^{-1}(\mathbf{c})$, then $f(\mathbf{p}) = \mathbf{c}$. If $\mathbf{x} \in \mathbb{R}^n$ is close to zero, then, since f is differentiable, we have

$$f(\mathbf{p} + \mathbf{x}) = f(\mathbf{p}) + f'(\mathbf{p}).\mathbf{x} + \epsilon(\mathbf{x})$$

where $\epsilon(\mathbf{x}) \to 0$ when $\mathbf{x} \to 0$ (see Dineen [112, p. 3, p. 12]). Because we wish to find $\mathbf{x}$ close to $\mathbf{0}$ such that $f(\mathbf{p} + \mathbf{x}) = \mathbf{c}$, we are considering points such that

$$f'(\mathbf{p}).\mathbf{x} + \epsilon(\mathbf{x}) = 0$$

and thus $f'(\mathbf{p}).\mathbf{x} \approx 0$ (where $\approx$ means approximately equal). Let us assume that $m \geq n$. Therefore, not surprisingly, we have something very close to the following system of linear equations

$$\frac{\partial f_1}{\partial x_1}(\mathbf{p})x_1 + \cdots + \frac{\partial f_1}{\partial x_m}(\mathbf{p})x_m = 0$$
$$\vdots \qquad\qquad (1.2)$$
$$\frac{\partial f_n}{\partial x_1}(\mathbf{p})x_1 + \cdots + \frac{\partial f_n}{\partial x_m}(\mathbf{p})x_m = 0,$$

whose matrix is the Jacobian Jf.

From linear algebra we know that

$$\begin{aligned}
\operatorname{rank} Jf = n &\iff n \text{ rows of } Jf \text{ are linearly independent} \\
&\iff n \text{ columns of } Jf \text{ are linearly independent} \\
&\iff Jf \text{ contains } n \text{ columns, and the associated} \\
&\quad\; n \times n \text{ matrix has nonzero determinant} \\
&\iff \text{the space of solutions of the system (1.2)} \\
&\quad\; \text{is } (m-n)\text{-dimensional}.
\end{aligned} \qquad (1.3)$$

Besides, if any of the conditions (1.3) are satisfied, and we select n columns that are linearly independent, then the variables concerning the remaining columns can be taken as a complete set of independent variables. If the conditions (1.3) are satisfied, we say that f has full or maximum rank at $\mathbf{p}$.

Example 1.20. Let us consider the following system of equations

$$\begin{aligned}
2x - y + z &= 0 \\
y \qquad\; - w &= 0,
\end{aligned}$$

whose matrix of coefficients is

$$\Lambda = \begin{bmatrix} 2 & -1 & 1 & 0 \\ 0 & 1 & 0 & -1 \end{bmatrix}.$$

The submatrix

$$\begin{bmatrix} 2 & -1 \\ 0 & 1 \end{bmatrix}$$

is obtained by taking the first two columns from A, and has determinant $2 \neq 0$. Thus, A has rank 2, or, equivalently, the two rows are linearly independent. So, the two variables z, w in the remaining two columns can be taken as the independent variables. In other words, $y = w$, $2x = y - z = w - z$, and hence $\{(\frac{w-z}{2}, w, z, w) : z \in \mathbb{R}, w \in \mathbb{R}\}$ is the solution set. Alternatively, the solution set can be written in the following form

$$\{(g(z,w), z, w) : (z,w) \in \mathbb{R}^2\}$$

where $g(z,w) = (\frac{w-z}{2}, w)$ is a mapping $g : \mathbb{R}^2 \to \mathbb{R}^2$. In this format, the solution space is the graph of g (defined in the next subsection).

Assuming that the rows of $Jf(\mathbf{p})$ are linearly independent is equivalent to supposing that the gradient vectors $\{\nabla f_1(\mathbf{p}), \ldots, \nabla f_n(\mathbf{p})\}$ are linearly independent in $\mathbb{R}^m$. The implicit mapping theorem states that with this condition we can solve the nonlinear system of equations (1.1) near $\mathbf{p}$ and apply the same approach to identify a set of independent variables. The hypothesis of a good linear approximation in the definition of differentiable functions implies that the equation systems (1.1) and (1.2) are very close to one another [112, p. 13]. Roughly speaking, this linear approximation is the tangent space to the solution set defined by the at $\mathbf{p}$.

Theorem 1.21. (Implicit Mapping Theorem, Munkres [292]) *Let $f : U \subset \mathbb{R}^m \to \mathbb{R}^n$ ($m \geq n$) be a differentiable mapping, let $\mathbf{p} \in U$ and assume that $f(\mathbf{p}) = \mathbf{c}$ and $\operatorname{rank} \mathbf{J}f(\mathbf{p}) = n$. For convenience, we also assume that the last n columns of the Jacobian are linearly independent. If $\mathbf{p} = (p_1, \ldots, p_m)$, let $\mathbf{p}_1 = (p_1, \ldots, p_{m-n})$ and $\mathbf{p}_2 = (p_{m-n+1}, \ldots, p_m)$ so that $\mathbf{p} = (\mathbf{p}_1, \mathbf{p}_2)$. Then, there exists an open set $V \subset \mathbb{R}^{m-n}$ containing $\mathbf{p}_1$, a differentiable mapping $g : V \to \mathbb{R}^n$, an open subset $U' \subset U$ containing $\mathbf{p}$ such that $g(\mathbf{p}_1) = \mathbf{p}_2$ and*

$$f^{-1}(\mathbf{c}) \cap U' = \{(\mathbf{x}, g(\mathbf{x})) : \mathbf{x} \in V\} = \operatorname{graph} g.$$

Therefore, locally every level set is a graph.

1.5.3 Graph of a Mapping

Definition 1.22. (Dineen [112, p. 6]) *Let U be open in $\mathbb{R}^m$. The **graph** of a mapping $f : U \subset \mathbb{R}^m \to \mathbb{R}^n$ is the subset of the product space $\mathbb{R}^{m+n} = \mathbb{R}^m \times \mathbb{R}^n$ defined by*

$$\operatorname{graph} f = \{(\mathbf{x}, \mathbf{y}) \,|\, \mathbf{x} \in U \text{ and } \mathbf{y} = f(\mathbf{x})\}$$

or

$$\operatorname{graph} f = \{(\mathbf{x}, f(\mathbf{x})) \,|\, \mathbf{x} \in U\}.$$

Example 1.23. Let us consider both mappings $f(t) = (\cos t, \sin t)$ and $g(t) = (\cos 2t, \sin 2t)$ of Example 1.10. They have the same image in $\mathbb{R}^2$, say a unit circle. However, their graphs are distinct point sets in $\mathbb{R}^3$. The graph of f is a circular helix $(t, \cos t, \sin t)$ in $\mathbb{R}^3$, Figure 1.1(b). But, although the graph of g is a circular helix with windings being around the same circular cylinder, those windings have half the pitch.

This suggests that there is a one-to-one correspondence between a mapping and its graph, that different mappings have distinct graphs. This leads us to think of a possible relationship between the smoothness of a mapping and the smoothness of its graph. In other words, the smoothness of a mapping determines the smoothness of its graph. This is corroborated by the following theorem.

Theorem 1.24. (Baxandall [35, p. 147]) *The graph of a C^1 mapping $f : U \subseteq \mathbb{R}^m \to \mathbb{R}^n$ is a smooth variety in $\mathbb{R}^m \times \mathbb{R}^n$.*

Proof. Consider the mapping $F : U \times \mathbb{R}^n \subseteq \mathbb{R}^m \times \mathbb{R}^n \to \mathbb{R}^n$ defined by

$$F(\mathbf{x}, \mathbf{y}) = f(\mathbf{x}) - \mathbf{y}, \quad \mathbf{x} \in U, \, \mathbf{y} \in \mathbb{R}^n.$$

The graph of f is the level set of F corresponding to the value $\mathbf{0}$, that is

$$\operatorname{graph} f = \{(\mathbf{x}, \mathbf{y}) \in \mathbb{R}^m \times \mathbb{R}^n \,|\, f(\mathbf{x}) - \mathbf{y} = \mathbf{0}\}.$$

To prove that graph f is a smooth variety in $\mathbb{R}^m \times \mathbb{R}^n$ we show that:

(i) F is a C^1 mapping.
(ii) $JF(\mathbf{x},\mathbf{y}) \neq (\mathbf{0},\mathbf{0})$ for all $\mathbf{x} \in U$, $\mathbf{y} \in \mathbb{R}^n$.

It follows from the definition of F above that for each $i = 1,\ldots,m$, $j = m+1,\ldots,m+n$ and each $\mathbf{x} \in U$, $\mathbf{y} \in \mathbb{R}^n$

$$\frac{\partial F}{\partial x_i}(\mathbf{x},\mathbf{y}) = \frac{\partial f}{\partial x_i}(\mathbf{x}) \quad \text{and} \quad \frac{\partial F}{\partial y_j}(\mathbf{x},\mathbf{y}) = -1.$$

Therefore the partial derivatives of F are continuous and so F is a C^1 mapping. Also, for any $\mathbf{x} \in U$, $\mathbf{y} \in \mathbb{R}^n$

$$JF(\mathbf{x},\mathbf{y}) = (Jf(\mathbf{x}),-1) \neq (\mathbf{0},\mathbf{0}).$$

This completes the proof.

Example 1.25. Let us consider the curves sketched in Figure 1.6. Figure 1.6(a) shows the curve $y = |x|$ in $\mathbb{R}^2$ that is not smooth. It is the graph of the function $f : \mathbb{R} \to \mathbb{R}$ that explicitly expresses y as a function of x, but f is not differentiable at $x = 0$. Nor is it the graph of (an inverse) function g expressing x as a function of y, because in the neighbourhood of $(0,0)$ the same value of y corresponds to two values of x.

Figure 1.6(b) shows another nonsmooth curve $xy = 0$ in $\mathbb{R}^2$, which is the union of the two coordinate axes, x and y. Any neighbourhood of $(0,0)$ contains infinitely many y values corresponding to $x = 0$, and infinitely many x values corresponding to $y = 0$. This means that the curve is not a graph of an explicit function $y = f(x)$, nor of a function $x = g(y)$. Incidentally, this curve can be regarded as a slice at $z = 0$ through the graph of $h : \mathbb{R}^2 \to \mathbb{R}$ where $h(x,y) = xy$, which defines the implicit curve $h(x,y)$ in $\mathbb{R}^2$.

Finally, the graph of the function $f(x) = x^{1/3}$, depicted in Figure 1.6(c), is a smooth curve. Note that the curve is smooth despite the function being not differentiable at $x = 0$. This happens because the curve is the graph of the function $x = f(y) = y^3$ that is differentiable.

From these examples, we come to the following conclusions:

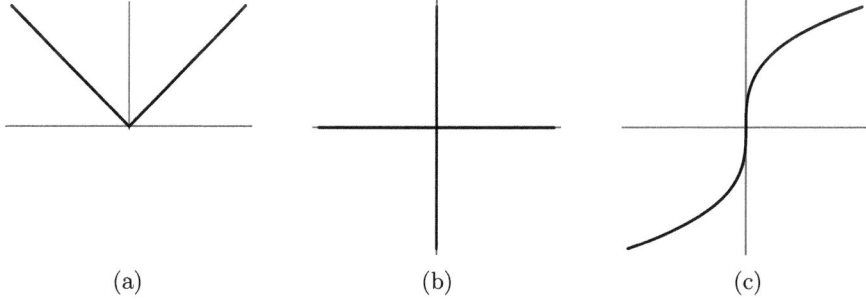

(a)   (b)   (c)

Fig. 1.6. Not all point sets in $\mathbb{R}^2$ are graphs of a mapping.

- Rewording Theorem 1.24, every point set that is the graph of a differentiable mapping is smooth.
- The fact that a mapping is not differentiable does not imply that its graph is not smooth; but if the graph is smooth, then it is necessarily the graph of a related function by changing the roles of the variables, possibly the inverse function. This is the case for the curve $x = y^3$ in Figure 1.6(c).
- The graph of a mapping that is not differentiable is possibly nonsmooth. This happens because of the differentiable singularities such as the cusp point in $y = |x|$, Figure 1.6.
- There are point sets in $\mathbb{R}^n$ that cannot be described as graphs of mappings, unless we break them up into pieces. For example, with appropriate constraints we can split $xy = 0$ (the union of axes in $\mathbb{R}^2$) into the origin and four half-axes, each piece described by a function. The origin is a cut point of $xy = 0$, that is, a topological singularity. The idea of partitioning a point set into smaller point sets by its topological singularities leads to a particular sort of stratification as briefly detailed in the next chapter. Another alternative to describe a point set that is not describable by a graph of a function is to describe it as a level set of a function.

The relationship between graphs and level sets plays an important role in the study of varieties. It is easy to see that *every graph is a level set*. Let us consider a mapping $f : U \subseteq \mathbb{R}^m \to \mathbb{R}^n$. We define $F : U \times \mathbb{R}^m \to \mathbb{R}^n$ by $F(\mathbf{x}, \mathbf{y}) = f(\mathbf{x}) - \mathbf{y}$. If $\mathbf{0}$ is the origin in $\mathbb{R}^n$, we have

$$(\mathbf{x}, \mathbf{y}) \in F^{-1}(\mathbf{0}) \iff F(\mathbf{x}, \mathbf{y}) = \mathbf{0}$$
$$\iff f(\mathbf{x}) - \mathbf{y} = \mathbf{0}$$
$$\iff (\mathbf{x}, \mathbf{y}) \in \text{graph } f.$$

Thus, $F^{-1}(\mathbf{0}) = \text{graph } f$ and every graph is a level set. This fact has been used to prove the Theorem 1.24. As a summary, we can say that:

- Not all varieties in some Euclidean space are graphs of a mapping.
- Every variety as a graph of a mapping is a level set.
- Every variety is a level set of a mapping.

This shows us why the study of algebraic and analytic varieties in geometry is carried out using level sets of mappings, i.e. point sets defined implicitly. The reason is a bigger geometric coverage of point sets in some Euclidean space. In addition to this, many (not necessarily smooth) varieties admit a global parametrisation, whilst others can only be partially (locally) and piecewise parametrised.

Example 1.26. Let $z = x^2 - y^2$ be a level set of a function $F : \mathbb{R}^3 \to \mathbb{R}$ defined by $F(x, y, z) = x^2 - y^2 - z$ corresponding to the value 0. It is observed that $\mathrm{J}F(x, y, z) = [2x \quad -2y \quad -1]$ is not zero everywhere. So $z = x^2 - y^2$ in $\mathbb{R}^3$ is smooth everywhere. It is a variety known as a *saddle surface*. Note that z is explicitly defined in terms of x and y. So, the saddle surface can be viewed as the graph of the function $f : \mathbb{R}^2 \to \mathbb{R}$ given by $f(x,y) = x^2 - y^2$. Consequently, the saddle surface can be given a global parametrisation $g : \mathbb{R}^2 \to \mathbb{R}^3$ defined by $g(x,y) = (x, y, x^2 - y^2)$.

Not all varieties can be *globally* parametrised, even when they are smooth. But, as proved later, every smooth level set can be always *locally* parametrised, i.e. every smooth level set is locally a graph. This fact is proved by the implicit mapping theorem.

Level sets correspond to implicit representations, say functions, on some Euclidean space, while graphs correspond to explicit representations. In fact, we have from calculus that

Definition 1.27. (Baxandall and Liebeck [35, p. 226]) *Let $f : X \subseteq \mathbb{R}^m \to \mathbb{R}$ be a function, where $m \geq 2$. If there exists a function $g : Y \subseteq \mathbb{R}^{m-1} \to \mathbb{R}$ such that for all $(x_1, \ldots, x_{m-1}) \in Y$,*

$$f(x_1, \ldots, x_{m-1}, g(x_1, \ldots, x_{m-1})) = 0,$$

*then the function g is said to be defined **implicitly** on Y by the equation*

$$f(x_1, \ldots, x_m) = 0.$$

Likewise, the graph of $g : Y \subseteq \mathbb{R}^{m-1} \to \mathbb{R}$ is the subset of $\mathbb{R}^m$ given by

$$\{(x_1, \ldots, x_{m-1}, x_m) \in \mathbb{R}^m \mid x_m = g(x_1, \ldots, x_{m-1})\}.$$

*The expression $x_m = g(\mathbf{x})$ is called the equation of the graph [35, p.100]. Hence, g is said to be **explicitly** defined on Y by the equation $x_m = g(x_1, \ldots, x_{m-1})$.*

Example 1.28. The graph of the function $f(x,y) = -x^2 - y^2$ has equation $-z = x^2 + y^2$. This graph is a 2-manifold in $\mathbb{R}^3$ called a paraboloid (Figure 1.7). The equation $-z = x^2 + y^2$ *explicitly* defines the paraboloid in $\mathbb{R}^3$.

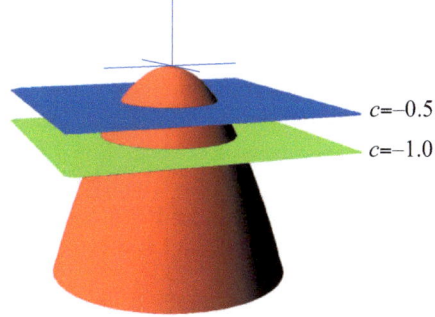

Fig. 1.7. The paraboloid $-z = x^2 + y^2$ in $\mathbb{R}^3$.

For $c < 0$ the plane $z = c$ intersects the graph in a circle lying below the level set $x^2 + y^2 = -c$ in the (x, y)-plane. The equation $x^2 + y^2 = -c$ of a circle (i.e. a 1-manifold) in $\mathbb{R}^2$ is said to define y *implicitly* in terms of x. This circle is said to be an implicit 1-manifold.

1.6 Rank-based Smoothness

Now, we are in position to show that the rank of a mapping gives us a general approach to check the C^r invertibility or C^r smoothness of a mapping, and whether or not a variety is smooth. This smoothness test is carried out independently of how a variety is defined, implicitly, explicitly or parametrically, i.e. no matter whether a variety is considered a level set, a graph, or an image of a mapping, respectively.

Definition 1.29. (Olver [313, p. 11]) *The **rank** of a mapping $f : \mathbb{R}^m \to \mathbb{R}^n$ at a point $\mathbf{p} \in \mathbb{R}^m$ is defined to be the rank of the $n \times m$ Jacobian matrix Jf of any local coordinate expression for f at the point $\mathbf{p}$. The mapping f is called **regular** if its rank is constant.*

Standard transformation properties of the Jf imply that the definition of rank is independent of the choice of local coordinates [313, p. 11] (see [58, p. 110] for a proof). Moreover, the rank of the Jacobian matrix (shortly rank Jf) provides us with a general algebraic procedure to check the smoothness of a submanifold or, putting it differently, to determine its singularities. It is proved in differential geometry that the set of points where the rank of f is maximal is an open submanifold of the manifold $\mathbb{R}^m$ (which is dense if f is analytic), and the restriction of f to this subset is regular. The subsets where the rank of a mapping decreases are *singularities* [313, p. 11]. The types and properties of such singularities are studied in singularity theory.

From linear algebra we have

$$\operatorname{rank} Jf = k \iff k \text{ rows of } Jf \text{ are linearly independent}$$
$$\iff k \text{ columns of } Jf \text{ are linearly independent}$$
$$\iff Jf \text{ has a } k \times k \text{ submatrix that has nonzero determinant.}$$

The fact that the $n \times m$ Jacobian matrix Jf has rank k means that it includes a $k \times k$ submatrix that is invertible. Thus, a necessary and sufficient condition for a k-variety to be smooth is that $\operatorname{rank} Jf = k$ at every point of it, no matter whether it is defined parametrically or implicitly by f. This is clearly a generalisation of Corollary 1.8, and is a consequence of a generalisation of the inverse mapping theorem, called the rank theorem:

Theorem 1.30. (Rank Theorem) *Let $U \subset \mathbb{R}^m$, $V \subset \mathbb{R}^n$ be open sets, $f : U \to V$ be a C^r mapping, and suppose that $\operatorname{rank} Jf = k$. If $\mathbf{p} \in U$ and*

$\mathbf{q} = f(\mathbf{p})$, there exists open sets $U_0 \subset U$ and $V_0 \subset V$ with $\mathbf{p} \in U_0$ and $\mathbf{q} \in V_0$, and there exists C^r diffeomorphisms

$$\phi : U_0 \to X \subset \mathbb{R}^m,$$

$$\psi : V_0 \to Y \subset \mathbb{R}^n$$

with X, Y open in $\mathbb{R}^m, \mathbb{R}^n$, respectively, such that

$$\psi \circ f \circ \phi^{-1}(X) \subset Y$$

and such that this mapping has the simple form

$$\psi \circ f \circ \phi^{-1}(p_1, \ldots, p_m) = (p_1, \ldots, p_k, 0, \ldots, 0).$$

Proof. See Boothby [58, p. 47].

This is a very important theorem because it states that a mapping of constant rank k behaves *locally* as a projection of $\mathbb{R}^m = \mathbb{R}^k \times \mathbb{R}^{m-k}$ to $\mathbb{R}^k$ followed by injection of $\mathbb{R}^k$ onto $\mathbb{R}^k \times \{0\} \subset \mathbb{R}^k \times \mathbb{R}^{n-k} = \mathbb{R}^n$.

1.6.1 Rank-based Smoothness for Parametrisations

The rank theorem for parametrisations is as follows:

Theorem 1.31. (Rank Theorem for Parametrisations) *Let U be an open set in $\mathbb{R}^m$ and $f : U \to \mathbb{R}^n$. A necessary and sufficient condition for the C^∞ mapping f to be a diffeomorphism from U to $f(U)$ is that it be one-to-one and the Jacobian Jf have rank m at every point of U.*

Proof. See Boothby [58, p. 46].

This is a generalisation of Corollary 1.8, with $m \leq n$. It means that the kernel[3] of the linear mapping represented by Jf is 0 precisely when the Jacobian matrix has rank m.

Let us review some simple examples of parametrised curves.

Example 1.32. We know that the bent curve in $\mathbb{R}^2$ depicted in Figure 1.6 and defined by the parametrisation $f(t) = (t, |t|)$ is not differentiable at $t = 0$, even though its rank is 1 everywhere.

Example 1.32 shows that the differentiability test should always precede the rank test in order to detect differentiable singularities.

---

[3] Let $F : X \to Y$ be a linear mapping of vector spaces. By the *kernel* of F, denoted by kernel F, is meant the set of all those vectors $\mathbf{x} \in X$ such that $F(V) = \mathbf{0} \in Y$, i.e. kernel $F = \{\mathbf{x} \in X : F(\mathbf{x}) = \mathbf{0}\}$ (see Edwards [128, p. 29]). In other words, the kernel of a linear mapping corresponds to the level set of a mapping.

Example 1.33. A parametrised curve that passes the differentiability test, but not the rank test, is the cuspidal cubic in $\mathbb{R}^2$ given by $f(t) = (t^3, t^2)$ (Figure 1.2(a)). The component functions are polynomials and therefore differentiable. However, the rank $Jf(t) = [3t^2 \quad 2t]$ is not 1 (i.e. its maximal value) at $t = 0$; in fact it is zero. This means that the parametrised cuspidal cubic is not smooth at $t = 0$, that is, it possesses a singularity at $t = 0$.

Example 1.34. Let us take the parametrised parabola in $\mathbb{R}^2$ given by $f(t) = (t, t^2)$ (Figure 1.2(b)). f is obviously differentiable, and its rank is 1 everywhere, so it is globally smooth.

Nevertheless, algorithmic detection of singularities of a parametrised variety fails for self-intersections, i.e. topological singularities. Let us see some examples.

Example 1.35. The curve parametrised by the differentiable mapping $f(t) = (t^3 - 3t - 2, t^2 - t - 2)$ is not smooth at $(0,0)$, despite the differentiability of f and its maximal rank. In fact, we get the same point $(0,0)$ on the curve for two distinct points $t = -1$ and $t = 2$ of the domain, that is, $f(-1) = f(2) = (0,0)$, and thus f is not one-to-one. These singularities are known as *self-intersections* in geometry or *topological singularities* in topology.

The problem with a parametrised self-intersecting variety is that its self-intersections are topological singularities for the corresponding underlying topological space, but not for the parametrisation. However, it is an easy task to check whether a non-self-intersecting point in a parametrised variety is singular or not. A non-self-intersecting point is singular if the rank of Jacobian at this point is not maximal.

Example 1.36. Let us consider a parametrisation $f(u,v) = (uv, u, v^2)$ of the Whitney umbrella without-handle (the negative z-axis) (Figure 1.4(b)). The effect of this parametrisation on $\mathbb{R}^2$ can be described as the 'fold' of the v-axis at the origin $(0,0)$ in order to superimpose negative v-axis and positive v-axis. The 'fold' is identified by the exponent 2 of the third component coordinate function. Thus, all points $(0, 0, v^2)$ along v-axis are double points and determine that all points on the positive z-axis are singularities or self-intersecting points in $\mathbb{R}^3$. However, this is not so apparent if we restrict the discussion to the Jacobian and try to determine where the rank drops below 2. In fact,

$$Jf(u,v) = \begin{bmatrix} v & u \\ 1 & 0 \\ 0 & 2v \end{bmatrix}$$

and we observe that the rank drops below 2 only at $(0,0)$. This happens because only $(0,0)$ is a differential singularity, that is, the tangent plane is not defined at $(0,0)$. Any other point on the positive z-axis has a parametrised neighbourhood that can be approximated by a tangent plane in relation to the parametrisation.

Example 1.37. Let $f : \mathbb{R}^2 \to \mathbb{R}^3$ be the mapping given by
$$f(x,y) = (\sin x, \, e^x \cos y, \, \sin y).$$
Then
$$\mathrm{J}f(x,y) = \begin{bmatrix} \cos x & 0 \\ e^x \cos y & -e^x \sin y \\ 0 & \cos y \end{bmatrix}$$
and hence
$$\mathrm{J}f(0,0) = \begin{bmatrix} 1 & 0 \\ 1 & 0 \\ 0 & 1 \end{bmatrix}$$
has rank 2, so that in a neighbourhood of $(0,0)$, the mapping f parametrises a subset of $\mathbb{R}^3$.

1.6.2 Rank-based Smoothness for Implicitations

The implicit function theorem is particularly useful for geometric modelling because it provides us with a computational tool to test whether an implicit manifold, and more generally a variety, is smooth in the neighbourhood of a point. Specifically, it gives us a local parametrisation for which it is possible to check the local C^r-invertibility by means of its Jacobian.

Before proceeding, let us see how C^r-invertibility and smoothness is defined for implicit manifolds and varieties.

Theorem 1.38. (Rank Theorem for Implicitations) *Let U be open in $\mathbb{R}^m$ and let $f : U \to \mathbb{R}$ be a C^r function on U. Let $(\mathbf{p}, q) = (p_1, \ldots, p_{m-1}, q) \in U$ and assume that $f(\mathbf{p}, q) = 0$ but $\frac{\partial f}{\partial x_m}(\mathbf{p}, q) \neq 0$. Then the mapping*
$$F : U \to \mathbb{R}^{m-1} \times \mathbb{R} = \mathbb{R}^m$$
given by
$$(\mathbf{x}, y) \mapsto (\mathbf{x}, f(\mathbf{x}, y))$$
is locally C^r-invertible at $(\mathbf{p}, q)$.

Proof. (See Lang [223, p.523]). All we need to do is to compute the derivative of F at $(\mathbf{p}, q)$. We write F in terms of its coordinates, $F = (F_1, \ldots, F_{m-1}, F_m) = (x_1, \ldots, x_{m-1}, f)$. Its Jacobian matrix is therefore
$$\mathrm{J}F(\mathbf{x}) = \begin{bmatrix} 1 & 0 & \cdots & 0 \\ 0 & 1 & \cdots & 0 \\ \vdots & & \ddots & \vdots \\ 0 & & \cdots & 1 & 0 \\ \frac{\partial f}{\partial x_1} & \frac{\partial f}{\partial x_2} & \cdots & & \frac{\partial f}{\partial x_m} \end{bmatrix}$$
and is invertible since its determinant is equal to $\frac{\partial f}{\partial x_m} \neq 0$ at $(\mathbf{p}, q)$. The inverse function theorem guarantees that F is locally C^r-invertible at $(\mathbf{p}, q)$.

As a corollary of this Theorem, we have the implicit function theorem for functions of several variables, which can be reworded as follows:

Theorem 1.39. (Multivariate Implicit Function Theorem) *Let U be open in $\mathbb{R}^m$ and let $f : U \to \mathbb{R}$ be a C^r function on U. Let $(\mathbf{p}, q) = (p_1, \ldots, p_{m-1}, q) \in U$ and assume that $f(\mathbf{p}, q) = 0$ but $\frac{\partial f}{\partial x_m}(\mathbf{p}, q) \neq 0$. Then there exists an open ball V in $\mathbb{R}^{m-1}$ centred at $\mathbf{p}$ and a C^r function*

$$g : V \to \mathbb{R}$$

such that $g(\mathbf{p}) = q$ and

$$f(\mathbf{x}, g(\mathbf{x})) = 0$$

for all $\mathbf{x} \in V$.

Proof. (See Lang [223, p. 524]). By Theorem 1.38 we know that the mapping

$$F : U \to \mathbb{R}^{m-1} \times \mathbb{R} = \mathbb{R}^m$$

given by

$$(\mathbf{x}, y) \mapsto (\mathbf{x}, f(\mathbf{x}, y))$$

is locally C^r-invertible at $(\mathbf{p}, q)$. Let $F^{-1} = (F_1^{-1}, \ldots, F_m^{-1})$ be the local inverse of F such that

$$F^{-1}(\mathbf{x}, z) = (\mathbf{x}, F_m^{-1}(\mathbf{x}, z)) \quad \text{for} \quad \mathbf{x} \in \mathbb{R}^{m-1}, \, z \in \mathbb{R}.$$

We let $g(\mathbf{x}) = F_m^{-1}(\mathbf{x}, 0)$. Since $F(\mathbf{p}, q) = (\mathbf{p}, 0)$ it follows that $F_m^{-1}(\mathbf{p}, 0) = q$ so that $g(\mathbf{p}) = q$. Furthermore, since F, F^{-1} are inverse mappings, we obtain

$$(\mathbf{x}, 0) = F(F^{-1}(\mathbf{x}, 0)) = F(\mathbf{x}, g(\mathbf{x})) = (\mathbf{x}, f(\mathbf{x}, g(\mathbf{x}))).$$

This proves that $f(\mathbf{x}, g(\mathbf{x})) = 0$, as shown by previous equality.

Note that we have expressed y as a function of $\mathbf{x}$ explicitly by means of g, starting with what is regarded as an implicit relation $f(\mathbf{x}, y) = 0$. Besides, from the implicit function theorem, we see that the mapping G given by

$$\mathbf{x} \mapsto (\mathbf{x}, g(\mathbf{x})) = G(\mathbf{x})$$

or writing down the coordinates

$$(x_1, \ldots, x_{m-1}) \mapsto (x_1, \ldots, x_{m-1}, g(x_1, \ldots, x_{m-1}))$$

provides a *parametrisation* of the variety defined by $f(x_1, \ldots, x_{m-1}, y) = 0$ in the neighbourhood of a given point $(\mathbf{p}, q)$. This is illustrated in Figure 1.8 for convenience. On the right, we have the surface $f(\mathbf{x}) = 0$, and we have also pictured the gradient $\operatorname{grad} f(\mathbf{p}, q)$ at the point $(\mathbf{p}, q)$ as in Theorem 1.39. Note that the condition $\frac{\partial f}{\partial x_m}(\mathbf{p}, q) \neq 0$ in Theorem 1.39 implies that the $\operatorname{grad} f(\mathbf{p}, q) = [\frac{\partial f}{\partial x_1} \, \frac{\partial f}{\partial x_2} \, \cdots \, \frac{\partial f}{\partial x_m}] \neq 0$.

An example follows to illustrate the implicit function theorem at work.

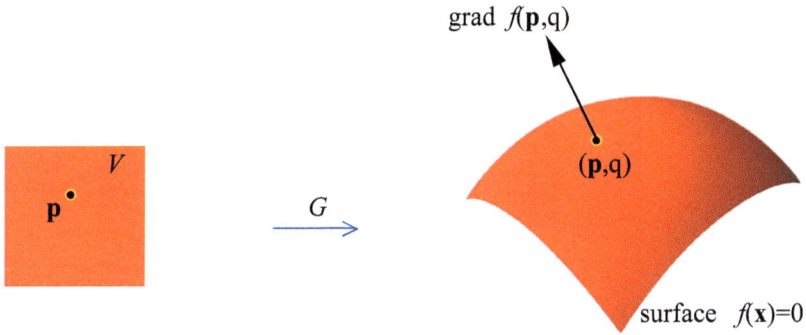

Fig. 1.8. Local parametrisation of an implicitly defined variety.

Example 1.40. The Whitney umbrella $x^2 - zy^2 = 0$ in $\mathbb{R}^3$ is the level set for the value 0 of the function $f : \mathbb{R}^3 \to \mathbb{R}$ given by $f(x,y,z) = x^2 - zy^2$. According to the Theorem 1.39, we have only to make sure that $\frac{\partial f}{\partial z} \neq 0$ in order to guarantee a regular neighbourhood for a point. But

$$\frac{\partial f}{\partial z} = -y^2 = 0 \quad \Rightarrow \quad y = 0$$

i.e. all points of $x^2 - zy^2 = 0$ with $y = 0$ are singular points. These singular points are then given by

$$\begin{cases} y = 0 \\ x^2 - zy^2 = 0 \end{cases} \Leftrightarrow \begin{cases} y = 0 \\ x = 0 \end{cases} \Leftrightarrow \{x = 0\} \cap \{y = 0\}$$

or, equivalently, the point set $\{(x,y,z) \in \mathbb{R}^3 : x = 0, y = 0\}$. That is, the singular set of the Whitney umbrella is the z-axis $0 \times 0 \times z$.

This result agrees with the fact that the Jacobian $\mathrm{J}f = [2x \quad 2yz \quad y^2]$ has maximal rank 1 for $(x,y,z) \neq (0,0,z)$. However, because the rank cannot fall below zero, we have no way to algorithmically detect via rank criterion any possible singularities in the z-axis. In fact, the z-axis is a smooth line, but we know that the origin is a special singularity of the Whitney umbrella provided that, unlike the points of the positive z-axis, it is a cut-point.[4]

The question now is whether or not there is any method to compute such singularities. An algorithm to determine the singularities of a variety is useful for many geometry software packages. For example, the graphical visualisation of the Whitney umbrella with-handle $x^2 - zy^2 = 0$ in $\mathbb{R}^3$ requires the detection of its singular set along the z-axis. Therefore, unless we use a parametric Whitney umbrella without-handle, such a point set cannot be visualised on

---

[4] In topology, a point of a connected space is a cut-point if its removal makes its space disconnected. For example, every point of a straight line is a cut-point because it splits the line into two; the same is not true for any circle point.

a display screen. This is an example amongst others that shows how much a stratification algorithm of varieties can be useful.

Amongst other applications of implicit function theorem, we can mention two:

- To prove the existence of smooth curves passing through a point on a surface [223, p. 525].
- To state the smoothness conditions when an implicit surface and a parametric surface are stitched along an edge.

The first refers a theorem of major importance because it allows the study of smoothness of higher-dimensional submanifolds via, for example, Taylor or Frénet approximations. The second is also important because it makes it possible to avoid the conversion of an implicit surface patch to its parametric representation, or vice-versa. So, in principle, it is possible to design a smooth surface composed of parametric and implicit patches.

1.7 Submanifolds

By definition, a submanifold is a subset of a manifold that is a manifold in its own right. In geometric modelling, manifolds are usually Euclidean spaces, and submanifolds are points, curves, surfaces, etc. in some Euclidean space of equal or higher dimension. Manifolds and varieties in an Euclidean space are usually defined by either the image, level set or graph associated with a mapping.

1.7.1 Parametric Submanifolds

As shown in previous sections, the smoothness characterisation of a submanifold clearly depends on its defining smooth mapping and its rank. We have seen that the notion of smooth mapping of constant rank leads to the definition of *smooth* submanifolds. In this respect, the rank theorem, and ultimately, the inverse function theorem, can be considered as the major milestones in the theory of smooth submanifolds. Notably, the smoothness of a mapping does not ensure the smoothness of a submanifold. In fact, not all smooth submanifolds, say parametric smooth submanifolds, can be considered as topological submanifolds, i.e. submanifolds equipped with the submanifold topology.

Extreme cases of mappings $f : M \to N$ of constant rank are those corresponding to maximal rank, that is, the rank is the same as the dimension of M or N.

Definition 1.41. *Let $f : M \to N$ be a smooth mapping with constant rank. Then, for all $\mathbf{p} \in M$, f is called:*

$$\begin{array}{rl} an \textbf{ immersion} & if \quad rank\, f = dim\, M, \\ a \textbf{ submersion} & if \quad rank\, f = dim\, N. \end{array}$$

Let us now concentrate on immersions, that is, mappings whose images are parametric submanifolds. To say that $f : M \to N$ is an immersion means that the differential $\mathrm{D} f(\mathbf{p})$ is injective at every point $\mathbf{p} \in M$. This is the same as saying that the Jacobian matrix of f has rank equal to $\dim M$ (which is only possible if $\dim M \leq \dim N$). Then by the rank theorem, we have

Corollary 1.42. *Let M, N be two manifolds of dimensions m, n, respectively, and $f : M \to N$ a smooth mapping. The mapping f is an* immersion *if and only if for each point $\mathbf{p} \in M$ there are coordinate systems (U, φ), (V, ψ) about $\mathbf{p}$ and $f(\mathbf{p})$, respectively, such that the composite $\psi f \varphi^{-1}$ is a restriction of the coordinate inclusion $\iota : \mathbb{R}^m \to \mathbb{R}^m \times \mathbb{R}^{n-m}$.*

Proof. See Sharpe [360, p. 15]. ∎

This corollary provides the canonical form for immersed submanifolds:

$$(x_1, \ldots, x_m) \mapsto (x_1, \ldots, x_m, 0, \ldots, 0).$$

Definition 1.43. *A smooth (analytic) m-dimensional* **immersed submanifold** *of a manifold N is a subset $M' \subset N$ parametrised by a smooth (analytic), one-to-one mapping $f : M \to M' \subset N$, whose domain M, the parameter space, is a smooth (analytic) m-dimensional manifold, and such that f is everywhere regular, of maximal rank m.*

Thus, an m-dimensional immersed submanifold M' is the *image* of an immersion $f : M \to M' = f(M)$. To verify that f is an immersion it is necessary to check that the Jacobian has rank m at every point. Observe that an immersed submanifold is defined by a *parametrisation*. Thus, an immersed submanifold is nothing more than a *parametrically defined submanifold*, or simply a **parametric submanifold**. Despite its smoothness, an immersed or parametric submanifold may include self-intersections. A submanifold with self-intersections is the image $M' = f(M)$ of an arbitrary regular mapping $f : M \to M' \subset N$ of maximal rank m, which is the dimension of the parameter space M. Examples of parametric submanifolds with self-intersections such as Bézier curves and surfaces are often found in geometric design activities. Immersed submanifolds constitute the largest family of parametric submanifolds. It includes the subfamily of parametric submanifolds without self-intersections, also known as *parametric embedded submanifolds*.

Definition 1.44. *An* **embedding** *is a one-to-one immersion $f : M \to N$ such that the mapping $f : M \to f(M)$ is a homeomorphism (where the topology on $f(M)$ is the subspace topology inherited from N). The image of an embedding is called an* **embedded submanifold**.

In other words, the topological type is invariant for any point of an embedded submanifold. This is why embedded submanifolds are often called simply submanifolds. Obviously, $f : M \to N$ considered as a smooth mapping is called an embedding if $f(M) \subset N$ is a smooth manifold and $f : M \to f(M)$ is a diffeomorphism [65, p. 10].

Parametric immersed submanifolds have been mainly used in computer-aided geometric design (CAGD) of parametric curves and surfaces, while embedded submanifolds are preferably used as "building blocks" of solids in solid geometric modelling, which usually embody mechanical parts and other engineering artifacts. This means that an eventual computational integration of these two research areas of geometric modelling becomes mandatory to reconcile immersed and embedded submanifolds.

Let us see first some examples of 1-dimensional immersed submanifolds that are not embedded manifolds.

Example 1.45. Let $f : \mathbb{R} \to \mathbb{R}^2$ an immersion given by $f(t) = (\cos 2\pi t, \sin 2\pi t)$. Its image $f(\mathbb{R})$ is the unit circle $\mathbb{S}^1 = \{(x,y) \,|\, x^2 + y^2 = 1\}$ in $\mathbb{R}^2$. This shows that an immersion need not be one-to-one into (injective) in the large, even though it is one-to-one locally. In fact, for example, all the points $t = 0, \pm 1, \pm 2, \ldots$ have the same image point $(0,1)$ in $\mathbb{R}^2$. Moreover, the circle intersects itself for consecutive unit intervals in $\mathbb{R}$, even though its self-intersections are not "visually" apparent. Thus, this circle is an immersed submanifold, but not an embedded submanifold in $\mathbb{R}^2$. The same holds if we consider the immersion $f : [0,1] \to \mathbb{R}^2$ because $f(0) = f(1)$. But, if we take the immersion $f :]0,1[\to \mathbb{R}^2$, its image is an embedded manifold, that is, a unit circle minus one of its points.

Example 1.46. Let $f :]-\infty, 2[\to \mathbb{R}^2$ be an immersion given by $f(t) = (-t^3 + 3t + 2, t^2 - t - 2)$. Its image $f(]-\infty, 2[)$ is an immersed 6-shaped submanifold of dimension 1 (Figure 1.9(a)). Although f is injective (say, injective globally, and consequently injective locally), that is, without self-intersections, its image is not an embedded manifold. This is so because $]-\infty, 2[$ and its image $f(]-\infty, 2[)$ are not homeomorphic. In fact the point $(0,0)$ in $f(]-\infty, 2[)$ is a cut point of $f(]-\infty, 2[)$, and hence the local topological type of such a

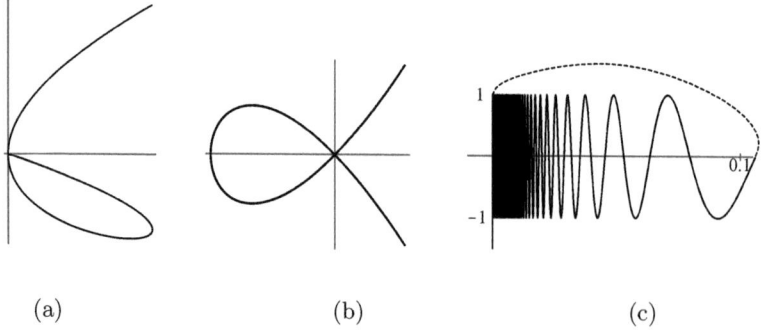

Fig. 1.9. Examples of immersed, but not embedded, submanifolds.

6-shaped submanifold is not constant. Note that the curve intersects itself at $t = -1$ and $t = 2$, but because $t = 2$ is not part of the domain, one says that the curve touches itself at the origin $(0,0)$.

Example 1.47. $f : \mathbb{R} \to \mathbb{R}^2$ defined by $f(t) = (t^2 - 1, t^3 - t)$ is an immersion (Figure 1.9(b)). It is not injective. However, it is injective when restricted to, say, the range $-1 < t < \infty$.

Example 1.48. A more striking example of a self-touching submanifold is given by the image of the mapping $f : \mathbb{R} \to \mathbb{R}^2$ so that

$$f(t) = \begin{cases} (\frac{1}{t}, \sin \pi t) & \text{for } 1 \leq t < \infty, \\ (0, t+2) & \text{for } -\infty < t \leq -1. \end{cases}$$

The result is a curve with a gap (Figure 1.9(c)). Let us connect the two pieces together smoothly by a dotted line as pictured in Figure 1.9(c). Then we get a smooth submanifold that results from the immersion of all of $\mathbb{R}$ in $\mathbb{R}^2$. This submanifold is not embedded because near $t = \infty$ the curve converges to the segment line $0 \times [-1, 1]$ in y-axis. In fact, while t converges to a point near ∞, its image converges to a line segment. Thus, the submanifold is not embedded because f is not a homeomorphism.

Embedded submanifolds are a subclass of immersed submanifolds that exclude self-intersecting submanifolds and self-touching submanifolds, that is, submanifolds that corrupt the local topological type invariance. Any other submanifold that keeps the same topological type everywhere in it is an embedded submanifold. Equivalently, a subset $f(M) \in N$ of a manifold N is called a smooth m-dimensional embedded submanifold if there is a covering $\{U_i\}$ of $f(M)$ by open sets (i.e. arbitrarily small neighbourhoods) of the ambient smooth manifold N such that the components of $U_i \cap f(M)$ are all connected open subsets of $f(M)$ of dimension m. Thus, there is no limitation on the number of components of an embedded submanifold in a chart of the ambient manifold; it may even be infinite [360, p. 19]. This means that, even with differential and topological singularities removed, a smooth embedded submanifold may be nonregular. Regular submanifolds intersect more neatly with coordinate charts of the ambient manifold; in particular, the family of components of this intersection do not pile up.

Definition 1.49. *An m-dimensional smooth submanifold $M \subset N$ is **regular** if, in addition to the regularity of the parametrising mapping, there is a covering $\{U_i\}$ of M by open sets of N such that, for each i, $U_i \cap M$ is a single open connected subset of M.*

By this definition, smooth regular submanifolds constitute a subclass of smooth embedded submanifolds. Let us see three counterexamples of regular submanifolds.

Example 1.50. Let $f:]1, \infty[\to \mathbb{R}^2$ be a mapping given by

$$f(t) = \left(\frac{1}{t} \cos 2\pi t, \frac{1}{t} \sin 2\pi t \right).$$

Its image (Figure 1.10(a)) in $\mathbb{R}^2$ is an embedded curve because the image of every point $t \in]1, \infty[$ is a point in $\mathbb{R}^2$; hence, f is a homeomorphism. Note that even near $t = \infty$, f is still a homeomorphism because its image is a point, the origin $(0,0)$. That is, a point and its image have the same dimension. (This is not true in Example 1.48.) However, the image of $]1, \infty[$ is not a regular curve because it spirals to $(0,0)$ as $t \to \infty$ and tends to $(1,0)$ as $t \to 1$, Figure 1.10(a). This happens because near (in a neighbourhood of) $t = \infty$ the relative neighbourhood in the image curve has several (possibly an infinite number of) components.

Example 1.51. Let us slightly change the previous mapping $f:]1, \infty[\to \mathbb{R}^2$ to be a mapping given by

$$f(t) = \left(\frac{t+1}{2t} \cos 2\pi t, \frac{t+1}{2t} \sin 2\pi t \right).$$

Its image (Figure 1.10(b)) in $\mathbb{R}^2$ is a nonregular embedded curve, now spiralling to the circle with centre at $(0,0)$ and radius $1/2$ as $t \to \infty$, Figure 1.10(b). It is quite straightforward to check that the Jacobian is always 1. In fact, it could be 0 if both derivatives of the component functions could vanish simultaneously on $]1, \infty[$; this would happen if and only if $\cos 2\pi t = -\tan 2\pi t$, an impossible equality.

Thus, every regular m-dimensional submanifold of an n-dimensional manifold locally looks like an m-dimensional subspace of $\mathbb{R}^n$. A trickier, but very important counterexample is as follows.

Example 1.52. Let us consider a torus $\mathbb{T}^2 = \mathbb{S}^1 \times \mathbb{S}^1$ with angular coordinates (θ, γ), $0 \leq \theta, \gamma < 2\pi$. The curve $f(t) = (t, kt) \bmod 2\pi$ is closed if k/t is a

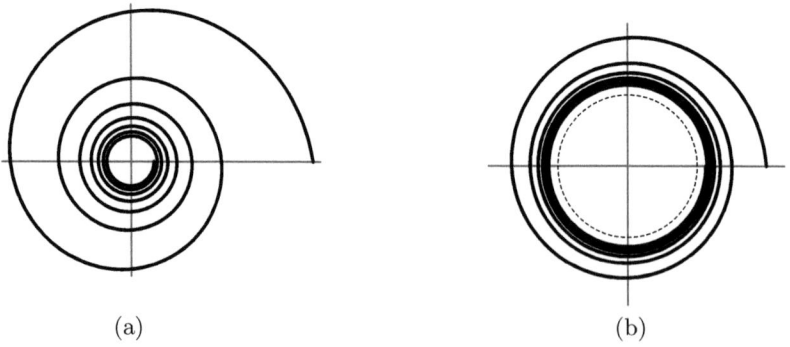

(a) (b)

Fig. 1.10. Counterexamples of regular submanifolds.

rational number, and hence a regular submanifold of $\mathbb{T}^2$, being $\mathbb{S}^1$ the parameter space. But, if k/t is irrational, the curve forms a dense subset of $\mathbb{T}^2$ and, consequently, is not a regular submanifold.

This example shows us that a regular submanifold such as a torus in $\mathbb{R}^3$ may include nonregular submanifolds. One should be careful to avoid irrational numbers in the representation and construction of submanifolds in a geometric kernel.

1.7.2 Implicit Submanifolds and Varieties

An alternative to the parametric approach for submanifolds is to define them *implicitly* as a common or intersecting level set of a collection of functions [313, p. 16]. We have seen this in Subsection 1.5.2, where the implicit mapping theorem was introduced. This theorem provides an immediate canonical form for regular manifolds as follows:

Theorem 1.53. (Olver [313, p. 14]) *A n-dimensional submanifold $N \subset \mathbb{R}^m$ is **regular** if and only if for each point $\mathbf{p} \in N$ there exist local coordinates $\mathbf{x} = (x_1, \ldots, x_m)$ defined on a neighbourhood U of $\mathbf{p}$ such that $U \cap N = \{\mathbf{x} : x_1 = \cdots = x_{m-n} = 0\}$.*

Therefore, every regular n-dimensional submanifold of an m-dimensional manifold locally looks like a n-dimensional subspace of $\mathbb{R}^m$. This means that all regular n-dimensional submanifolds are locally equivalent. They are the basic constituents of some space decompositions introduced in Chapter 2.

Let us now see how all this works for varieties. They are generalisations of implicit submanifolds, and thus they are defined by submersions. In general, the variety $V_{\mathcal{F}}$ determined by a family of real-valued functions $\mathcal{F}$ is defined by the subset where they simultaneously vanish, that is,

$$V_{\mathcal{F}} = \{\mathbf{x} \mid f_i(\mathbf{x}) = 0 \text{ for all } f_i \in \mathcal{F}\}.$$

In particular, when these functions $\{f_i\}$ are components of a mapping $f : \mathbb{R}^m \to \mathbb{R}^n$, the variety $V_f = \{f(\mathbf{x}) = 0\}$ is just the set of solutions to the simultaneous system of equations $f_1(\mathbf{x}) = \cdots = f_n(\mathbf{x}) = 0$.

It is clear that the notion of rank has a natural generalisation to (infinite) families of smooth functions.

Definition 1.54. *Let $\mathcal{F}$ be a family of smooth real-valued functions $f_i : M \to \mathbb{R}$, with $M, \mathbb{R}$ smooth manifolds. The **rank** of $\mathcal{F}$ at a point $\mathbf{p} \in M$ is the dimension of the space spanned by their differentials. The family is **regular** if its rank is constant on M.*

Definition 1.55. *A set $\{f_1, \ldots, f_k\}$ of smooth real-valued functions on a manifold M with a common domain of definition is called **functionally dependent** if, for each $\mathbf{p} \in M$, there is a neighbourhood U and a smooth* function $H(y_1, \ldots, y_k)$, not identically zero on any subset of $\mathbb{R}^k$, such that $H(f_1(\mathbf{x}), \ldots, f_k(\mathbf{x})) = 0$ for all $\mathbf{x} \in U$. The functions are called **functionally independent** if they are not functionally dependent when restricted to any open subset of M.

Example 1.56. The functions $f_1(x,y) = x/y$ and $f_2(x,y) = xy/(x^2+y^2)$ are functionally dependent on the upper half-plane $\{y > 0\}$ because the second can be written as a function of the first, $f_2 = f_1/(1+f_1^2)$.

Thus, for a regular family of functions, the rank gives us the number of functionally independent functions it contains. So, we obtain an implicit function family theorem generalising the implicit mapping theorem as follows.

Theorem 1.57. (Implicit Function Family Theorem) *If a family of functions $\mathcal{F}$ is regular of rank n, there exists n functionally independent functions $f_1, \ldots, f_n \in \mathcal{F}$ in the neighbourhood of any point, with the property that any other function $g \in \mathcal{F}$ can be expressed as a function thereof, $g = H(f_1, \ldots, f_n)$.*

Proof. See Olver [313, p.13].

Thus, if $f_1, \ldots, f_r$ is a set of functions whose $m \times r$ Jacobian matrix has maximal rank r at $\mathbf{p} \in M$, they also have, by continuity, the same rank r in a neighbourhood of $U \subset M$ of $\mathbf{p}$, and hence are functionally independent near $\mathbf{p}$. As expected, Theorem 1.57 also implies that, locally, there are at most m functionally independent functions on any m-dimensional manifold M.

Definition 1.58. *A variety (or system of equations) $V_\mathcal{F}$ is **regular** if it is not empty and the rank of $\mathcal{F}$ is constant.*

Clearly, the rank of $\mathcal{F}$ is constant if $\mathcal{F}$ itself is a regular family. In particular, regularity holds if the variety is defined by the vanishing of a mapping $f: N \to \mathbb{R}^r$ which has maximal rank r at each point $\mathbf{x} \in V_\mathcal{F}$, or equivalently, at each solution $\mathbf{x}$ to the system of equations $f(\mathbf{x}) = 0$ [313, p. 16]. The implicit function family theorem 1.57, together with Theorem 1.53, shows that a regular variety is a regular submanifold, as stated by the following theorem.

Theorem 1.59. *Let $\mathcal{F}$ be a family of functions defined on an m-dimensional manifold M. If the associated variety $V_\mathcal{F} \subset M$ is regular, it defines a regular submanifold of dimension $m - r$.*

Proof. See Olver [313, p. 17].

As for parametric submanifolds, to say that an implicit submanifold is regular means that it is smooth. However, a smooth parametric submanifold is not necessarily regular. But, for implicit submanifolds, regularity and smoothness coincide. This is so because, unlike a parametric submanifold, regularity of an implicit submanifold is completely determined by the regularity of its defining family of functions.

Thus, Theorem 1.59 gives us a simple criterion for the smoothness of a submanifold described implicitly.

1.7 Submanifolds   37

Example 1.60. Let $f : \mathbb{R}^3 \to \mathbb{R}$ be a function given by $f(x, y, z) = x^2 + y^2 + z^2 - 1$. Its Jacobian matrix $[2x \ \ 2y \ \ 2z]$ has rank 1 everywhere except at the origin, and hence its variety (the unit sphere) is a regular 2-dimensional submanifold of $\mathbb{R}^3$.

Example 1.61. The function $f : \mathbb{R}^3 \to \mathbb{R}$ given by $f(x, y, z) = xyz$ is not regular, and its variety (the union of the three coordinate planes) is not a submanifold.

The fact that regularity and smoothness coincide for implicit submanifolds suggests that we may have an algorithm to determine singularities on a variety via the Jacobian matrix. Let us define regular points and singular points before providing some examples that illustrate the computation of such singularities.

Definition 1.62. *Let $f : U \subset \mathbb{R}^m \to \mathbb{R}^r$ be a smooth mapping. A point $\mathbf{p} \in \mathbb{R}^m$ is a **regular point** of f, and f is called a submersion at $\mathbf{p}$, if the differential $Df(\mathbf{p})$ is surjective. This is the same as saying that the Jacobian matrix of f at $\mathbf{p}$ has rank r (which is only possible if $r \leq m$). A point $\mathbf{q} \in \mathbb{R}^r$ is a **regular value** of f if every point of $f^{-1}(\mathbf{q})$ is regular.*

Instead of 'nonregular' we can also say *singular* or *critical*. In general, we have:

Definition 1.63. *Let $f : U \subset \mathbb{R}^m \to \mathbb{R}^r$ be a smooth mapping. A point $\mathbf{p} \in \mathbb{R}^m$ is a **singular point** of f if the rank of its Jacobian matrix falls below its largest possible value $\min(m, r)$. Likewise, a **singular value** is any $f(\mathbf{p}) \in \mathbb{R}^r$ where $\mathbf{p}$ is a singular point.*

Recall that a singular point of an immersion determines a singular point in a parametric submanifold, but its self-intersections are not determined by the singular points of its associated function. This happens because the regularity of an immersion at a given point is necessary but not sufficient to guarantee the regularity of its image. But, for implicit submanifolds and varieties, the regularity of functions is necessary and sufficient to ensure their regularity.

Example 1.64. Let $f : \mathbb{R} \to \mathbb{R}$ given by $f(x) = x^2$. Then any $c \neq 0$ is a regular value of f. Its Jacobian $[2x]$ has rank 1 iff $x \neq 0$; hence $x = 0$ is the only singular point of f. This corresponds to the minimum point of the graph of f (i.e. the vertex of a parabola), but here we are concerned with implicit submanifolds that are defined by level sets, not graphs.

Example 1.65. Let $f : \mathbb{R}^2 \to \mathbb{R}$ given by $f(x, y) = 2x^2 + 3y^2$. Its Jacobian $[4x \ \ 6y]$ has rank 1 unless $x = y = 0$. So any $c \neq 0$ is a regular value of f. For $c > 0$, $f^{-1}(c)$ is an ellipse in the plane.

Example 1.66. Let $f : \mathbb{R}^2 \to \mathbb{R}$ given by $f(x, y) = x^3 + y^3 - xy$. The maximal possible rank for its Jacobian $[3x^2 - y \ \ 3y^2 - x]$ is 1, and we can find all

points where this fails, i.e. all singular points, by solving the system $\partial f/\partial x = \partial f/\partial y = 0$, that is,
$$\begin{cases} 3x^2 - y = 0 \\ 3y^2 - x = 0. \end{cases}$$

This yields the points $(0,0)$ and $(\frac{1}{3}, \frac{1}{3})$ as the only singular points of f. Since $f(0,0) = 0$ and $f(\frac{1}{3}, \frac{1}{3}) = -\frac{1}{27}$ it follows that any c other than 0 or $-\frac{1}{27}$ is a regular value of f. Also, 0 is a regular value of restrictions $f|(\mathbb{R}^2 - \{(0,0)\})$ and $-\frac{1}{27}$ is a regular value of $f|(\mathbb{R}^2 - \{(\frac{1}{3}, \frac{1}{3})\})$. This is because the singular points $(0,0)$, $(\frac{1}{3}, \frac{1}{3})$ do not belong to the domain of the restrictions of f, say $f|(\mathbb{R}^2 - \{(0,0)\})$, $f|(\mathbb{R}^2 - \{(\frac{1}{3}, \frac{1}{3})\})$, respectively.

Figure 1.11 illustrates $f^{-1}(c)$ for some values of c. For $c = 0$ we have the well-known folium of Descartes (Figure 1.11(a)). The folium of Descartes is the variety $x^3 + y^3 - xy = 0$ which self-intersects at the singular point $(0,0)$, i.e. the level set defined by $f(x,y) = 0$. The level set defined by $f(x,y) = -\frac{1}{27}$ is the variety $x^3 + y^3 - xy = -\frac{1}{27}$ (Figure 1.11(c)) whose singular point is the isolated point $(\frac{1}{3}, \frac{1}{3})$. For $c = -\frac{1}{54}$, we have the regular variety $x^3 + y^3 - xy = -\frac{1}{54}$ (Figure 1.11(b)).

Example 1.67. Let $f : \mathbb{R}^3 \to \mathbb{R}$ be given by $f(x,y,z) = x^2 - zy^2$. The associated variety has dimension $m - r = 3 - 1 = 2$, but the maximal possible rank of its Jacobian $[2x \ -2zy \ -y^2]$ is 1. Its singular points are the solutions of the following system of equations:
$$\begin{cases} 2x = 0 \\ -2zy = 0 \\ -y^2 = 0 \end{cases} \iff \begin{cases} x = 0 \\ zy = 0 \\ y = 0 \end{cases}.$$

The expressions $x = 0$ and $y = 0$ denote the two coordinate planes in $\mathbb{R}^3$, whose intersection is the z-axis. That is, the Jacobian vanishes along the

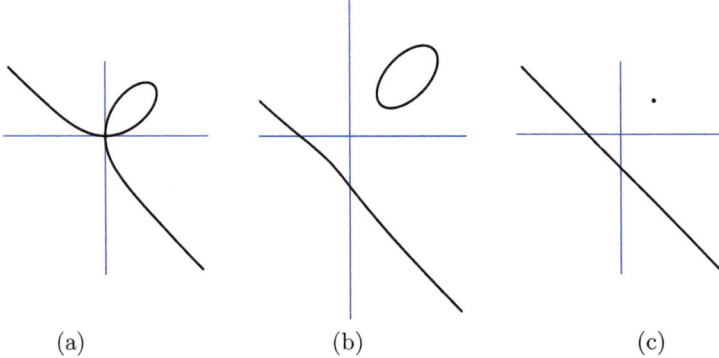

(a)   (b)   (c)

Fig. 1.11. Varieties as level sets $x^3 + y^3 - xy = c$.

z-axis, or, equivalently, Each point in the z-axis is a singular point. Since $f(0,0,z) = 0$ it follows that any c other than 0 is a regular value of f. Also, 0 is a regular value of $f|(\mathbb{R}^3 - \{(0,0,z)\})$. Figure 1.12(a) illustrates $f^{-1}(0)$, the Whitney umbrella with-handle (already seen in Figure 1.4(a)).

Example 1.68. Let $f : \mathbb{R}^3 \to \mathbb{R}$ be given by $f(x,y,z) = y^2 - z^2x^2 + x^3$. As for the previous example, the Jacobian $(-2z^2x + 3x^2 \quad 2y \quad -2zx^2)$ vanishes precisely on the z-axis. The z-axis is the line of "double points" where the surface intersects itself at $c = 0$. This surface is depicted in Figure 1.12(b).

Example 1.69. Let $f : \mathbb{R}^3 \to \mathbb{R}^2$ be the mapping given by $f(x,y,z) = (xy, xz)$. The Jacobian of f is
$$\begin{pmatrix} y & x & 0 \\ z & 0 & x \end{pmatrix}$$
which has rank 2 unless all 2×2 minors are zero, i.e. unless $xz = xy = x^2 = 0$, which is equivalent to $x = 0$. Since $f(0,y,z) = (0,0)$, any point of $\mathbb{R}^2$ other than $(0,0)$ is a regular value. This variety (the union of the x-axis and the plane $x = 0$) has dimension 2 and is the intersection of two 2-dimensional varieties defined by the levels sets of the components functions of f. The first level set is the union of the planes $x = 0$ and $y = 0$, while the second level set is the union of the the planes $x = 0$ and $z - 0$ in $\mathbb{R}^3$.

In short, the implicit function theorem and its generalisations allow us to determine the singular set of an implicit variety. In the particular case of an implicit surface $f(x,y,z) = 0$, the singular set is a 0- or 1-dimensional set at which all the partial derivatives simultaneously vanish. Therefore, in essence, a k-dimensional smooth (or differentiable) submanifold can be approximated by a k-dimensional subspace of $\mathbb{R}^n$ at each of its points. In particular, this the same as saying that a smooth curve in $\mathbb{R}^2$ can be approximated by a tangent line at each one of its points, a smooth surface by its tangent plane, etc. It is

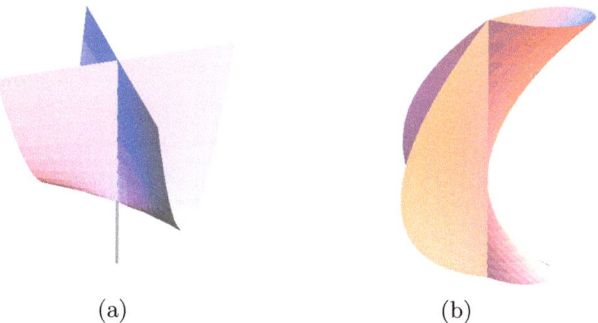

(a) (b)

Fig. 1.12. (a) Whitney umbrella with-handle as a level set $x^2 - zy^2 = 0$; (b) the surface $y^2 - z^2x^2 + x^3 = 0$.

clear that such an approximation is not possible at (differential) singularities; for example, a tangent plane flips at any corner and along any edge of the surface of cube.

1.8 Final Remarks

In this chapter, we have seen that manifolds can be either smooth or nonsmooth. Nonsmooth manifolds are in principle piecewise smooth manifolds. This leads us to the idea of partitioning a n-dimensional manifold into smooth k-dimensional submanifolds ($k \leq n$). The family of smooth submanifolds of dimension less than n are singularities of such a n-dimensional manifold. This simple idea is based on the pioneering work of two mathematicians, Whitney and Thom, nowadays known as Thom-Whitney stratification theory. They shows us that there is a close relationship between the concepts of differentiability and stratificability of manifolds. Notably, both concepts are related even when they are applied to more general geometric point sets such as algebraic, analytic or even semianalytic varieties.

The essential key for having a smooth manifold is the concept of diffeomorphism, that is, a differentiable mapping with a differentiable inverse. The differentiability of a mapping is not enough to guarantee the smoothness of a manifold; its inverse must be also differentiable. As noted in [132, p. 106], smoothness and differentiability do not agree. Smoothness means that the mapping which defines a submanifold is a diffeomorphism.

Only a diffeomorphism (i.e. a smooth mapping with smooth inverse) ensures the smoothness of a parametric curve or surface. Thus, the smoothness of a submanifold depends more on the properties of the mapping used to define it than on its associated geometric invariants (e.g. curvature and torsion). The use of a geometric invariant may be not conclusive to ensure smoothness on a submanifold, as a topological invariant (e.g. Betti numbers) is not sufficient to characterise the continuity of a subspace.

The relationship between the invertibility and smoothness of a mapping has led us to its algebraic counterpart, that is, the relationship between the invertibility of the Jacobian and smoothness of a submanifold. We have shown that this relationship is independent of whether we treat submanifolds as level sets, images, or graphs of mappings, i.e. it is representation-independent. So, we have shown that C^1 smoothness can be determined by the rank-based criterion. This suggests that we can determine the singularities of a submanifold by observing where the rank is not constant.

2
Spatial Data Structures

This chapter presents an overview of several spatial decomposition techniques, as well as their associated data structures. We assume that the reader is familiar with some basic concepts of set theory, topology and geometry.

Spatial decompositions apply to both ambient spaces and their subspaces. In this textbook, we will focus on particular subspaces, say implicit curves and surfaces. Spatial decompositions of these subspaces are here called *object decompositions*. For example, the resolution of singularities of a level set (e.g. implicit surface in $\mathbb{R}^3$) gives rise to its decomposition into manifolds. These object decompositions are particularly useful for rendering implicit curves and surfaces through continuation algorithms (see Chapter 6 for further details).

Decompositions that cover all the ambient space (e.g. a bounding box or even the whole $\mathbb{R}^n$) containing an embedded object are called *space decompositions*. These decompositions are used by space-partitioning algorithms for implicit objects (see Chapter 7 for more details).

2.1 Preliminary Notions

Let X be a topological space.[1] There are many ways to decompose X. Such decompositions can be grouped together in two families: *coverings* and *partitions*. A covering of X is a collection $\{X_i\}$ of subsets of X such that $\bigcup X_i = X$.

---

[1] A topology $\mathcal{T}$ is a collection of sets U_i that satisfies the following two axioms: (i) the union of any (may be infinite) number of those sets also belongs to such a collection; (ii) the intersection of a finite number of those sets also belongs to such a collection. That is, a topology is closed in respect to the union and intersection of its sets. Furthermore, from the axiom (i) we can say that the set $X = \bigcup \{U_i \in \mathcal{T}\}$ is necessarily in $\mathcal{T}$ because $\mathcal{T}$ is a subcollection of itself, and every set U_i of $\mathcal{T}$ is a subset of X. The set X is called the *space* of the topology $\mathcal{T}$ and $\mathcal{T}$ is a topology for X. Besides, the pair $(X, \mathcal{T})$ is a *topological space* if an additional axiom is satisfied: (iii) $\varnothing, X \in \mathcal{T}$. Thus, $\mathcal{T}$ is a topology on X iff $(X, \mathcal{T})$ is a topological space.

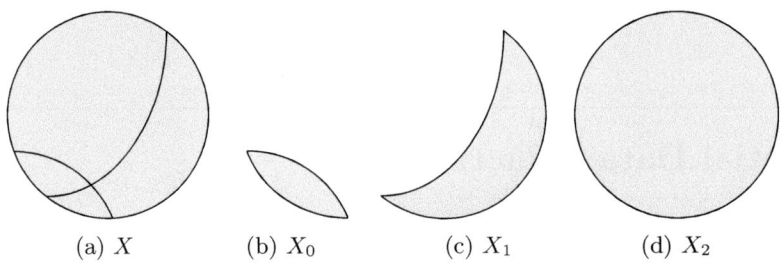

Fig. 2.1. A covering of a point set X.

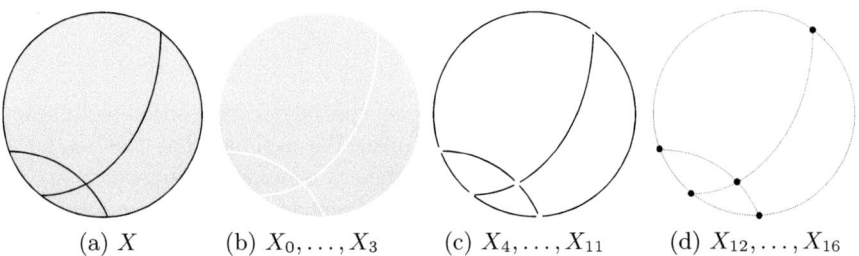

Fig. 2.2. A partition of a point set X into faces, edges and vertices.

If a subset of a covering of X still covers X, it is said to be a *subcovering*. This is illustrated in Figure 2.1, where $X = X_0 \cup X_1 \cup X_2$, with X_2 the only subset that covers X totally.

But, if the subsets X_i are all disjoint, we say that such a space decomposition is a *partition*, i.e. $X_i \cap X_j = \varnothing$ for any $i \neq j$. For example, in Figure 2.2, we have the partition $X = \bigcup_{i=0}^{16} X_i$, where $X_0, \ldots, X_3$ are faces, $X_4, \ldots, X_{11}$ are edges, and $X_{12}, \ldots, X_{16}$ are vertices.

This chapter focuses mainly on partitions. There are two main families of polygonisation algorithms for implicit curves and surfaces: continuation algorithms, and space-partitioning algorithms. Both of these are based on partitioning methods, in that both create a partitioned implicit object after finding a finite collection of its points. They differ from each other through the manner in which sampling points are found.

Continuation algorithms do not require the partitioning of the ambient space. It is the implicit object itself (e.g. a curve or surface) that is sampled directly, partitioned and approximated by a polyline or a triangular mesh, respectively. Thus, continuation algorithms use *object partitionings* (see Chapter 6 for further details).

By contrast, in order to sample an implicit object, space decomposition algorithms do partition the ambient space. This partitioning of the ambient space into convex cells allows us to sample an implicit object against the edges of those cells. Therefore, we can say that the resulting polyline that approximates a curve, or mesh that approximates a surface, is obtained indirectly from the partitioning of the ambient space. That is, space decomposition algorithms for implicit curves and surfaces use *space partitionings* (see Chapter 7 for more details).

2.2 Object Partitionings

2.2.1 Stratifications

Stratifications have been extensively studied in mathematics mainly since the 1970s just to pave the way for the resolution of singularities of algebraic, semi-algebraic, semi-analytic, and sub-analytic varieties (see, for example, Whitney [412, 413], Thom [385, 386], Lojasiewicz [241, 242] and Hironaka [188, 189]). Middleditch et al. [271, 162] introduced them in geometric modelling in the late 90's, in part by influence of the development of the Djinn project [21].

A *stratification* is a partition of a subset of $\mathbb{R}^n$ into manifolds (called strata), thus providing a structure for point sets, regardless of whether they are manifold or not [272]. Such a subset can be stratified in many ways. For example, the partition of the manifold X in Figure 2.2 is a stratification because all resulting subsets X_i are manifolds, i.e. they are all locally homeomorphic to $\mathbb{R}^k$, with $k = 0, 1, 2$. A counterexample appears in Figure 2.3(a), where the partitioned set $X = \bigcup_{i=0}^{1} X_i$ in (a) is not a stratification because X_0 is not a manifold (it self-intersects). But, the sets $X = \bigcup_{i=0}^{2} X_i$ in (b) and $X = \bigcup_{i=0}^{4} X_i$ in (c) are both stratifications of the same point set.

Figure 2.3 shows that there are many ways of partitioning and stratifying point sets. Evidently, not all are of interest in geometric modelling. For example, in solid modelling, boundary representations (B-Reps) represent geometric objects which are Whitney stratifications. These stratifications are

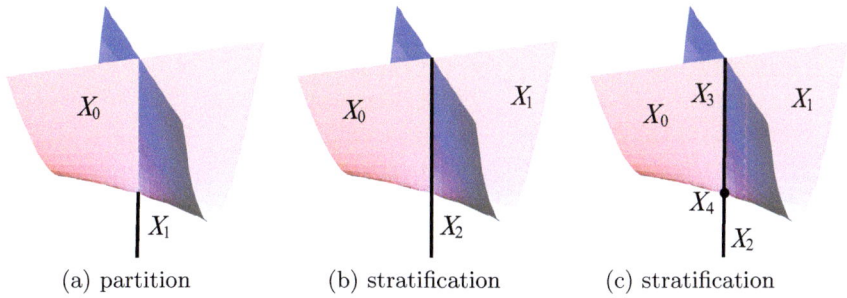

Fig. 2.3. (a) A partition that is not a stratification; (b) a stratification; (c) another stratification of the same point set.

occasionally called "cell complexes" in geometric modelling literature, but the term is inadequate because of the essential difference between local and global topological properties, as described in this section.

Whitney stratifications enjoy two main properties: (i) local finiteness; (ii) local topological invariance. Local finiteness means that in the neighbourhood of each stratum point there are only a finite number of other strata. Local topological invariance means that the topological type of the neighbourhood of any point of a stratum is the same for every point of such a stratum. Neighbourhoods with the same topological type is a way of saying that they are topologically equivalent. For example, both stratifications (b) and (c) in Figure 2.3 have a finite number of strata, so they are globally, and consequently locally, finite; but, only (c) is a Whitney stratification.

In fact, not all points of the stratum X_2 belonging to the stratification (b) have topologically equivalent neighbourhoods. Intuitively, this is so because the top part of X_2 bounds simultaneously X_0 and X_1, while its bottom part bounds no strata; there are three topological types along the z-axis: one along the positive subaxis, one along the negative one, and one around the origin. For a more comprehensive study on stratifications, the reader is referred to Shiota [365], Middleditch et al. [272], and Gomes [161].

Whitney stratifications: a data structure. With the advent of solid modelling in the 1970s, and until the end of 1990s, general data structures for representing geometric objects were proposed in the literature in order to cope with the requirements imposed by computer-aided design and manufacturing (CAD/CAM) systems and applications. The "building blocks" of CAD geometric kernels were and still are manifolds or strata, while those traditionally used in computational geometry, and now widely used in computer graphics, are cells and simplexes.

Let us then outline a general dimension-independent topological data structure for *finite* Whitney stratifications:

```
typedef map<int, vector<Stratum*>*> Skeleton;

class Stratum {
  int id;                   // stratum id
  int dim;                  // stratum dimension
  vector<Stratum*> *as      // adjacent strata
  vector<Stratum*> *is;     // incident strata
  geometry *g;              // geometry
}

class Object {
  int id;                   // object id
  Skeleton *sk;             // map of n-sleketa
}
```

The containers `vector` and `map` are appliances provided by the STL (standard template library) of C++. Note that this data structure is dynamically dimension independent, in that when we need to add an n-stratum to it but the corresponding object n-skeleton does not exist yet, we have only to create a new entry in `sk` that maps the dimension n onto a vector of n-dimensional strata, say the n-skeleton. By definition, the *n-skeleton* of an object is the set of n- and lower-dimensional strata.

This data structure reinforces both the local finiteness condition and the frontier condition by representing both the finite set `is` of incident $(n+1)$-dimensional strata and the finite set `as` of adjacent $(n-1)$-dimensional strata for each n-dimensional stratum. Recall that the frontier condition states that the frontier of each cell is given by the union of a subset of the lower-dimensional cells.

Finally, despite its simplicity, this data structure is prepared to host manifold and nonmanifold geometric objects partitioned into strata, as required in geometric and solid modelling. Nevertheless, a more comprehensive description of this data structure appears in [161], where it is called DiX data structure.

2.2.2 Cell Decompositions

A *cell decomposition* can be defined as a partition of the space into cells. By definition, an n-dimensional *cell* is homeomorphic to $\mathbb{R}^n$. For example, the subsets $X_0, \ldots, X_{16}$ of X depicted in Figure 2.2 are all cells. They do not need be convex. But, they need to be simply connected, i.e. without homotopic holes.

A counterexample is given by the 1-sphere $\mathbb{S}^1 = \{\mathbf{p} \in \mathbb{R}^2 : x_\mathbf{p}^2 + y_\mathbf{p}^2 = 1\}$. In fact, $\mathbb{S}^1$ admits a Whitney stratification consisting of a single 1-stratum X_0 (Figure 2.4(a)), but its cell decomposition requires at least two cells, i.e. a 0-cell X_0 and a 1-cell X_1 (Figure 2.4(b)).

Another counterexample is given by an annulus without bounding vertices (Figure 2.4(c)). As known, an annulus possesses two 1-dimensional boundaries. Its simplest Whitney stratification consists of a 2-stratum X_2 (i.e. a face with

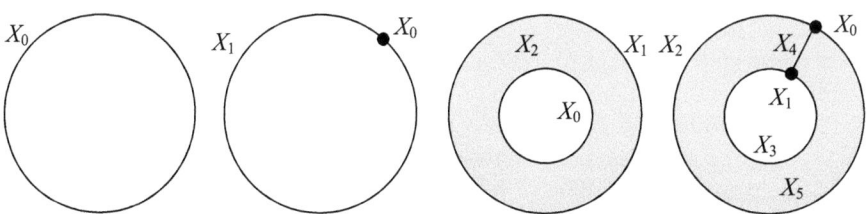

(a) stratification   (b) cell decomposition   (c) stratification   (d) cell decomposition

Fig. 2.4. Four stratifications, two of which are also cell decompositions.

a through hole) bounded by two 1-strata X_0 and X_1, and no 0-strata. The corresponding cell decomposition requires at least two 0-strata, X_0 and X_1, bounding distinct 1-strata, X_2 and X_3, respectively, with a dummy 1-stratum X_4 connecting those two 0-strata; all these 0- and 1-strata form the frontier of a 2-stratum X_5 (Figure 2.4(d)).

Thus, unlike strata which may possess zero or more boundaries, a cell is a manifold with exactly one nonempty boundary. As noted by Middleditch et al., it is no coincidence that classical boundary representations of solid objects use artificial vertices and edges as a way to facilitate their cellular partitioning; for example, this was important for the pioneer boundary representations to guarantee that the resulting complexes would satisfy the Euler-Poincaré formula.

A particular cell decomposition is the cell complex. A *cell complex* is a collection of cells together with their boundaries, as well as further information describing how the cells fit together. Like a Whitney stratification, a cell complex also satisfies the frontier condition. However, a cell complex may possess an infinite number of cells, so that mathematicians often use a more restricted cell complex, called CW complex [410, 411]. A *CW complex* satisfies two important conditions:

- *Closure finiteness.* This is the C condition, i.e. the frontier of each cell is the *finite* union of lower-dimensional cells.
- *Weak topology.* This the W condition, i.e. the closed subsets are exactly those sets that have a closed intersection with the closure of each cell.

The C condition is equivalent to say that the closure of each cell is contained in a finite subcomplex. It imposes a finiteness restriction on the frontier condition, i.e. the union of lower-dimensional cells bounding a given cell must be finite. It is worth nothing that closure finiteness is not the same as local finiteness. A complex is locally finite if each of its points is in a finite number of cell closures. Closure finiteness neither implies nor is implied by local finiteness, but both conditions are satisfied when a complex has a finite number of cells. Thus, finite cell complexes are inherently CW-complexes that are also locally finite. For example, the decomposition of the 2-disk $\mathbb{D}^2 = \{\mathbf{p} \in \mathbb{R}^2 : x_\mathbf{p}^2 + y_\mathbf{p}^2 \leq 1\}$ into a 2-cell (its interior $\mathbb{B}^2 = \{\mathbf{p} \in \mathbb{R}^2 : x_\mathbf{p}^2 + y_\mathbf{p}^2 < 1\}$) and an infinite number of 0-cells bounding it (its frontier $\mathbb{S}^1 = \{\mathbf{p} \in \mathbb{R}^2 : x_\mathbf{p}^2 + y_\mathbf{p}^2 = 1\}$) is not a CW complex because, despite its local finiteness, it does not satisfy the closure finiteness condition.

The W condition defines a unique topology on the cell complex called the W-topology or weak topology. This condition also imposes a finiteness restriction, but now on the number of sets of the topology covering the complex. In fact, a weaker topology is one that has fewer closed sets. The weak topology on a cell complex is the smallest collection of subsets such that the intersection with each cell closure is closed within such a cell closure. Therefore, the weak topology $\mathcal{W} = \{W_i\}_{i=0,\ldots,n}$ consists of the following sets:

(i) the sets formed from the closure of each cell;
(ii) the sets resulting from the union of any number of sets in $\mathcal{W}$.
(iii) the whole set X of cells;
(iv) the empty set $\varnothing$ of cells;

For example, the cell complex shown in Figure 2.4(d) is a CW complex. It satisfies the closure finiteness because the closure of any cell has a finite number of cells. Its weak topology consists of the following collection of sets:

(i) *Closures of cells.* These sets of cells are $W_0 = \overline{X_0} = \{X_0\}$, $W_1 = \overline{X_1} = \{X_1\}$, $W_2 = \overline{X_2} = \{X_0, X_2\}$, $W_3 = \overline{X_3} = \{X_1, X_3\}$, $W_4 = \overline{X_4} = \{X_0, X_1, X_4\}$, and $W_5 = \overline{X_5} = \{X_0, X_1, X_2, X_3, X_4, X_5\}$.
(ii) *Unions of sets.* By combining the previous sets through the set-theoretic union, we obtain the following sets for the weak topology: $W_6 = \overline{X_0} \cup \overline{X_1} = \{X_0, X_1\}$, $W_7 = \overline{X_0} \cup \overline{X_3} = \{X_0, X_1, X_3\}$, $W_8 = \overline{X_1} \cup \overline{X_2} = \{X_0, X_1, X_2\}$, $W_9 = \overline{X_2} \cup \overline{X_3} = \{X_0, X_1, X_2, X_3\}$, $W_{10} = \overline{X_2} \cup \overline{X_4} = \{X_0, X_1, X_2, X_4\}$, and $W_{11} = \overline{X_3} \cup \overline{X_4} = \{X_0, X_1, X_3, X_4\}$. Note that the remaining unions are already in the weak topology. For example, the union $\overline{X_0} \cup \overline{X_2} = \{X_0, X_2\}$ is precisely the set W_2. Recall that, according to set theory, there are no repeated elements in a set.
(iii) *The whole set of cells.* In this particular CW complex, the whole set X of cells is just the set W_5.
(iv) *The empty set of cells.* The empty set $\varnothing$ of cells is also part of this CW topology. It is necessary to guarantee the closeness of the set-theoretic intersection in the CW topology.

In fact, as for any topology, the intersection between two subsets of a weak topology is always one of its subsets. For the example above, it is easy to see that the intersection of any two subsets of the weak topology $\mathcal{W}$ is also a subset of $\mathcal{W}$. The same is true for the union of any two subsets of $\mathcal{W}$. In short, both the closure-finite and weak-topology conditions are satisfied for the closed cells of that CW complex.

A CW-complex can be built up by attaching cells of increasing dimensions. Informally speaking, attaching an n-cell to a CW-complex is carried out by identifying the boundary of the cell with the finite union of a subset of $(n-1)$-cells in the complex. Therefore, by using this attachment rule, and starting off with the empty set, a CW-complex X can be inductively constructed out by gluing the 0-cells, 1-cells, 2-cells, and so forth; this originates a filtration $X^{(-1)} \subseteq X^{(0)} \subseteq X^{(1)} \subseteq X^{(2)} \subseteq \cdots$ of X such that $X = \bigcup_{i \geq -1} X^{(i)}$, with $X^{(-1)} = \varnothing$. The set $X^{(i)}$ obtained from $X^{(i-1)}$ by attaching the collection of i-cells is nothing more than the i-skeleton of X. Note that this definition of CW complex does not allow us to attach i-cells before $(i-1)$- and lower-dimensional cells. Although some authors allow this (i.e. relaxation of the attachment order), it seems to be common practice to restrict CW complexes to the definition given above, and to call a space built up by attaching cells with unrestricted order of dimensions a cell complex. Apart from these subtleties, we can say that CW complexes are *finite* cell complexes, which match the memory storage limitations of modern computers. A more comprehensive study on cell and CW complexes can be found in Lundell and Weingram [250]. CW complexes were introduced in mathematics by Whitehead [410, 411].

Finite cell complexes: data structures. In computational geometry there are data structures for finite cell complexes and CW complexes. The *cell-tuple data structure* due to Brisson [64] is a well-known data structure to represent finite cell complexes. It consists of a set of tuples of incident cells of increasing dimension: (vertices, edges, faces). For example, the cell complex depicted in Figure 2.4(d) can be represented by the following set of four cell-tuples:

$$(X_0, X_2, X_5)$$
$$(X_0, X_4, X_5)$$
$$(X_1, X_3, X_5)$$
$$(X_1, X_4, X_5)$$

To understand better the incidence scheme underlying this data structure, let us consider the first tuple (X_0, X_2, X_5). This tuple tell us that the edge X_2 is incident at the vertex X_0, and the face X_5 is incident on the edge X_2; conversely, X_0 is adjacent to X_2, which in turn is adjacent to X_5. As for stratifications, the term "is adjacent" means "bounds" or "is in the boundary of." Thus, the incidence and adjacency relations between cells are defined by the order of such cells in each tuple.

Note that the data structure described above for finite Whitney stratifications may also be used to represent finite cell complexes because every n-cell is an n-stratum (but not vice-versa). For 2-dimensional cellular objects in $\mathbb{R}^3$, such a data structure would be as follows:

```
class Vertex {
  int id;                   // vertex id
  vector<Edge*> *ie;        // incident edges
  Point *p;                 // geometry
}

class Edge {
  int id;                   // edge id
  Vertex *v1, *v2;          // adjacent vertices
  vector<Face*> *if;        // incident faces
}

class Face {
  int id;                   // face id
  vector<Edge*> *ae;        // adjacent edges
  Point *nf;                // face normal
}
```

```
class Object {
  int id;                  // object id
  vector<Vextex*> *vv;     // vector of vertices
  vector<Edge*> *ev;       // vector of edges
  vector<Face*> *fv;       // vector of faces
}
```

This data structure was proposed by Silva and Gomes in [369], and called AIF (adjacency and incidence framework) data structure. Unlike traditional B-rep data structures, it does not include any topologically oriented cells (e.g. half-edges of the half-edge data structure) [256]. Nevertheless, the AIF data structure is geometrically oriented because every face includes a normal vector nf as appears defined in the class Face. The consistent orientation of such normal vectors on the object surface is acquired through an inducing mechanism similar to the one described in [369] and [59]. Such an inducing mechanism requires to traverse the frontier of each face in the same manner, i.e. either clockwise or counterclockwise. From this induced topological orientation for all faces, we are able to generate a geometric orientation (i.e. a normal vector) for each face of the object.

2.2.3 Simplicial Decompositions

In a space of dimension at least n, an *n-simplex* (plural simplexes or simplices) is an n-dimensional manifold with boundary whose interior is topologically equivalent to $\mathbb{R}^n$, i.e. a n-cell. In geometric terms, a n-simplex is the convex hull of a set of (n + 1) affinely independent points in some n- or higher-dimensional Euclidean space. Therefore, a n-simplex is a linear, convex, closed n-cell; some examples in $\mathbb{R}^3$ are depicted in Figure 2.5.

A *simplicial complex* is a space decomposed into a collection of simplices, sometimes also called a *triangulation*. Simplicial complexes are a particular case of CW complexes in that closures of cells are simplices (or simplexes). In more formal terms, a simplicial complex K in $\mathbb{R}^n$ is a collection of simplices that satisfy the following conditions [291]:

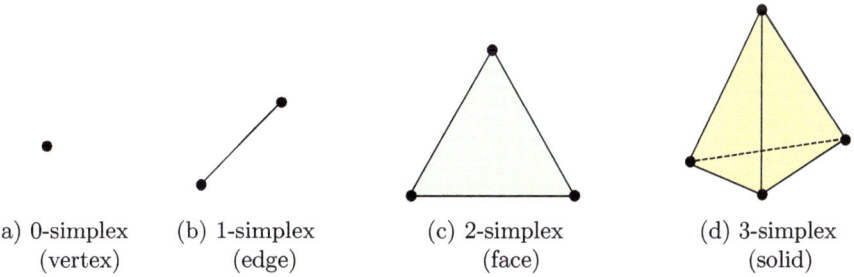

(a) 0-simplex (vertex)   (b) 1-simplex (edge)   (c) 2-simplex (face)   (d) 3-simplex (solid)

Fig. 2.5. Simplices in $\mathbb{R}^3$.

(i) Every frontier simplex of a simplex of K is in K, and
(ii) The intersection of any two simplices of K is a frontier simplex of each of them.

Alternatively, a n-simplicial decomposition can be viewed as a union of k-simplices ($0 < k \leq n$) that is closed under intersection, and such that the only time that one simplex is contained in another is as a boundary simplex. This is a *constructive* view of looking at simplicial complexes, i.e. simplices can be used as building blocks to construct simplicial complexes by gluing simplices together through their boundary simplices (or their intersecting simplices).

Simplicial complexes: data structures. In the computational geometry literature, we easily find data structures for simplicial complexes (and cell complexes) because Delaunay triangulations have been often applied to explore and to study the properties and subtleties of subspaces in $\mathbb{R}^n$. But, triangulations have also become very popular in geometric modelling and computer graphics in last decade, mainly due to the emergence of multiresolution meshes and compression techniques as a consequence of the Internet and Web revolution. Besides, graphics cards are optimised to process triangles rapidly.

The incidence graph data structure [126] is a well-known dimension-independent data structure for simplicial complexes. This data structure can be viewed as a particular case of those described above for Whitney stratifications and cell complexes for two reasons:

- It also hosts the boundary and co-boundary of each d-simplex. The boundary of a d-simplex is roughly the set of $(d-1)$-simplexes which are adjacent to it, while its co-boundary is given by the set of $(d+1)$-simplexes which are incident on it.
- A d-simplex, roughly speaking, is a particular case of a d-cell.

A very compact, yet less general, data structure for 2D cell and simplicial complexes is the star-vertices data structure [205]. It is as follows:

```
class Neighbour
{
   Vertex *vtx; // the neighbour vertex
   int nxt;    // index to find next vertex of the face
};

class Vertex
{
   float x, y, z; // geometry coordinates
   int num_nb;    // the number of neighbours to this vertex
   Neighbour *nb; // pointer to the array of neighbours
};
```

```
class Mesh
{
   vector<Vertex> *aov; // pointer to array of vertices
};
```

The star of vertices around a given vertex is represented by the field nb in the class Vertex. The index nxt works as a pointer to the next vertex of the star, which endows the data structure with a topological orientation. This data structure is quite compact because it only encodes vertices and their neighbours explicitly, not edges and faces. This fact may slow down geometric algorithms involving traversal and reasoning algorithms. The star-vertex data structure is particularly useful for encoding triangulations in $\mathbb{R}^2$.

In the literature, many other data structures that represent geometric objects have been proposed in recent decades, namely: the winged-edge [33], the half-edge [256], the DCEL [288], the quad-edge [170], the lath [203], the corner-table [342], the facet-edge [113], the handle-face [246], the cell-tuple [64], the nG-map [238] and the TCD graph [130], amongst others. For a more detailed study of simplicial complexes and triangulations, the reader is referred to Floriani and Hui [104] and Hjelle and Dæhlen [191].

2.3 Space Partitionings

These decompositions primarily stress on the decomposition of the ambient space itself, instead of its embedded geometric objects (e.g. implicit curves and surfaces). In this case, the partitioning of any object is a consequence of the partitioning of its ambient space. Some examples of these space partitionings include quadtrees, octrees, and BSP trees.

Partitioning the ambient space normally results in a collection of convex cells. However, there is no unique way to partition the space into similar cells— hence the nonexistence of a unique representation for a given object. Any unambiguous partitioning is valid, although for a given model some partitions are better than others, depending on the *problem* we intend to solve.

For sampling implicit curves (respectively, surfaces), we normally use some kind of space partitioning of a rectangular region into cells in $\mathbb{R}^2$ (respectively, $\mathbb{R}^3$). Some examples of space partitionings are shown in Figure 2.6. The cell occupancy can be described either sequentially (Figure 2.6(a)), or in a hierarchical way (Figure 2.6(b), (c) and (d)).

The sequential enumeration (Figure 2.6(a)), also known as *exhaustive enumeration*, partitions a rectangular region into axially aligned cells with the same size such that the resulting rectilinear grid is easily represented as an n-dimensional array (where n is the dimension of the space). This technique has applications in fields such as digital image processing where the data is obtained from 2D image scanning devices), computer tomography (CT), magnetic resonance imaging (MRI), and other scanning devices capable of

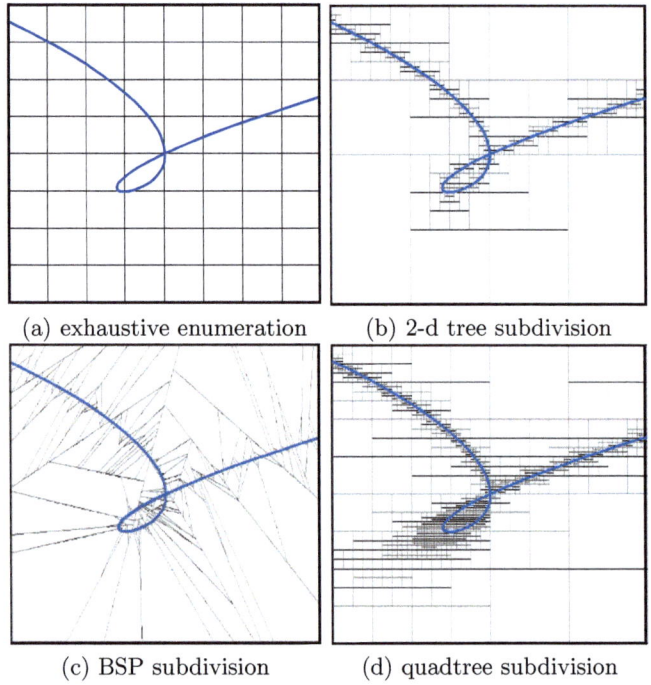

Fig. 2.6. Some space partitionings for implicit curves.

processing 3D data. A popular example of this technique is provided by the marching cubes algorithm due to Lorensen and Cline [247], which will be described later in Chapter 7.

By contrast, in a hierarchical structure the relationship between cells is granted by the very way in which the space is partitioned: smaller cells are derived from larger ones, and can be arranged in a tree structure. This technique is also known as *subdivision*, as the larger cells are further away from the curve or surface, whereas the finer cells tend to gather around the curve or surface and adapt to its shape (Figure 2.6(b),(c) and (d)). Thus, subdivision is a recursive partition of space into cells that altogether cover the rectangular region of interest where the curve or surface lies in. The corresponding hierarchical data structures thus generated are 2^n-trees, which particularise to bintrees ($n = 1$), quadtrees ($n = 2$) and octrees ($n = 3$) in 1D, 2D and 3D, respectively, or higher dimensions.

2.3.1 BSP Trees

Binary space partitioning (BSP) recursively splits the space into convex subspaces by hyperplanes (i.e. a higher-dimensional generalisation of the concepts of a point in a straight line, a straight line in a plane, a plane in a 3D space, and so forth). Each hyperplane divides a space into two convex subspaces. This recursive partition of the space is usually encoded into a binary tree or bintree, irrespective of the number of dimensions. This is a major advantage from the point of view of implementation because binary trees are a common data structure, easy to store and to browse. This matching between the BSP and a bintree results in a BSP tree, as illustrated in Figure 2.7. Each internal node of a BSP tree is associated with a splitting hyperplane. In Figure 2.7, these hyperplanes are the bisection lines $l_1, \ldots, l_5$.

BSP trees find applications in many science and engineering fields. In the context of the present textbook, they are particularly useful to sample implicit curves and surfaces; for example, the curve shown in Figure 2.6(c) was polygonised after sampling some of its points against the bisection lines of a binary space partitioning of a rectangular subspace of $\mathbb{R}^2$. This approach was first proposed by Fuchs et al. in 3D computer graphics to determine visible surfaces of polyhedra, without the need for a z-buffer, and to increase the rendering efficiency [148]. In fact, the membership test of an arbitrary point in space against a BSP tree is well known. Since then, BSP trees have found some other applications, including geometric operations of CSG (constructive solid geometry) shapes in CAD systems, ray tracing, collision detection in robotics and 3D FPS (first-person shooter) computer games involving navigation through indoor environments, as well as other applications that involve handling of complex spatial scenes.

Doom was probably the first computer game to use a BSP data structure. This is so because the scenery in computer games is usually built up using polyhedral shapes, which can be easily represented accurately with this technique. This includes convex as well as concave polyhedra. In a BSP representation of a polyhedral object, each face is contained in a splitting plane of the ambient space. In this case, the normal of the plane is assumed to point towards the empty half-space, for illumination purposes.

Conversion algorithms from other representations of geometric objects into BSP and vice versa are well known in geometric modelling community. In particular the conversion from a B-rep model into a BSP tree has been studied by Thibault and Naylor [298, 384]).

BSP: data structure. Many variants of the BSP tree data structure appear in the literature, so that sometimes it is not easy to distinguish application-dependent from application-independent data. On the contrary, the following BSP tree data structure clearly splits such data:

```
class Node
{
   Hyperplane *hp;      // splitting hyperplane
   Space *s;            // intersection of half-spaces
   Data *data;          // application-dependent data
};
```

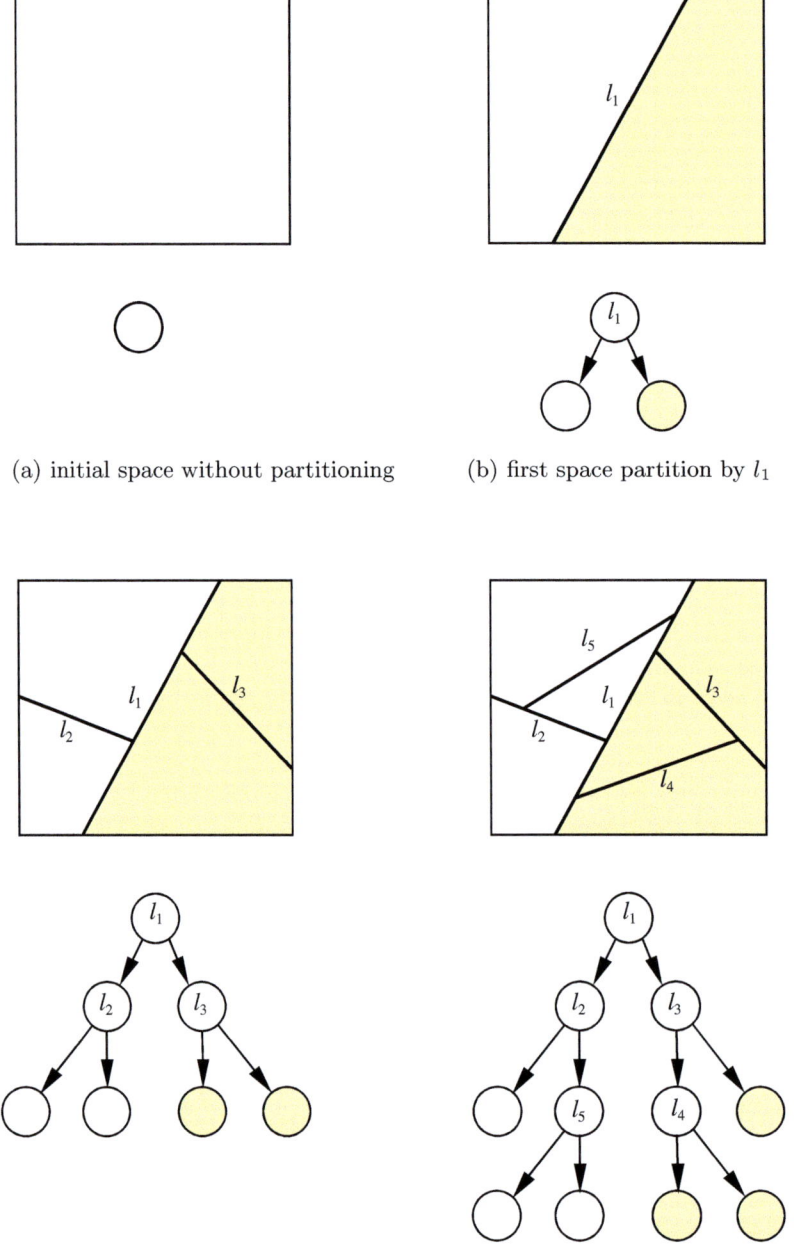

Fig. 2.7. Binary space partitioning of a space $\Omega \in \mathbb{R}^2$ and its corresponding bintree.

```
class BSPTree
{
    Node *node;              // the top node
    BSPTree *subtree[2];     // its two subtrees
}
```

The class `BSPTree` encodes the hierarchical structure of a binary tree, i.e. it is application independent. Each BSP node keeps not only the geometry of the its associated space `s` and its bisection hyperplane `hp`, but also any specific `data` required by the application.

2.3.2 K-d Trees

K-d trees (short for k-dimensional trees) are a particular case of BSP trees, where k denotes the dimension of the ambient space to be subdivided. The extra restriction applied to k-d trees is that the space is divided by planes that are always mutually perpendicular, and parallel to the coordinate axes. The resulting subspaces are boxes, also known as hyper-rectangles. Even though the division is always performed at right angles, it can be uneven, in that a box need not be split into two equal sub-boxes.

The k-d tree is a multidimensional binary search tree for points in the ambient k-dimensional space. Thus, given a set of n points in a k-dimensional box, we can construct a binary tree which represents a binary partition of the space into axially aligned subboxes by hyperplanes such that each point is contained in its own region. This illustrated in Figure 2.8, where the insertion of points A, B, C, D, E and F in the ambient space Ω by lexicographic order have caused a decomposition of Ω into smaller boxes. The resulting space partition and respective 2-d tree appears in Figure 2.8(f). A black square node ■ of the 2-d tree indicates that a box in Ω contains a given point, while a white square node □ denotes an empty box inside Ω; a circle node enclosing the letter identifier of a point means that either the x-coordinate or y-coordinate of such a point defines a splitting line of the space.

In a unidimensional binary search tree, nodes have a single key. In a k-d tree, nodes cycle through a set of k keys. As for a traditional binary search tree, nodes are inserted and retrieved using $<$ and $\geq$. But, the key that determines the subtree to follow (i.e. left or right) depends on the the level of the tree. At level l, the key number is given by $l \bmod k + 1$. That is, the first key is used at level 0 (root), the second key at level 1, and so forth until all keys have been used. Then, we recycle through the keys for the remaining nodes down in the tree. These keys usually represent the orthogonal axes x and y in 2D, x, y and z in 3D, and so on.

Let us look again at the 2-d tree in Figure 2.8. The two possible splitting directions are x and y. The area of interest is first split into two with a line parallel to the y-axis, being the key given by $x = 10$ (i.e. the x-coordinate of the point A). The point B is then inserted into the left node of A because its

56   2 Spatial Data Structures

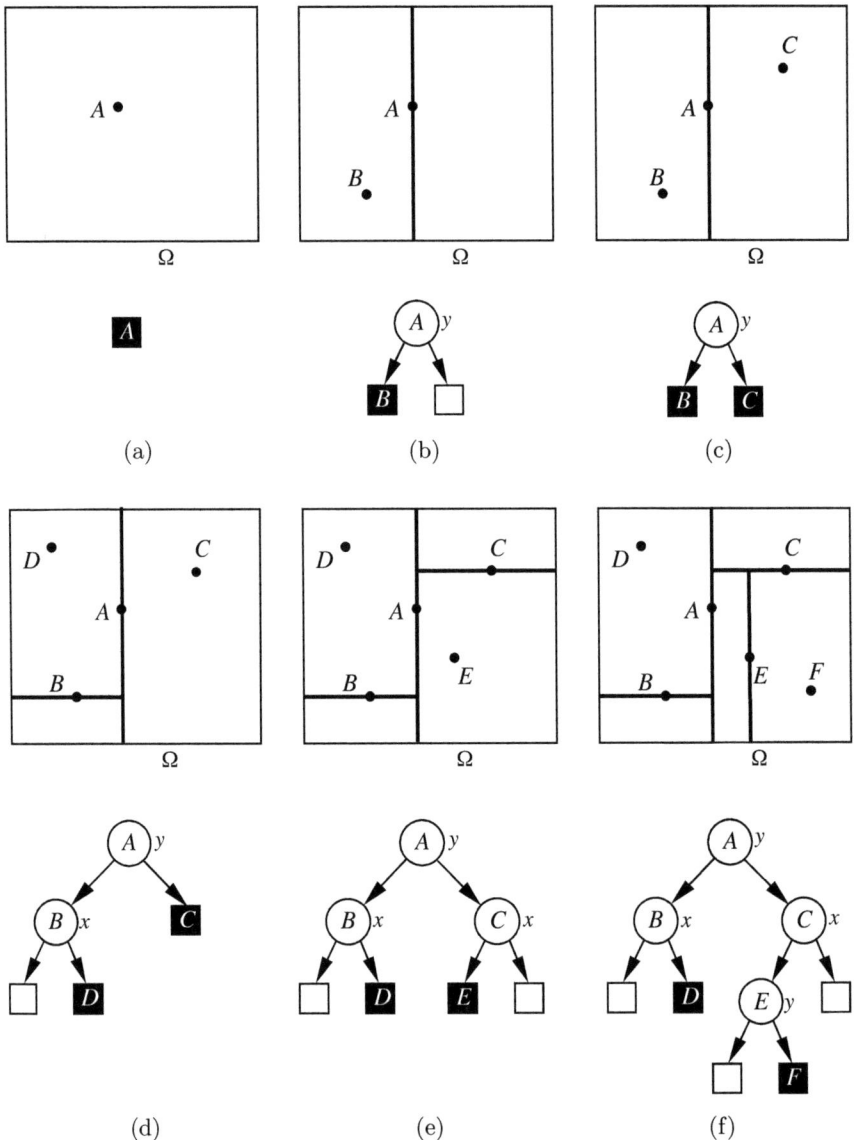

Fig. 2.8. A 2-d tree and its associated space decomposition.

x-coordinate satisfies $x < 10$, while the point C is associated with the right node of A provided that its x-coordinate satisfies $x > 10$. Then, at the level 1, each of the subboxes generated through B and C are cut parallel to the x axis. In short, the splitting lines are parallel to x-axis at the odd levels, and parallel to y-axis at the even levels.

K-d trees are constructed by partitioning point sets. Figure 2.8 illustrates a construction algorithm of a k-d tree by inserting points incrementally and partitioning the corresponding cell accordingly. But, other flavours of algorithms to construct k-d trees can be devised. For example, given a discrete point set, each tree node is defined by a splitting hyperplane that partitions the set of points into subsets (left and right subsets), each with half the points of the parent node. In this case, we get a balanced k-d tree.

The correspondence between a k-d tree and a binary partition of the space makes it well suited to support spatial queries on point sets. In computational geometry, k-d trees are used to carry out the nearest neighbour point search, the point location inside a polygon, and the orthogonal range search (to find all the points that lie within a given box).

In geometric modelling, 2-d trees can be used to carry out a binary space partitioning that progressively approximates an implicit curve or surface. This is illustrated in Figure 2.6(b) for a curve. But, it can be also used to decompose a filled geometric figure into boxes adaptively, depending on the local curvature of its frontier, as usual in engineering analysis and finite element methods. In this case, the recursive space decomposition of the figure stops when the sub-boxes are either entirely empty, or entirely full, or too small to divide further.

The k-d tree data structure can be encoded as follows:

```
class Node
{
   Box *box;   // its associated box
   Data *data; // application-dependent data
};

class kdTree
{
   Node *node;          // the top node
   kdTree *subtree[2]; // its two subtrees
}
```

Most of that structure is a standard binary tree, with links pointing upwards (if desired) and downwards. Additionally, there needs to be a convention for encoding a direction of space along which the node is being subdivided; it is appropriate to use the same field in order to mark a leaf node. A char field has been suggested for this purpose, though the information is merely two-bit wide. The coordinate of the division is the absolute offset, in the direction direction, of a division plane whose normal is along that same direction.

At each stage of the division, it is also worth storing the coordinates of the current node's box (so as to avoid having to recalculate them each time). Since the box is expected to be axially aligned, the box type can be, for example, a collection of three intervals for the ranges in x, y and z.

```
class box {
    interval xi, yi, zi; // intervals for ranges in x, y, z
}
```

One sensitive detail is the fullness indicator. Depending on the reasons for which the subdivision is being carried out, the `full` flag may represent one bit of information indicating whether a box is full or empty. The subdivision needs to stop once the box size has reached a threshold, in which case some leaf nodes are "approximated" to one of the two values even if they contain some surface and could, theoretically, be subdivided further. (This is dealt with in a similar manner in the case of quadtrees and octrees, introduced below.)

Alternatively, a third kind of value may be stored (as suggested in the example above), whereby a box can be flagged as "partially full." In other words, this box contains surface and further algorithms may process the information pertaining to the surface patch in each partially full leaf node, for example for the purpose of rendering. More details on k-d trees can be found in de Berg et al. [97] and Samet [347].

2.3.3 Quadtrees

A *quadtree* is a tree data structure so that each node has up to four children, i.e. a particular case of the 2^n-tree, with $n = 2$. In other words, each node of a 2^n-tree splits along n dimensions giving rise to 2^n children. Once again, the directions in which a region is split are axially aligned. Finkel and Bentley were who first proposed the quadtree data structure in 1974 [137].

When associated to a recursive partition of a 2D space, it splits a region of interest along the two axial directions from which four quadrants or regions are obtained, each stored into one of the four tree nodes. Note that such regions may have arbitrary shapes, i.e. they are not necessarily square or rectangular. Usually, a quadtree appears associated to a squared box which is partitioned into subsidiary boxes recursively until some stopping condition be satisfied.

There are many types of quadtrees in the literature. They may be classified according to the type of data they represent, namely points, lines, curves, areas, etcetera. For example, a curve is approximated by a collection of edges, whereas a volumetric object is approximated by a collection of voxels—both stored in a treelike fashion.

The *point quadtree* is similar to the point 2-d tree previously described in that the subdivision of the space always occurs on a point.

Another type is given by the *edge quadtree*, which stores lines rather than points. This is illustrated in Figure 2.6(d), where an implicit curve is approximated by adaptively partitioning boxes to an adequate sampling resolution. Some boxes appear unnecessarily subdivided in Figure 2.6(d), but that does depend on the subdivision criterion, not the quadtree itself.

In geometric modelling, the *region quadtree* is the most familiar quadtree data structure, which is used to approximately represent a point set (i.e. a shape) in the plane. In this case, a quadtree can be viewed as a particular type of space-partitioning tree. The recursive division process stops when all the current sub-quadrants are either full or empty, or when they become too small to be subdivided. Partially full quadrants are divided further in order to establish finer features of the region being studied. When these partially full quadrants become very small, they end up being also classified as full or empty. A single bit in each leaf indicates which is the case. Figure 2.9 illustrates a region quadtree of a shape in the plane and its tree data structure in which each node has four children. Every node in the three (Figure 2.9(b)) corresponds to a quadrant or squared box in the region of interest Ω; hence, the quadrants NE (northeast), NW (northwest), SW (southwest) and SE (southeast).

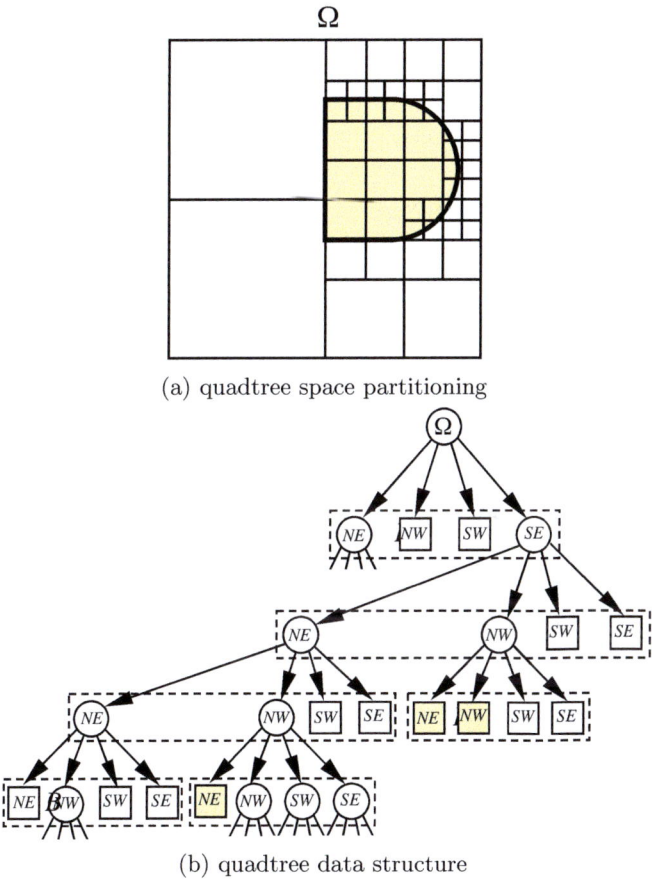

(a) quadtree space partitioning

(b) quadtree data structure

Fig. 2.9. Planar quadtree.

2 Spatial Data Structures

Therefore, quadtrees can be encoded as follows:

```
class Node
{
   Box *box;              // box it represents in the plane
   Data *data;            // application-dependent data
}

class Quadtree
{
   Node *node;            // top node
   Quadtree *quadrant[4];// the quadrants: NE, NW, SW and SE
}
```

Beyond their applicability in computational geometry, geometric modelling and computer graphics, quadtrees are also widely used in image processing (e.g. image representation and storing of raster data) and spatial information analysis (e.g. spatial indexing in databases) as needed in numerous further applications, from computer vision, geographical information systems, astronomy and cartography, etc.

2.3.4 Octrees

An *octree* is a 2^3-tree, i.e. it is the 3-dimensional analogue of a quadtree, i.e. an octree has eight children instead of four (see Figure 2.10). Therefore, octrees are good candidate data structures for representing 3D embedded geometric objects in memory because an octree breaks the space of interest (i.e. the initial bounding box) into eight boxes, called *octants*, by three hyperplanes (usually, axis-aligned planes). These boxes are then recursively partitioned into eight sub-boxes. Similar to the quadtree, this process continues until a box is sufficiently homogenous (either full or empty) that it can be represented

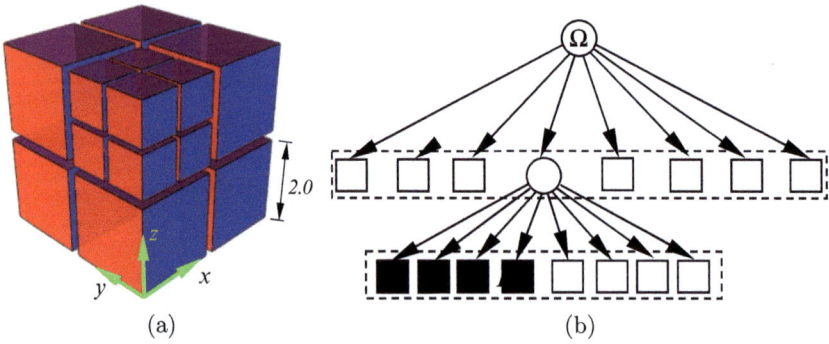

Fig. 2.10. Some space partitionings for implicit curves.

by a single node. Recall that, in octree jargon, while an octant is of any size, *voxels* (for "volume element") refer to smallest octants (i.e. those lying in the leaf nodes of the tree).

Looking at Figure 2.10, we easily conclude that the octree data structure can be written as follows:

```
class Node
{
    Box *box;          // octant it represents
    Data *data;        // application-dependent data
}

class Octree
{
   Node *node;         // top node
   Octree *octant[8];// array of octants or 3D boxes for node
}
```

The field `data` is capable of storing arbitrary information, but octrees are commonly used for representing surfaces or volumes in 3D. For example, implementations often use six bits in each octant to indicate whether any of the octant's six faces is on the surface of the volume. Extra face information (e.g. colour) requires a collection of up to six indices or pointers to an auxiliary face structure. Additionally, each leaf node denotes whether the space in its corresponding octant is either empty or full, while interior nodes represent partially full octants. That is, partially full octants have eight child octants (hence the prefix "oct").

Octrees are a data structure for storing information about curves, surfaces or volumes in a 3D space. Octrees have particular advantages over other representations when 3D spaces contain *blobs* or volumes which are highly connected (e.g. a human body). In [237], Libes uses octrees in modelling dynamic surfaces. They are also useful in collision detection, as usual when we need to compute robot paths. In fact, when a robot interacts with the geometric objects existing in the ambient space, the octrees allow us to detect intersections between them.

However, octrees also have disadvantages, in particular in dynamic environments where the geometric objects can grow and move without predefined constraints. For a discussion about the advantages and disadvantages of octrees and a running/space-time analysis, the reader is referred to Meagher [264, 265]. Navazo and Brunet developed a solid geometric modeller based on octrees, whose representation was even extended to non-manifold domain [69, 296, 297]. Octrees also find applications in ray tracing and computer graphics [154, 212, 421]. For a more comprehensive survey of octrees and related spatial representations of geometric objects, the reader is referred to Samet [346, 347].

2.4 Final Remarks

As seen above, there are many different ways of partitioning spaces. Some partitions directly decompose geometric objects as subspaces of the ambient space, while others do the same indirectly by first decomposing the ambient space. Different partitions normally lead to distinct spatial data structures. This chapter just explores the relation between space partitions (including object partitions) and their possible spatial data structures. It would interesting to observe how different space decompositions determine not only different data structures, but also different algebraic structures, but that is not in the scope of the current textbook. For that, the reader is referred to Shapiro [356, 357] for further details.

Part II

Sampling Methods

Overview

Sampling an implicit curve or surface is a discretisation process that outputs a finite set of points of such an implicit object. Usually, sampling is the first stage of the rendering pipeline, say *sampling* → *polygonisation* → *rendering*. Sampling an implicit object is carried out against line segments. For 2D curves, these segments split up the subspace, where the curve lies in, into two. For 3D surfaces, these segments are, for example, the voxel edges of an octree space decomposition.

Each sampling point is computed by using some root finding method. Thus, sampling an implicit curve or surface can be reduced to the problem of root finding. This problem has been much studied and, as known, there are many ways of approaching it. For polynomials, root finding involves three major steps:

- Removal of the repeated roots.
- Root isolation.
- Root approximation.

The first two steps may be computed in an algebraic or symbolic manner, avoiding this way eventual problems of numerical instability. The first step is achieved by determining the square-free polynomial having the same roots as the original polynomial. In science and engineering applications, a square-free polynomial is also known as a *polynomial without repeated roots*. More details about this root cleaning-up step are given in the beginning of the first chapter of Part II.

The motivation for the second step lies in the fact that an interval may contain several real roots of a polynomial. Isolation of these real roots is a process of finding a number of real subintervals such that each subinterval contains exactly one real root of the polynomial. Thus, this intermediate step outputs a set of subintervals, each one of which surely contains a single real root, as required input for the third step. There are various root isolation techniques, some of which appear described in the next two chapters of Part II.

The third step involves the numerical computation of each root previously isolated inside a subinterval. For that, we can use an 1-point or 2-point numerical method. An 1-point numerical method uses the current estimate to compute the next one in the subinterval towards a curve or surface. For example, the Newton-Raphson method is an 1-point numerical method. A 2-point numerical method computes the next estimate from the two current estimates that bracket the root, as it is the case of, for example, the false position method. These 2-point methods assume that if two distinct points have function values with opposite signs, then there exists a point between them at which the function evaluates to zero. In geometric terms, this intermediate point is just a sampled point of an implicit curve or surface. Some root-finding algorithms are described in the third chapter of Part II.

Therefore, this part of the book deals with different computational methods to sampling implicit curves and surfaces. In particular, we are interested in getting a deeper understanding of such methods, their advantages and disadvantages, and how they are applied to functions in 2D and 3D spaces.

3
Root Isolation Methods

This chapter deals with bounding and isolating the zeros (i.e. roots) of a polynomial function in a given region of interest. The function is assumed to be in its *implicit* form $f(\mathbf{x}) = 0$, although some of the theory is also relevant to explicit and parametric functions. If the function has only one variable, the region of interest is an interval. The procedure of root isolation is normally preceded by procedures that find bounds for the interval where the polynomial's roots are likely to lie. Together, bounding and isolating are known as *root location* methods [372]. Root location methods are important to guarantee the correctness of sampling methods for curves and surfaces.

The theory of root location methods is easiest to explain for univariate polynomial functions defined on a finite interval domain. This is because, historically, the algebraic methods for root location were developed before their use in geometric modelling became obvious. It is, however, possible to extend the definition of "roots" to refer to zeros of multivariate functions. In 2D, for example, the zeros of an bivariate polynomial lie on a curve, and isolating such zeros in $\mathbb{R}^2$ is equivalent to locating the points of the curve in the Euclidean plane. Similarly, in 3D one can talk about the skin or surface of an object defined by a trivariate polynomial. Generalisations to higher dimensions follow from there.

3.1 Polynomial Forms

An implicit polynomial function can be represented in an infinity of ways, all of which are equivalent modulo algebraic manipulation. Some of the most significant forms depend on the bases available on the ring of polynomials with real coefficients, others depend on particular ways of arranging the terms. Let us list and discuss a few:

- Power form
- Factored form
- Bernstein form

3.1.1 The Power Form

The *power* form is the expanded polynomial in monomials, after removal of superfluous terms (e.g. $x - x = 0$), and reordering in descending order of exponents. Therefore, a degree-n polynomial (also called n-order polynomial) in the power basis is written as follows:

$$f(x) = \sum_{i=0}^{n} a_i x^i \qquad (3.1)$$

where $a_i \in \mathbb{R}$ and $a_n \neq 0$.

For example, the following power form polynomial

$$f(x) = x^2 + x - 6 \qquad (3.2)$$

is a 2-order polynomial because its highest power of x is 2. It is also *monic* because its leading coefficient, say the coefficient of x^2, is 1. To find the roots of this 2-order polynomial, we can use the well-known quadratic formula

$$x = \frac{-b \pm \sqrt{b^2 - 4ac}}{2a}. \qquad (3.3)$$

After general algebraic formulæ for the cubic and quartic equations [67] had been found in the 16th century, Abel (in 1824) gave the first accepted proof of the insolubility of the quintic. Later on, in 1831, Galois proved that no formula exists for polynomials of degrees equal to or greater than five. This motivated the appearance and development of the field of numerical analysis.

3.1.2 The Factored Form

By the fundamental theorem of algebra, any degree-n polynomial has exactly n roots (or zeros), which are real or complex. For the 2-order polynomial $f(x)$ in Equation (3.2) above, let us assume its roots are both real and denote them by α_0 and α_1. This means that $f(\alpha_0) = 0$ and $f(\alpha_1) = 0$, so that we can write

$$f(x) = (x - \alpha_0)(x - \alpha_1)$$

This is known as the *factored form* of the monic polynomial $f(x)$. For a nonmonic polynomial, in order to make it monic, it is possible to divide all coefficients by the coefficient of the highest power term. This division by a nonzero constant does not change the polynomial's zeros. By multiplying out the symbolic factored form, we obtain

$$f(x) = (x - \alpha_0)(x - \alpha_1) = x^2 - (\alpha_0 + \alpha_1)x + \alpha_0 \alpha_1$$

Comparing this form with the original power form of the polynomial $f(x)$ in Equation (3.2), we come to the following nonlinear system of two equations in two unknowns

$$\begin{cases} -(\alpha_0 + \alpha_1) = 1 \\ \alpha_0 \alpha_1 = -6 \end{cases}$$

which yields the solution zeros $\alpha_0 = 2$ and $\alpha_1 = -3$. Thus, the conversion of a factored form polynomial into its power form is clearly easier than the reverse process. This is even more the case with higher-order polynomials.

3.1.3 The Bernstein Form

In CAGD, the Bernstein polynomials are often used to define free-form *parametric* surfaces (and curves), namely: Bézier, B-spline or NURBS surfaces [132].

On the contrary, in this book, we use the Bernstein form polynomials to define *implicit* curves and surfaces. In fact, similar to the power form polynomials, a 2D implicit curve can be defined as the zero set of a bivariate Bernstein form polynomial in $\mathbb{R}^2$, while an implicit surface is defined as the zero set of a trivariate Bernstein form polynomial in $\mathbb{R}^3$.

But, for the time being, let us concentrate on the univariate case. The univariate Bernstein polynomial basis is normally defined for a variable x that varies in the interval $[0,1]$. However, it is possible to remove the constraint $x \in [0,1]$ and to extend the domain of the Bernstein polynomials to the generic interval $[a,b]$.

For a given $n \in \mathbb{N}$ there are $n+1$ univariate degree-n Bernstein polynomials. By definition, the univariate Bernstein basis functions of degree n on the interval $[a,b]$ (see also Lorentz [248]) are defined by:

$$B_i^n(x) = \binom{n}{i} \frac{(x-a)^i (b-x)^{n-i}}{(b-a)^n}, \quad \forall x \in [a,b], \quad i = 0, 1, \ldots, n. \quad (3.4)$$

The fact that the polynomial set $(B_i^n)_{i=0,n}$ forms a basis for the ring of degree-n polynomials means that any univariate power form polynomial of degree n or lower can be represented on the interval $[a,b]$ using its equivalent Bernstein form as follows:

$$f(x) = \underbrace{\sum_{i=0}^n a_i x^i}_{power\ form} = \underbrace{\sum_{i=0}^n b_i^n B_i^n(x)}_{Bernstein\ form} \quad (3.5)$$

where b_i^n are the Bernstein coefficients corresponding to the degree-n base. Both univariate representations (3.5) are equivalent on the interval $[a,b]$ and conversion between them is fairly straightforward.

The Unit Interval [0,1]

On the unit interval $[0,1]$ for example, the univariate Bernstein coefficients are easily computed in terms of the power coefficients (also shown by Farouki and Rajan [133, 134]):

$$b_i^n = \sum_{j=0}^{i} \frac{\binom{i}{j}}{\binom{n}{j}} a_j \qquad (3.6)$$

Therefore, the formula (3.6) can be used to design an algorithm of conversion between the power and the Bernstein form of an univariate polynomial. This is shown in the following example.

Example 3.1. (Univariate Bernstein form polynomial in $[0,1]$) Given the polynomial $p(x)$ in the power form

$$p(x) = x^3 - 5x^2 + 2x + 4$$

its equivalent Bernstein form (valid for $x \in [0,1]$) is obtained using the Formula (3.6) above:

$$p(x) = 4B_0^3 + \frac{14}{3}B_1^3 + \frac{11}{3}B_2^3 + 2B_3^3$$
$$= 4(1-x)^3 + \frac{14}{3}3x(1-x)^2 + \frac{11}{3}3x^2(1-x) + 2x^3$$

where B_i^3 are the Bernstein polynomials of degree 3, namely:

$$B_0^3 = (1-x)^3$$
$$B_1^3 = 3x(1-x)^2$$
$$B_2^3 = 3x^2(1-x)$$
$$B_3^3 = x^3$$

In order to generalise the Formula 3.6 for polynomials with more than one variable, it is convenient to express polynomials as matrix products. The formulae below show how to calculate the desired set of Bernstein coefficients in the matrix $\mathbf{B}$ in terms of the power coefficients given in matrix $\mathbf{A}$.

Let us then first rewrite both polynomial representations in the matrix notation:

$$f(x) = \sum_{i=0}^{n} a_i x^i = \begin{bmatrix} 1 & x & \cdots & x^n \end{bmatrix} \begin{bmatrix} a_0 \\ a_1 \\ \vdots \\ a_n \end{bmatrix} = \mathbf{XA} \qquad (3.7)$$

$$f(x) = \sum_{i=0}^{n} b_i^n B_i^n(x) = \begin{bmatrix} B_0^n(x) & B_1^n(x) & \cdots & B_n^n(x) \end{bmatrix} \begin{bmatrix} b_0 \\ b_1 \\ \vdots \\ b_n \end{bmatrix} = \mathbf{B_X B} \qquad (3.8)$$

Now, by expanding the elements of $\mathbf{B}_X$, we have:

$$\begin{aligned}
\mathbf{B}_X &= \begin{bmatrix} B_0^n(x) & \cdots & B_n^n(x) \end{bmatrix} \\
&= \begin{bmatrix} \binom{n}{0}(1-x)^n & \cdots & \binom{n}{n}x^n \end{bmatrix} \\
&= \begin{bmatrix} \binom{n}{0}\left(1 + \binom{n}{1}(-x) + \cdots + \binom{n}{n}(-x)^n\right) & \cdots & \binom{n}{n}x^n \end{bmatrix} \\
&= \underbrace{\begin{bmatrix} 1 & x & \cdots & x^n \end{bmatrix}}_{\mathbf{X}} \underbrace{\begin{bmatrix} 1 & 0 & \cdots & 0 \\ \binom{n}{0}\binom{n}{1}(-1)^1 & \binom{n}{1}\binom{n-1}{0}(-1)^0 & \cdots & 0 \\ \vdots & \vdots & \ddots & 0 \\ \binom{n}{0}\binom{n}{n}(-1)^1 & \binom{n}{1}\binom{n-1}{n-1}(-1)^{n-1} & \cdots & \binom{n}{n}\binom{n-n}{0}(-1)^0 \end{bmatrix}}_{\mathbf{C}_X}
\end{aligned}$$

Thus,
$$f(x) = \mathbf{B}_X \mathbf{B} = \mathbf{X} \mathbf{C}_X \mathbf{B}$$

So, by equating (3.7) and (3.8), we now obtain the Bernstein coefficients matrix $\mathbf{B}$ in terms of the power coefficients matrix $\mathbf{A}$ as follows:

$$\mathbf{X}\mathbf{A} = \mathbf{X}\mathbf{C}_X\mathbf{B}$$

or, equivalently,
$$\mathbf{B} = (\mathbf{C}_X)^{-1}\mathbf{A}$$

where $(\mathbf{C}_X)^{-1}$ is the inverse matrix of $\mathbf{C}_X$.

A General Interval $[a, b]$

The constraint $x \in [0, 1]$ can be relaxed by extending the domain of the Bernstein polynomials to a generic interval $[a, b]$, as given by (3.4) and rewritten here as follows:

$$B_i^n(x) = \binom{n}{i}\left(\frac{x-a}{b-a}\right)^i\left(1 - \frac{x-a}{b-a}\right)^{n-i}, \quad \forall x \in [a, b]. \tag{3.9}$$

As before, a Bernstein form polynomial $f(x)$ is written as:

$$f(x) = \mathbf{B}_X \mathbf{B}, \quad \forall x \in [a, b], \tag{3.10}$$

where $\mathbf{B}_X$ is the vector of Bernstein polynomials and $\mathbf{B}$ is the Bernstein coefficients matrix.

Using an analogous sequence of steps as above, $\mathbf{B}_X$ can be expressed as follows:

$$\begin{aligned}
\mathbf{B}_X &= \begin{bmatrix} B_0^n(x) & B_1^n(x) & \cdots & B_n^n(x) \end{bmatrix} \\
&= \begin{bmatrix} 1 & \frac{x-a}{b-a} & \cdots & \left(\frac{x-a}{b-a}\right)^n \end{bmatrix} \mathbf{C}_X
\end{aligned}$$

or, equivalently,

$$\mathbf{B}_X = \begin{bmatrix} 1 & x-a & \cdots & (x-a)^n \end{bmatrix} \underbrace{\begin{bmatrix} \frac{1}{(b-a)^0} & & 0 \\ & \ddots & \\ 0 & & \frac{1}{(b-a)^n} \end{bmatrix}}_{\mathbf{D}_X} \mathbf{C}_X$$

$$= \begin{bmatrix} 1 & x-a & \cdots & (x-a)^n \end{bmatrix} \mathbf{D}_X \mathbf{C}_X$$

$$= \begin{bmatrix} \sum_{k=0}^{0} \binom{0}{k} x^k (-x)^{0-k} & \sum_{k=0}^{1} \binom{1}{k} x^k (-x)^{1-k} & \cdots & \sum_{k=0}^{n} \binom{n}{k} x^k (-x)^{n-k} \end{bmatrix} \mathbf{D}_X \mathbf{C}_X$$

$$= \underbrace{\begin{bmatrix} 1 & x & \cdots & x^n \end{bmatrix}}_{\mathbf{X}} \underbrace{\begin{bmatrix} 1 & \binom{1}{0}(-x)^1 & \cdots & \binom{n}{0}(-x)^n \\ 0 & \binom{1}{1}(-x)^{1-1} & \cdots & \binom{n}{1}(-x)^{n-1} \\ \vdots & \vdots & \ddots & \vdots \\ 0 & 0 & \cdots & \binom{n}{n}(-x)^{n-n} \end{bmatrix}}_{\mathbf{E}_X} \mathbf{D}_X \mathbf{C}_X$$

or
$$\mathbf{B}_X = \mathbf{X}\mathbf{E}_X\mathbf{D}_X\mathbf{C}_X. \tag{3.11}$$

Therefore, the Bernstein coefficients matrix $\mathbf{B}$ for a generic interval $[a, b]$ can be determined from:
$$\mathbf{X}\mathbf{A} = \mathbf{X}\mathbf{E}_X\mathbf{D}_X\mathbf{C}_X\mathbf{B}$$

that is,
$$\mathbf{B} = (\mathbf{C}_X)^{-1}(\mathbf{D}_X)^{-1}(\mathbf{E}_X)^{-1}\mathbf{A}.$$

3.2 Root Isolation: Power Form Polynomials

A real root isolation algorithm is an algorithm that, given a univariate real function (e.g. a polynomial function), computes a sequence of disjoint intervals each containing exactly one of the function's distinct real roots.

In this section, the focus is on the root isolation for power form polynomials. (Similar root isolation algorithms for Bernstein form polynomials are dealt with in the next section.) When the function is a power form polynomial, the most common methods for isolating its real roots are:

- Descartes' rule of signs
- Sturm sequences
- Interval arithmetic

The first two are algebraic procedures that compute the total number of roots that a power form polynomial has in a given interval. Descartes' rule only provides an *upper bound*, i.e. the maximum number of positive and negative real roots of a polynomial. Sturm's method computes the polynomial's *exact* number of roots.

Isolation by interval arithmetic has a major advantage, in that its use is not confined to polynomials—it applies to general continuous functions. This method only requires the function to be evaluated at the extremities of the interval for which it is being studied. However, when the function has more than one variable, this method presents a number of problems, mainly to do with singularities, multiple roots and other special points. Some of these are tackled in Milne's thesis [273], but his theoretical solutions remain difficult to implement. For example, it may be thought that repeated roots can be removed through factoring the polynomial and removing the factors that appear more than once. This is not a practical option. Also, transcendental functions return values that have no finite representation within the discrete range of data formats. Consequently, there is no guarantee of correctness of the results, even using multiple precision arithmetic libraries for storing floating point numbers. In this case, the function evaluation is said to be *ill-conditioned*.

A thorough discussion of Interval Arithmetic and its uses in root isolation can be found in Chapter 4.

3.2.1 Descartes' Rule of Signs

Descartes' rule of signs provides an upper bound on the number of positive and negative roots of a power form polynomial. For positive roots, it states this number does not exceed the number of sign changes of the nonzero coefficients of the power form of the polynomial, arranged in the order of exponents. More precisely, the number of positive roots of the polynomial is:

- either equal to the number of sign changes between consecutive nonzero coefficients,
- or less than it by a multiple of 2.

More formally, we have:

Theorem 3.2. (Descartes' Rule of Signs). *The number of positive roots of a power form polynomial does not exceed the number of sign changes of its coefficients and differs from it by a multiple of two.*

Proof. See Krandick and Mehlhorn [217].

Note that a zero coefficient is not counted as a sign change, and multiple roots are counted separately.

For negative roots, we use a corollary of the previous theorem, which is as follows:

Corollary 3.3. *The number of negative roots of a power form polynomial $f(x)$ is equal to the number of positive roots of $f(-x)$.*

Proof. See Levin [231].

In other words, the number of negative roots is given by the number of sign changes after replacing $-x$ for x in $f(x)$, or less than it by a multiple of 2.

Example 3.4. The power form polynomial

$$f(x) = x^3 - 3x^2 + 4$$

has two sign changes; the first between the first and second terms, while the second occurs between the second and third terms. Therefore it has exactly two positive roots. Negating the odd-power terms, we obtain

$$f(-x) = -x^3 - 3x^2 + 4$$

i.e. a polynomial with one sign change, so the original polynomial has exactly one negative root. The polynomial easily factors as

$$f(x) = x^3 - 3x^2 + 4 = (x-2)^2(x+1)$$

so the roots are 2 (twice) and -1.

Note that Descartes' rule of signs, first described by René Descartes in his work *La Geometrie* (an appendix of the famous masterpiece entitled "Discours de la Méthode" written in 1637), provides a bound to the number of roots of a power-form polynomial, not a bound to the interval where they lie in. For a more recent mathematical discussion about Descartes' rule of signs, the reader is referred to Anderson et al. [16], Grabiner [166] and Levin [231].

Some well-known polynomial real root isolation algorithms based on Descartes' rule of signs are found in the literature. Uspensky's 1948 book presents an early version of these algorithms for a square-free polynomial with real coefficients [394]. However, the worst-case complexity of this algorithm grows exponentially with the number of digits in the coefficients. This fact motivated Collins and Akritas to propose the modified Uspensky algorithm in order to guarantee polynomial complexity [89]. In [202], Johnson and Kandrick present a powerful and fast method which can be applied to polynomials with both integer and real algebraic number coefficients, including the pseudo-code of the algorithms. Rouillier and Zimmermann [343] bring up-to-date (or at least to 2004) the various improvements of root isolation methods based on Descartes' rule of signs.

3.2.2 Sturm Sequences

By contrast with numerical algorithms, algebraic algorithms do not try to evaluate the roots in the first instance, but rather study the existence of roots in a given interval. Of course, in the cases where root counting techniques are available, a divide-and-conquer approach can subsequently help to isolate each of the roots in a separate subinterval. This subinterval provides tighter lower and upper bounds for each root, from which it can be determined by using an appropriate numerical method (e.g. Newton-Raphson method).

Sturm sequences are part of a root isolation method established by Sturm in 1829. His theorem provides the number of real roots of a univariate polynomial in a given interval [375]. Uspensky [394], Davenport et al. [95], and

Bronstein and Semendjajew [67] also describe Sturm's theory, the main features of which are given below.

Let $f \in \mathbb{R}_n[x]$ be a n-order univariate polynomial with real coefficients and no multiple roots (i.e. f and its first derivative f' are relatively prime). Let $s_0, \ldots, s_k$ be the sequence of polynomials such that:

$$\begin{aligned} s_0 &= f \\ s_1 &= f' \\ &\vdots \\ s_i &= -\mathrm{mod}(s_{i-2}, s_{i-1}) \end{aligned} \qquad (3.12)$$

with $i = 2, \ldots, k$ and $s_k \in \mathbb{R} \setminus \{0\}$, where mod means the remainder from the division of two polynomials of $\mathbb{R}[x]$. (The recurrence rule given above is a valid particularisation of Sturm's more general formulation: $s_{i-2} = s_{i-1} q_i - s_i$, with $\deg(s_i) < \deg(s_{i-1})$.)

In fact, only the *sign* of the evaluation of the elements of the Sturm sequence (3.12) are needed for the root finding. The algorithm for the construction of Sturm's sequence is similar to an application of Euclid's algorithm to f and f'. Since the polynomials f and f' are supposed to be relatively prime, and the terms of the sequence are polynomials of decreasing degree, ultimately a constant is obtained. The null terms appearing after the constant term are ignored, so Sturm's sequence is always finite.

Sign Variation. If $a \in \mathbb{R}$ is *not* one of the roots of $f(x)$, denote by $V(a)$ the number of sign changes in the sequence $s_0(a), s_1(a), \ldots, s_k(a)$. Note that a sign change is counted whenever $s_i(a) s_j(a) < 0$ $(j \geq i+1)$ and $s_l(a) = 0$ $(i < l < j)$. In other words, the zeros in the evaluated sequence $s_0(a), s_1(a), \ldots, s_k(a)$ are ignored.

Theorem 3.5. (Sturm's Theorem). *Let $a, b \in \mathbb{R}$, $a < b$ such that neither is a root of $f(x)$. Then the number of the roots that $f(x)$ has in the interval $]a, b[$ equals $V(a) - V(b)$.*

Proof. See Sturm [375] or Uspensky [394].

Corollary 3.6. *Let $f(x)$ in $\mathbb{R}[x]$ without multiple roots and $s_0, \ldots, s_k$ be a sequence of polynomials defined in Sturm's theorem. Let us denote by $V(+\infty)$ (respectively, $V(-\infty)$) the number of sign changes in the sequence formed with the leading coefficients of $s_0, \ldots, s_k$ (respectively, $s_0(x), \ldots, s_k(x)$). Then, the total number of real roots of $f(x)$ equals $V(-\infty) - V(+\infty)$.*

Proof. See Uspensky [394].

Theorem 3.7. (Cauchy's Theorem). *Let $f(x) = a_n x^n + \cdots + a_0 \in \mathbb{R}[x]$ with $a_n \neq 0$. Let $M = 1 + |\frac{a_{n-1}}{a_n}| + \cdots + |\frac{a_0}{a_n}|$. Then $f(x)$ has no roots on $[M, +\infty[$ (respectively, on $]-\infty, -M]$) and its sign is the same as the one of a_n (respectively of $(-1)^n a_n$).*

3 Root Isolation Methods

Cauchy's theorem provides the very useful interval $]-M, M[$ containing all the roots of the polynomial $f(x)$, i.e. their bounds.

Example 3.8. Let $f(x) = (x-1)(x-2)(x-3)(x+1)(x+2)$ a polynomial in the factored form such that its roots are explicitly given in its expression. The steps of the algorithm that determines the Sturm sequence for $f(x)$ is as follows:

s_0 :
$$s_0 = (x-1)(x-2)(x-3)(x+1)(x+2)$$
$$= x^5 - 3x^4 - 5x^3 + 15x^2 - 12$$

s_1 :
$$s_1 = f' = 5x^4 - 12x^3 - 15x^2 + 30x + 4$$

s_2 :
$$-\text{mod}(s_0, s_1) = \frac{86x^3 - 180x^2 - 170x + 288}{25}$$
$$s_2 = 86x^3 - 180x^2 - 170x + 288$$

s_3 :
$$-\text{mod}(s_1, s_2) = \frac{15{,}400x^2 - 18{,}900x - 16{,}900}{1849}$$
$$s_3 = 15{,}400x^2 - 18{,}900x - 16{,}900$$

s_4 :
$$-\text{mod}(s_2, s_3) = \frac{282{,}897}{1694}x - \frac{49{,}923}{242}$$
$$s_4 = 282{,}897x - 349{,}461$$

s_5 :
$$-\text{mod}(s_3, s_4) = \frac{4{,}840{,}000}{289}$$
$$s_5 = 4{,}840{,}000$$

The sequence stops here because s_5 is a constant. Also, note that multiplying a polynomial by a nonzero constant does not affect its set of roots, so that the terms of a Sturm sequence can be freely multiplied by convenient numbers.

According to Sturm's theorem, the number of roots will be given by the number of sign changes encountered. Let us then evaluate the terms of the Sturm sequence for several values of x, and study the sign variations on several intervals. This is illustrated in Table 3.1.

For example, given the sequence $s_0(-\frac{5}{2}), s_1(-\frac{5}{2}), \ldots, s_5(-\frac{5}{2})$, we observe that the number $V(-\frac{5}{2})$ of sign changes in such a sequence is equal to 5; analogously, we get $V(-\frac{3}{2}) = 4$, $V(0) = 3$, $V(\frac{3}{2}) = 2$, $V(\frac{5}{2}) = 1$, and $V(\frac{7}{2}) = 0$.

3.2 Root Isolation: Power Form Polynomials

Table 3.1. Evaluation of Sturm's sequence at several values of x and the corresponding sign variations.

s_i \ x	$-\frac{5}{2}$	$-\frac{3}{2}$	0	$\frac{3}{2}$	$\frac{5}{2}$	$\frac{7}{2}$
s_0	−	+	−	+	−	+
s_1	+	−	+	+	−	+
s_2	−	−	+	−	+	+
s_3	+	+	−	−	+	+
s_4	−	−	−	+	+	+
s_5	+	+	+	+	+	+
$V(s_i)$	5	4	3	2	1	0

Algorithm 1 Sturm sequence algorithm

1: **procedure** STURM($f(x), L[\,]$)
2: $\quad i \leftarrow 0$ and $s_i \leftarrow f(x)$
3: $\quad L[\,] \leftarrow \text{add}(p_i)$ ▷ first polynomial
4: $\quad d \leftarrow \text{degree}(s_i)$ ▷ find out its degree
5: $\quad$ **if** $d \geq 1$ **then**
6: $\quad\quad s_{i+1} \leftarrow \text{diff}(s_i)$ ▷ next term = first derivative
7: $\quad\quad$ **while** degree(s_{i+1}) > 0 **do**
8: $\quad\quad\quad L[\,] \leftarrow \text{add}(s_{i+1})$ ▷ append next polynomial
9: $\quad\quad\quad r \leftarrow -s_i \bmod s_{i+1}$
10: $\quad\quad\quad s_{i+2} \leftarrow \text{numerator}(r)$ ▷ remove denominators
11: $\quad\quad\quad s_i \leftarrow s_{i+1}$
12: $\quad\quad\quad s_{i+1} \leftarrow s_{i+2}$
13: $\quad\quad$ **end while**
14: $\quad\quad L[\,] \leftarrow \text{add}(s_{i+1})$ ▷ append the last (constant) term
15: $\quad$ **end if**
16: $\quad$ **return** $L[\,]$ ▷ Sturm's sequence as a list
17: **end procedure**

Therefore, from Table 3.1 it follows that:

$$V\left(-\frac{5}{2}\right) - V(0) = 5 - 3 = 2 \text{ roots in the interval } \left]-\frac{5}{2}, 0\right[$$

$$V\left(-\frac{3}{2}\right) - V\left(\frac{3}{2}\right) = 4 - 2 = 2 \text{ roots in the interval } \left]-\frac{3}{2}, \frac{3}{2}\right[$$

$$V\left(-\frac{3}{2}\right) - V\left(\frac{7}{2}\right) = 4 - 0 = 4 \text{ roots in the interval } \left]-\frac{3}{2}, \frac{7}{2}\right[\text{ and so on.}$$

Sturm Sequence Algorithm. The terms of the Sturm sequence are computed according to Formula (3.12) and are successively appended to a list L (steps 7–13 of Algorithm 1). When a degree-zero polynomial is found, the algorithm appends such a polynomial to the list (step 14) and returns the list as a result (step 16).

3.3 Root Isolation: Bernstein Form Polynomials

Bernstein-form polynomials have become increasingly popular for CAGD applications because of their *numerical stability*. This term is borrowed from the field of numerical analysis, where the numerical stability of an algorithm [186] is meant to express the extent to which approximation errors in the terms of a calculation, as well as the order in which commutative operations are being carried out, affect the accuracy of the final result.

Similarly, the *numerical stability of a polynomial* amounts to the numerical stability of its corresponding evaluation algorithm. In that sense, it is interesting to study the effect of a (small) perturbation in one of the polynomial's coefficients onto the number and location of its roots, onto the location of its graph curve, or onto the extent to which it interpolates a given set of points.

Polynomials stored in the Bernstein form are more numerically stable than their equivalent power form. This theoretical result, established by Farouki and Rajan [133, 134], has now found its way into many geometric modelling packages. Unsurprisingly, the robustness of the evaluation algorithms is achieved at the expense of extra computations, since the Bernstein forms of polynomials have significantly more terms and hence more operations need to be performed for each evaluation.

For low-degree polynomials, the advantage of using the Bernstein form polynomials is not immediately obvious. However, in axially aligned areas which are further away from the origin, high-degree polynomials in the power form usually operate with large powers of large numbers. This means that any small errors in the coordinates of a point can cause a significant change in the value of the polynomial at that point. On the contrary, since the Bernstein base is more numerically stable than the power base, minor perturbations introduced in the coefficients tend not to affect the accuracy of the polynomial evaluation.

Let us then briefly review some relevant real root isolation methods relevant to polynomials in Bernstein form:

- Variation diminishing methods
- Hull approximation methods
- Descartes' rule of signs

All these methods use recursive subdivision as the basic technique behind the root isolation. If they are used for Bernstein-form polynomials, then a new (equivalent) for of the polynomial is recomputed at each step of the subdivision, according to the subinterval being studied at that step.

The difference between these three methods lies in the computations they each perform during the intermediate steps in order to isolate the real roots. For example, it is possible to sample only the ends of the interval, to evaluate the function's gradient, or to draw conclusions from an evaluation of the function's value set.

Variation Diminishing Methods

Lane and Riesenfeld introduced this technique in 1981 (see [222]). Since then several variants have been proposed by other researchers, namely Schneider [354] and Spencer [372]. This technique works with a polynomial defined on a finite interval. It repeatedly subdivides the interval domain of the polynomial into two parts by its midpoint, with a view to isolating (and ultimately approximating) the polynomial's real roots. This recursive subdivision of the interval stops when either the root is approximated to the desired precision, or it is established that no root exists in one of the subintervals, whereupon that subinterval is eliminated. The Lane-Riesenfeld algorithm combines recursive bisection with the variation diminishing property of the Bernstein polynomials to know whether or not a root exists in the subinterval. Binary subdivision involves $\mathcal{O}(n^2)$ steps and provides one bit of accuracy for each step.

Hull Approximation Methods

Instead of using the variation diminishing property, hull approximation methods exploit the convex hull property of the Bernstein polynomials. This is in order to isolate, as well as to approximate the real roots of a polynomial. Rajan-Klinkner-Farouki's method [332] is well-known in this category. It uses parabolic hulls to isolate and approximate simple real roots of a Bernstein-form polynomial. A parabolic hull is a parabolic generalisation of the convex hull property of the Bernstein-form polynomial. This method possesses cubic convergence when approximating a root, which makes it a very fast root-finding method even for high degree polynomials (examples up to degree 2048). In his thesis [372], Spencer also describes a method of this type to isolate and approximate real roots for Bernstein-form polynomials.

Descartes' Rule of Signs

As Eigenwillig et al. refer in [129], root isolation based on Descartes' rule of signs was cast into its modern form by Collins and Akritas [89] for polynomials in the power form. An analogous formulation for polynomials in the Bernstein form was first described by Lane and Riesenfeld [222], and more recently by Mourrain et al. [284] (see also [32] and [286]).

Proposition 3.9. *Let $f(x) = \sum_{i=0}^{n} b_i^n B_i^n(x)$ be a Bernstein form polynomial of degree n on the interval $]a, b[$. Let $V(b)$ be the number of sign changes in the list of Bernstein coefficients $b = b_0^n, \ldots, b_n^n$ and N the number of roots of $f(x)$ in $]a, b[$ counted with multiplicities. Then*

(i) $V(b) \geq N$,
(ii) $V(b) - N$ *is even.*

Proof. See Mourrain et al. [284].

This claim can be viewed as Descartes' rule of signs for Bernstein form polynomials. In other words, it provides the number of roots of a Bernstein form polynomial.

Nevertheless, $V(b)$ only provides us an upper bound for the number N of roots of $f(x)$ in $]a,b[$. However, under some circumstances, $V(b)$ yields the exact number of roots. This is stated by the following theorems:

Proposition 3.10. (One-Circle Theorem) *The open disc bounded by the circle centred at the midpoint of $[a,b]$ does not contain any root of $f(x)$ if and only if $V(b) = 0$.*

Proof. See Mourrain et al. [284] or Krandick and Mehlhorn [217].

Proposition 3.11. (Two-Circle Theorem) *The union of two open discs bounded by the circumcircles of two equilateral triangles sharing $[a,b]$ as one of their edges contains precisely one simple root of $f(x)$ (which is then necessarily a real root) if and only if $V(b) = 1$.*

Proof. See Mourrain et al. [284] or Krandick and Mehlhorn [217].

Both one-circle and two-circle theorems provide the stopping conditions of a recursive algorithm (see steps 4–9 of Algorithm 2) that subdivides the interval $[a,b]$ into subintervals to isolate roots therein. Taking into account that the coefficients of a Bernstein form polynomial depend on the interval being considered, we have to have an algorithm capable of computing the Bernstein coefficients of $f(x)$ on the subintervals $[a,c]$ and $[c,b]$ from those on $[a,b]$ (step 12 of Algorithm 2). This algorithm is known as de Casteljau's algorithm (see Algorithm 3).

Algorithm 2 Real Root Isolation

1: **procedure** BERNSTEINROOTISOLATION($f(x), [a,b], L$)
2:     $b_{[a,c]} \leftarrow$ list of Bernstein's coefficients for $[a,c]$
3:     $V(b_{[a,c]}) \leftarrow$ number of sign variations of Bernstein's coefficients for $[a,c]$
4:     **if** $V(b_{[a,c]}) = 0$ **then**   ▷ one-circle theorem
5:        **return** L ▷ stopping condition: $[a,c]$ is not inserted into L
6:     **end if**
7:     **if** $V(b_{[a,c]}) = 1$ **then**   ▷ two-circle theorem
8:        $L \leftarrow L \cup [a,c]$ ▷ stopping condition: $[a,c]$ is inserted into L
9:        **return** L
10:    **end if**
11:    **if** $V(b_{[a,c]}) > 1$ **then**   ▷ de Casteljau subdivision algorithm
12:        CASTELJAU($f(x), b_{[a,b]}, b_{[a,c]}, b_{[c,b]}$)
13:        BERNSTEINROOTISOLATION($f(x), [a,c], L$)
14:        BERNSTEINROOTISOLATION($f(x), [c,b], L$)
15:    **end if**
16: **end procedure**

Algorithm 3 de Casteljau's algorithm

1: **procedure** CASTELJAU($f(x), b_{[a,b]}, b_{[a,c]}, b_{[c,b]}$)
2:     $\alpha \leftarrow \frac{c-a}{b-a}$ and $\beta \leftarrow \frac{b-c}{b-a}$ ▷ first and second weights
3:     $b_i^{(0)} \leftarrow b_i, i = 0, \ldots, p$ ▷ initialisation of Bernstein's coefficients
4:     **for** $i = 1, \ldots, p$ **do**
5:         **for** $j = 0, \ldots, p - i$ **do**
6:             $b_j^{(i)} \leftarrow \alpha b_j^{(i-1)} + \beta b_{j+1}^{i-1}$ ▷ triangle of Bernstein coefficients
7:         **end for**
8:     **end for**
9:     **return** $l_{[a,c]} \leftarrow b_0^{(0)}, \ldots, b_0^{(j)}, \ldots, b_0^{(p)}$ ▷ Bernstein's coefficients for $[a, c]$
10:    **return** $l_{[c,b]} \leftarrow b_0^{(p)}, \ldots, b_j^{(p-j)}, \ldots, b_p^{(0)}$ ▷ Bernstein's coefficients for $[c, b]$
11: **end procedure**

The Bernstein coefficients computed in step 7 of de Casteljau's algorithm form a triangle as follows:

$$
\begin{array}{ccccccc}
b_0^{(0)} & b_1^{(0)} & \ldots & & \ldots & b_{p-1}^{(0)} & b_p^{(0)} \\
& b_0^{(1)} & \ldots & & \ldots & b_{p-1}^{(1)} & \\
& & \ldots & & \ldots & & \\
& & & \ldots & & & \\
& & b_0^{(p-1)} & & b_1^{(p)} & & \\
& & & b_0^{(p)} & & &
\end{array}
$$

The Bernstein coefficients on the interval $[a, b]$ appear on the top side of the triangle, while those on the subintervals $[a, c]$ and $[c, b]$ appear on the left and right sides of the triangle, respectively.

For more details about the Descartes root isolation for univariate polynomials in both power and Bernstein forms, the reader is referred to Mourrain et al. [284]. Another recommended reference is Eigenwillig et al. [129] in which a basis-free or unified approach for Descartes' method is described.

3.4 Multivariate Root Isolation: Power Form Polynomials

Root isolation methods are not easily extendible to multivariate polynomials. They are still a topic of active research.

3.4.1 Multivariate Descartes' Rule of Signs

A possible first attempt to come to a multivariate version of Descartes' rule was due to Itenberg and Roy [200], in 1996. But in 1998 Li and Wang [235] gave a counterexample to their conjecture.

3.4.2 Multivariate Sturm Sequences

The generalisation of Sturm's theorem is not immediate, but was made possible through the work of Milne [273, 274]. His generalisation relies heavily on resultants, and does not deal with singularities, nor with multiple roots. This makes it possible (though not easy) to use in 2D, but increasingly difficult to adjust for higher dimensions. An implementation that relies on exact arithmetic is given in Voiculescu's thesis [401].

Milne's theory is meant to generalise the Sturm technique to n dimensions. However, applications for $n > 2$ are somewhat difficult to implement, the main inconvenience being finding suitable starting terms for the sequence. For instance, the mere case $d = 3$ requires the initial term in the Sturm sequence to be the product of two polynomial resultants. The problem with this is that such a product will introduce "spurious roots" at the intersections of the two resultants. These roots' coordinates are such that some components are roots of one resultant and some others are roots of another resultant. Although they do not make the initial polynomials vanish, they do make the polynomial product vanish. The elimination of these "spurious" points in the root counting technique is not straightforward and makes the algorithm almost impracticable.

In order to deal with this impediment Milne introduced the so-called "volume function" in the calculation of which Gröbner bases are essential—yet notoriously difficult to compute. Gröbner bases were introduced by Buchberger in his PhD thesis [70]. Other references introducing the theory of Gröbner bases are the books of Cox et al. [91] and Becker and Weispfenning [38].

In his thesis [321], Pedersen describes similar algebraic root-counting methods. In [322] he attempts a generalisation of the Sturm theory, based on ideas expressed by Hermite [184]. Pedersen's investigations were contemporary to Milne's. Their results are comparable. See also Gonzalez-Vega and Trujillo [164] for more details.

3.5 Multivariate Root Isolation: Bernstein Form Polynomials

Section 3.3, gave an overview of the way in which univariate root isolation through Bernstein-form polynomials can be based on Descartes' rule of signs. It is easy to assume that the absence of a Descartes-like rule for polynomials of more than one variable might undermine the generalisation the isolation method to multivariate polynomials.

However, this is not the case: since the Bernstein coefficients can be seen as a tensor, we have only to use the univariate de Casteljau subdivision n times. This subdivision can be performed independently for each variable [236]. This technique was recently developed by Mourrain and Pavone [283] and can be viewed as a follow-up of the interval projected polyhedron algorithm proposed by Sherbrooke and Patrikalakis [364].

Evidently, this requires a preliminary algorithm capable of converting a multivariate power form polynomial into a multivariate Bernstein polynomial such as those outlined earlier in Section 3.5.1.

3.5.1 Multivariate Bernstein Basis Conversions

The multivariate Bernstein form polynomial $f(\mathbf{x})$, with $\mathbf{x} = (x_0, \ldots, x_{n-1})$, of maximum degree $\mathbf{d} = (d_0, \ldots, d_{n-1})$ can be obtained by rewriting Definition 3.5 in the form of tensor products as follows:

$$f(\mathbf{x}) = \sum_{k_0=0}^{d_0} \cdots \sum_{k_n=0}^{d_n} b_{k_1,\ldots,k_n} B_{k_0}^{d_0}(x_0) \ldots B_{k_n}^{d_n}(x_n) \qquad (3.13)$$

The Bernstein coefficients $b_{k_1,\ldots,k_n}$ can be seen as a tensor of dimension n.

Methods for converting a multivariate power form polynomial into a multivariate Bernstein polynomial have been proposed by Berchtold et al. [40] and by Garloff [425], both outlined below.

3.5.2 Bivariate Case

Berchtold et al. [40] note that the implicit expression of a bivariate polynomial in the power basis can also be rewritten in terms of matrix multiplication:

$$f(x,y) = a_{00} + a_{10}x + a_{01}y + a_{11}xy + \cdots + a_{mn}x^m y^n = \mathbf{X}\mathbf{A}\mathbf{Y}$$

where

$$\mathbf{X} = \begin{pmatrix} 1 & x & \cdots & x^m \end{pmatrix} \qquad \mathbf{Y} = \begin{pmatrix} 1 \\ y \\ \vdots \\ y^n \end{pmatrix} \qquad \mathbf{A} = \begin{pmatrix} a_{00} & \cdots & a_{0n} \\ \vdots & & \vdots \\ a_{m0} & \cdots & a_{mn} \end{pmatrix}$$

By analogy with the univariate case,

$$f(x,y) = \mathbf{X}\mathbf{A}\mathbf{Y} = \mathbf{B}_X \mathbf{B} \mathbf{B}_Y$$

where $\mathbf{B}_X$ and $\mathbf{B}_Y$ are Bernstein vectors in the variables $x \in [\underline{x}, \overline{x}]$ and $y \in [\underline{y}, \overline{y}]$. These vectors can be decomposed as shown in Section 3.1.3.

In the case of the Bernstein vector corresponding to the variable y the factors $\mathbf{C}$, $\mathbf{D}$ and $\mathbf{E}$ in Equation (3.11) will appear in reverse order. This happens because $\mathbf{B}_Y$ is a column vector (as opposed to $\mathbf{B}_X$ which is a row vector).

Hence, $\forall x \in [\underline{x}, \overline{x}]$, $\forall y \in [\underline{y}, \overline{y}]$, by equating the power form and Bernstein-form polynomials we obtain:

$$\mathbf{X A Y} = \mathbf{X E}_X \mathbf{D}_X \mathbf{C}_X \ \mathbf{B} \ \mathbf{C}_Y \mathbf{D}_Y \mathbf{E}_Y \mathbf{Y} \tag{3.14}$$

where

$$\mathbf{B} = (\mathbf{C}_X)^{-1}(\mathbf{D}_X)^{-1}(\mathbf{E}_X)^{-1} \ \mathbf{A} \ (\mathbf{E}_Y)^{-1}(\mathbf{D}_Y)^{-1}(\mathbf{C}_Y)^{-1}$$

Example 3.12. (Bivariate Bernstein form polynomial in $[0,1] \times [0,1]$) Given the equation of a circle centred at $(\frac{1}{2}, \frac{1}{2})$ and of radius $\frac{2}{5}$ in either the canonical or the expanded power form

$$\begin{aligned} f(x,y) &= \left(x - \frac{1}{2}\right)^2 + \left(y - \frac{1}{2}\right)^2 - \left(\frac{2}{5}\right)^2 \\ &= x^2 - x + \frac{17}{50} + y^2 - y \end{aligned}$$

the conversion algorithm based on Formula (3.14) finds the following equivalent Bernstein form in $[0,1] \times [0,1]$:

$$\begin{aligned} b_f(x,y) = & \left(\frac{17}{50}(1-x)^2 - \frac{8}{25}x(1-x) + \frac{17}{50}x^2\right)(1-y)^2 \\ & + 2\left(-\frac{4}{25}(1-x)^2 - \frac{33}{25}x(1-x) - \frac{4}{25}x^2\right)y(1-y) \\ & + \left(\frac{17}{50}(1-x)^2 - \frac{8}{25}x(1-x) + \frac{17}{50}x^2\right)y^2 \end{aligned}$$

3.5.3 Trivariate Case

It is possible to generalise this formula further, to 3D and higher dimensions. Its trivariate version is rather difficult to write in linear form because the order and direction in which the tensor products of the matrices involved is essential for the correctness of the calculation. This method was given jointly by Berchtold [41] and Voiculescu [401].

By analogy with the univariate and bivariate cases, the implicit expression of a trivariate polynomial in the power basis can also be rewritten in terms of matrix multiplication:

$$\begin{aligned} f(x,y,z) = \ & a_{000} \\ & + a_{100}x + a_{010}y + a_{001}z \\ & + a_{110}xy + a_{101}xz + a_{011}yz \\ & + \cdots + a_{mnl}x^m y^n z^l \\ = \ & \mathbf{Y} \otimes_y (\mathbf{X} \otimes_x \mathbf{A}) \otimes_z \mathbf{Z} \end{aligned}$$

where $\mathbf{A}_{m \times n \times l}$ is the three-dimensional coefficient tensor, and $\mathbf{X}$, $\mathbf{Y}$ and $\mathbf{Z}$ are chosen such that the tensor multiplications are well-defined.

The following types of tensor multiplication have been chosen :

3.5 Multivariate Root Isolation: Bernstein Form Polynomials

$$\otimes_x : Q_{q\times m} \otimes_x \mathbf{A}_{m\times n\times l} = R_{q\times n\times l}$$
$$\otimes_y : Q_{q\times n} \otimes_y \mathbf{A}_{m\times n\times l} = R_{m\times q\times l}$$
$$\otimes_z : \mathbf{A}_{m\times n\times l} \otimes_z Q_{l\times q} = R_{m\times n\times q}$$

If $\mathbf{B}_X$, $\mathbf{B}_Y$ and $\mathbf{B}_Z$ are Bernstein vectors in the respective variables, the Bernstein form of the polynomial $f(x,y,z)$ is:

$$f(x,y,z) = \mathbf{Y} \otimes_y (\mathbf{X} \otimes_x \mathbf{A}) \otimes_z \mathbf{Z} = \mathbf{B}_Y \otimes_y (\mathbf{B}_X \otimes_x \mathbf{B}) \otimes_z \mathbf{B}_Z$$

The Bernstein vectors can be decomposed as shown in Equation 3.11. When the power form is made equal to the Bernstein form, the following relation is obtained:

$$\mathbf{Y} \otimes_y (\mathbf{X} \otimes_x \mathbf{A}) \otimes_z \mathbf{Z} = \begin{array}{c} \mathbf{YE}_Y \mathbf{D}_Y \mathbf{C}_Y \\ \otimes_y \\ (\mathbf{X} \otimes_x (\mathbf{E}_X \otimes_x (\mathbf{D}_X \otimes_x (\mathbf{C}_X \otimes_x \mathbf{B})))) \\ \otimes_z \\ \mathbf{C}_Z \mathbf{D}_Z \mathbf{E}_Z \mathbf{Z} \end{array}$$

In this equation the three-dimensional tensor $\mathbf{B}$ is being multiplied consecutively by each of the two-dimensional factors. At each stage another three-dimensional tensor is produced. After the $\otimes_x$-multiplication with the vector $\mathbf{X}$, the three-dimensional tensor is reduced to two dimensions. The rest of the multiplications are the usual two-dimensional ones.

Hence, $\forall x \in [\underline{x}, \overline{x}], \forall y \in [\underline{y}, \overline{y}], \forall z \in [\underline{z}, \overline{z}]$, the Bernstein coefficients tensor $\mathbf{B}$ can be calculated by:

$$\mathbf{B} = (\mathbf{C}_Y)^{-1} \otimes_y (\mathbf{D}_Y)^{-1} \otimes_y (\mathbf{E}_Y)^{-1}$$
$$\otimes_y ((\mathbf{C}_X)^{-1} \otimes_x (\mathbf{D}_X)^{-1} \otimes_x \underbrace{(\mathbf{E}_X)^{-1} \otimes_x \mathbf{A}}_{\leftarrow})$$
$$\otimes_z (\mathbf{E}_Z)^{-1} \otimes_z (\mathbf{D}_Z)^{-1} \otimes_z (\mathbf{C}_Z)^{-1}$$

It is essential in this equation that the order of the multiplications is starting from the tensor $\mathbf{A}$ outwards (according to the orientation of the arrows).

Example 3.13. (Trivariate Bernstein form polynomial in $[2,3] \times [6,7] \times [4,5]$) The Bernstein form of the polynomial $f(x,y,z) = x^3 y^2 z^7$ in the 3D box specified above is:

$$\begin{aligned}b_f(x,y,z) = &\left(288\,(3-x)^3\,z^7 + 1296\,(x-2)\,(3-x)^2\,z^7\right.\\ &\left.+1944\,(x-2)^2\,(3-x)\,z^7 + 972\,(x-2)^3\,z^7\right)(7-y)^2\\ &+2\left(336\,(3-x)^3\,z^7 + 1512\,(x-2)\,(3-x)^2\,z^7\right.\\ &\left.+2268\,(x-2)^2\,(3-x)\,z^7 + 1134\,(x-2)^3\,z^7\right)(y-6)\,(7-y)\\ &+\left(392\,(3-x)^3\,z^7 + 1764\,(x-2)\,(3-x)^2\,z^7\right.\\ &\left.+2646\,(x-2)^2\,(3-x)\,z^7 + 1323\,(x-2)^3\,z^7\right)(y-6)^2\end{aligned}$$

3.5.4 Arbitrary Number of Dimensions

Zettler and Garloff [425] give an equivalent formula for the calculation of the coefficients for an n-variate Bernstein form polynomial.

Let $l \in \mathbb{N}$ be the number of variables and $\mathbf{x} = (x_1, \ldots, x_l) \in \mathbb{R}^l$. A multi-index I is defined as $I = (i_1, \ldots, i_l) \in \mathbb{N}^l$. For two given multi-indices $I, J \in \mathbb{N}^l$ the following conventions are made:

Notation. Write $I \leq J$ for the case where $0 \leq i_1 \leq j_1, \ldots, 0 \leq i_l \leq j_l$.

Notation. Denote the product $\binom{i_1}{j_1} \cdots \binom{i_l}{j_l}$ by $\binom{I}{J}$.

Notation. Denote by the product $x_1^{i_1} \cdots x_l^{i_l}$ $\mathbf{x}^I$.

Let $p(\mathbf{x})$ be a multivariate polynomial in l variables with real coefficients.

Definition 3.14. $D = (d_1, \ldots, d_l)$ is the tuple of maximum degrees so that d_k is the maximum degree of x_k in $p(\mathbf{x})$, for $k = 1, \ldots, l$.

Definition 3.15. The set $S = \{I \in \mathbb{N}^l : I \leq D\}$ contains all the tuples from $\mathbb{R}^l$ which are 'smaller than or equal to' the tuple D of maximum degrees.

Then an arbitrary polynomial $p(\mathbf{x})$ can be written as:

$$p(\mathbf{x}) = \sum_{I \in S} a_I \mathbf{x}^I$$

where $a_I \in \mathbb{R}$ represents the corresponding coefficient[1] to each $\mathbf{x}^I \in \mathbb{R}^l$.

As in equation 3.5, a univariate Bernstein polynomial of degree n on the unit interval $[0, 1]$ is defined by:

$$B_k^n(x) = \binom{n}{k} x^k (1-x)^{n-k} \qquad k = 0, \ldots, n;\ x \in [0,1].$$

---

[1] Note that some of the a_I may be 0.

For the multivariate case consider, without loss of generality, a unit box $U = [0,1]^l$ and the Ith Bernstein polynomial of degree D is defined by:

$$B_I^D(\mathbf{x}) = B_{i_1}^{d_1}(x_1) \times \cdots \times B_{i_l}^{d_l}(x_l) \qquad \mathbf{x} \in \mathbb{R}^l.$$

The Bernstein coefficients $B_I(U)$ of p over the unit box $U = [0,1]^l$ are given by:

$$B_I(U) = \sum_{J \leq I} \frac{\binom{I}{J}}{\binom{D}{J}} a_J \qquad I \in S.$$

And so the Bernstein form of a multivariate polynomial p is defined by:

$$p(\mathbf{x}) = \sum_{I \in S} B_I(U) B_I^D(\mathbf{x}).$$

For the uni-, bi- and trivariate polynomials in the examples above, this formula and the alternative formulae (by Berchtold et al. [40]) given earlier in this chapter generate the same Bernstein form polynomial.

3.6 Final Remarks

This chapter has given an overview of some of the most significant root isolation techniques for real functions. In particular, real root isolation of univariate integer polynomials is a classical and well studied problem, so a variety of algorithms can be found in the literature. We have merely scratched the surface of the existing literature of this topic in the area of algebraic and symbolic computation.

Interval subdivision-based algorithms for real root isolation are based either on Descartes' rule of signs or on Sturm sequences. In general terms, the idea behind these two approaches consists of partitioning a given interval containing all the real roots into disjoint subintervals such that distinct roots are assigned distinct subintervals. For that, Descartes' approach repeatedly transforms the original polynomial and counts the sign variations of the coefficients, while Sturm's approach constructs a signed remainder sequence of polynomials and evaluates them over the interval of interest.

Besides, as recently proven, both Descartes' (either power basis or Bernstein basis) and Sturm's approaches achieve the same bit complexity bound [117, 129]. For an alternative to the subdivision-based algorithms, the interested reader is referred to a recent paper due to Tsigaridas and Emiris [391] and Sharma [359] (and the bibliography therein), where the continued fractions-based algorithms for root isolation are approached.

4
Interval Arithmetic

4.1 Introduction

The fundamental idea behind the interval arithmetic (IA) is that the values of a variable can be expressed as ranging over a certain interval. If one computes a number A as an approximation to some unknown number X such that $|X - A| \leq B$, where B is a precise bound on the overall error in A, we will know for sure that X lies in the interval $[A - B, A + B]$, no matter how A and B are computed. The idea behind IA was to investigate computations with intervals, instead of simple numbers.

In fact, when we use a computer to make real number computations, we are limited to a finite set of floating-point numbers imposed by the hardware. In these circumstances, there are two main options for approximating a real number. One is to use a simple floating point approximation of the number and to propagate the error of this approximation whenever the number is used in a calculation. The other is to bind the number in an interval (whose ends may also be floating point values) within which the number is guaranteed to lie. In the latter case, any calculation that uses the number can just as well use its interval approximation instead. This chapter deals with computations involving two floating-point numbers as intervals—the subject covered by interval arithmetic. Approximations carried out with a single floating-point number are studied in the next chapter.

Interval arithmetic, also known as interval mathematics, interval analysis, or interval computation, has been developed by mathematicians and computer scientists since the late 1950s and early 1960s as an approach to putting bounds on rounding errors in arithmetic computations. In this respect, Ramon Moore's PhD thesis [278], as well as his book [279] and other papers published *a posteriori*, played an important role in the development of interval arithmetic. Interval analysis is now a field of study in itself, widely used in numerical analysis and geometric modelling, as well as many other computation processes which require some guarantee in the results of calculations.

Other relevant references in interval arithmetic can be found in the literature. For example, Alefeld and Herzberger [4] propose using intervals in Newton-like methods for finding the roots of univariate functions and also in solving systems of equations. Neumaier [299] takes the concept of intervals further and develops distance definitions and topological properties for intervals. Methods for finding enclosures for the range of a function are given, as well as interval-based methods for solving systems of equations.

Apart from the classical way of looking at intervals, other approaches exist whereby the interval is regarded as an approximation of its centre. Ratschek and Rockne [335], as well as Neumaier [299], discuss the use of *centred-form intervals*. Comba and Stolfi [90] and Andrade et al. [17] take this approach even further in their *affine arithmetic*. Affine arithmetic (AA) still regards the interval as an approximation of the number at its centre, but at the same time keeps track of the various levels of error affecting the computed quantity at different steps of the evaluation of an expression. Their results are quite encouraging, in that they are tighter than the ones produced by the traditional IA. But as expected, there is a tradeoff between accuracy and computation time cost.

IA and AA are also used in research areas such as computer graphics and geometric modelling. At our best knowledge, Suffern and Fackerell [379] and Snyder [370] were who first introduced interval arithmetic in these research areas. For example, Snyder [370] explains the advantages of using IA in geometric modelling as opposed to approaching global problems by finding roots of polynomials. The main point he makes is that IA controls the approximation errors during the floating-point computation by computing bounds rather than exact values. The other major advantage of using interval methods is that they are exhaustive and can give information about the whole region of interest. Other references in scientific computing, computer graphics and geometric modelling include de Figueiredo and Stolfi [102], Heidrich et al. [181], Cusatis et al. [99], Voiculescu [401], Martin et al. [259], Bowyer et al. [62, 63], Bülher and Barth [72], Michelucci [270], Bülher [71], Shou et al. [367], Figueiredo et al. [103], Fang et al. [131], Paiva et al. [315], and Miyajima and Kashiwagi [275].

In this chapter, we look at the IA and AA rules, and describe how they can be used in geometric modelling. This is important because geometric modelling not only involves high precision calculations, but also uses intervals in order to denote and study regions of space, regardless of whether they contain implicit curves, surfaces or solids. For example, a point can be approximated by the intervals that give bounds for its coordinates. Hence a neighbourhood in the shape of a box describes the region of space where that point is guaranteed to lie. Evaluating the function at that point (or some similar potential value for the whole box of coordinate ranges) has geometrical meaning: it is a measure of how far away the point (or the box) is from the surface represented by the function. This measurement is only relative, as the function value merely indicates which of several points is closer to a given surface but it does not actually help evaluate the distance from a point to the surface.

4.2 Interval Arithmetic Operations

The execution of an automatic computation usually involves the propagation of inaccuracies and rounding errors, because floating point values are merely rational approximations of real numbers. If interval ranges are used instead of a single approximation, then an automatic computation results in a range of possible values for the final solution. This solution is generally described by means of an interval. Once again, this is only one intuitive motivation for using intervals and introducing arithmetic operations on the set of intervals. The exact way in which intervals are used in geometric modelling is explained later.

4.2.1 The Interval Number

Owing to Moore's work, the mathematical concept of number has been generalised to the ordered pair of its approximations—*the interval number*. An interval number $\mathbf{x}$ is denoted as the ordered pair of reals $[\underline{x}, \bar{x}]$, $\underline{x} \leq \bar{x}$, which defines the set of real numbers

$$[\underline{x}, \bar{x}] = \{x \mid \underline{x} \leq x \leq \bar{x}\}$$

When one of the extremities of the interval needs to be excluded from the interval set, variations of the following notation are used: $]\underline{x}, \bar{x}] = \{x \mid \underline{x} < x \leq \bar{x}\}$. Either or both extremities of an interval can be excluded from the set by using the appropriate inequalities. This particular notation has the advantage of distinguishing between the open interval $]\underline{x}, \bar{x}[$ and the pair of numbers $(\underline{x}, \bar{x})$.

4.2.2 The Interval Operations

The rules of arithmetic can be redefined so that they apply to interval numbers. If $\mathbf{x} = [\underline{x}, \bar{x}]$ and $\mathbf{y} = [\underline{y}, \bar{y}]$, and the operator $\odot \in \{+, -, \times, /\}$ then the four elementary arithmetic operations will follow the scheme:

$$\mathbf{x} \odot \mathbf{y} = \{x \odot y : x \in \mathbf{x}, \ y \in \mathbf{y}\}$$

An interval operation must produce a new interval containing all the possible results that can be obtained by performing the operation in question on any element of the argument intervals. This template produces simpler specific rules for each of the arithmetic operators (see also Higham [186]):

4 Interval Arithmetic

Addition:
$$\mathbf{x} + \mathbf{y} = [\underline{x} + \underline{y}, \overline{x} + \overline{y}] \tag{4.1}$$

Subtraction:
$$\mathbf{x} - \mathbf{y} = [\underline{x} - \overline{y}, \overline{x} - \underline{y}] \tag{4.2}$$

Multiplication:
$$\mathbf{x} \times \mathbf{y} = [\min\{\underline{x}\underline{y}, \underline{x}\overline{y}, \overline{x}\underline{y}, \overline{x}\overline{y}\}, \max\{\underline{x}\underline{y}, \underline{x}\overline{y}, \overline{x}\underline{y}, \overline{x}\overline{y}\}] \tag{4.3}$$

Division:
$$\mathbf{x}/\mathbf{y} = [\min\{\underline{x}/\underline{y}, \underline{x}/\overline{y}, \overline{x}/\underline{y}, \overline{x}/\overline{y}\}, \max\{\underline{x}/\underline{y}, \underline{x}/\overline{y}, \overline{x}/\underline{y}, \overline{x}/\overline{y}\}] \tag{4.4}$$

Depending on the circumstances in which interval division is used, it may be appropriate to declare division by an interval containing zero as *undefined* or to express it as a union of two semi-infinite intervals.

Interval division can also be written as follows:

$$\mathbf{x}/\mathbf{y} = \mathbf{x} \times \frac{1}{\mathbf{y}} \tag{4.5}$$

where

$$\frac{1}{\mathbf{y}} = \begin{cases} \left[\frac{1}{\overline{y}}, \frac{1}{\underline{y}}\right] & \text{if } \underline{y} > 0 \text{ or } \overline{y} < 0 \\ \left]-\infty, \frac{1}{\underline{y}}\right] \cup \left[\frac{1}{\overline{y}}, \infty\right[& \text{if } \underline{y} \leq 0 \leq \overline{y} \end{cases}$$

The addition and multiplication operations are commutative, associative and subdistributive. The subdistributivity property comes from that fact that the set $\mathbf{x}(\mathbf{y} + \mathbf{z})$ is a subset of $\mathbf{xy} + \mathbf{xz}$.

An additional operation is the exponentiation of an interval. Interestingly, it is defined differently from number exponentiation as follows:

Exponentiation:

$$[\underline{x}, \overline{x}]^{2n+1} = [\underline{x}^{2n+1}, \overline{x}^{2n+1}] \tag{4.6}$$

$$[\underline{x}, \overline{x}]^{2n} = \begin{cases} [\underline{x}^{2n}, \overline{x}^{2n}] & \text{if } 0 \leq \underline{x} < \overline{x} \\ [0, M^{2n}] & \text{if } \underline{x} < 0 \leq \overline{x} \\ [\overline{x}^{2n}, \underline{x}^{2n}] & \text{if } \underline{x} < \overline{x} < 0 \end{cases} \tag{4.7}$$

where n is any natural number and $M = \max\{|\underline{x}|, |\overline{x}|\}$.

In particular, for even values of k,

$$[\underline{x}, \overline{x}]^k \neq \underbrace{[\underline{x}, \overline{x}] \times \cdots \times [\underline{x}, \overline{x}]}_{k}$$

which is proved by a simple counterexample:

$$[-1, 2] \times [-1, 2] = [-2, 4]$$
$$[-1, 2]^2 = [0, 4]$$

The interval resulting from an even power exponentiation is always *entirely positive* (even when the interval which is being raised to the even power contains negative numbers).

4.3 Interval Arithmetic-driven Space Partitionings

Interval arithmetic is especially useful in geometric modelling when objects (e.g. points, curves, surfaces, and solids) are represented by implicit functions and are categorised by means of space partitioning. As seen above, an interval can be regarded as an entity which gives lower and upper approximations of a number. Since a point in Euclidean space is a pair of real coordinates in 2D (respectively, a triplet in 3D), it can be naturally approximated by a pair (respectively, triplet) of intervals, i.e. an axially aligned *box*. Thus, the classical point membership testing method used in geometric modelling can be extended to a *box testing method*.

We are then able to combine interval arithmetic with axially aligned space partitionings to locate objects defined implicitly. This is illustrated in Figure 4.1, where combining interval arithmetic and a 2-d tree space partitioning allows us to locate the following curve defined implicitly as follows:

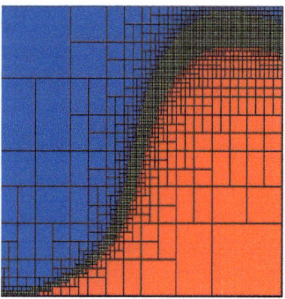

Fig. 4.1. An implicit curve specified by the power-form polynomial $p(x, y)$ defined below, in the ambient space $[0, 1] \times [0, 1]$ and using a minimum box size of $\frac{1}{2^7} \times \frac{1}{2^7}$.

$$p(x,y) = \frac{9446}{10{,}000}\,xy - \frac{700{,}443{,}214}{100{,}000{,}000}\,x^3y^2 + \frac{764{,}554}{100{,}000}\,x^4y^3 + \frac{564}{1000}\,y^4 - x^3$$

The curve is somewhere in the region of the green boxes, i.e. those boxes in which p evaluates to an interval that straddles zero. The red boxes denote entirely negative boxes, i.e. boxes in which p evaluates negative everywhere. The blue boxes identify entirely positive boxes, i.e. boxes in which p evaluates positive everywhere.

Here is an example that illustrates this box classification. Given the axially aligned box $[\underline{x}, \overline{x}] \times [\underline{y}, \overline{y}]$, the two variables of the curve expression x and y are replaced by the two interval coordinates $[\underline{x}, \overline{x}]$ and $[\underline{y}, \overline{y}]$, respectively. This substitution produces an interval expression which is then evaluated by applying IA rules. This evaluation results in an interval. For example, let us consider the box $[\frac{1}{2}, \frac{5}{8}] \times [0, \frac{1}{8}]$ in Figure 4.1. Substituting x and y by $[\frac{1}{2}, \frac{5}{8}]$ and $[0, \frac{1}{8}]$, respectively, in the expression of p, we obtain

$$\begin{aligned}
p_{rat}\left(\left[\frac{1}{2},\frac{5}{8}\right],\left[0,\frac{1}{8}\right]\right) &= \frac{9446}{10{,}000}\left[\frac{1}{2},\frac{5}{8}\right]\left[0,\frac{1}{8}\right] - \frac{700{,}443{,}214}{100{,}000{,}000}\left[\frac{1}{2},\frac{5}{8}\right]^3\left[0,\frac{1}{8}\right]^2 \\
&+ \frac{764{,}554}{100{,}000}\left[\frac{1}{2},\frac{5}{8}\right]^4\left[0,\frac{1}{8}\right]^3 + \frac{564}{1000}\left[0,\frac{1}{8}\right]^4 - \left[\frac{1}{2},\frac{5}{8}\right]^3 \\
&= \left[-\frac{27{,}086{,}029{,}053}{10^{11}}, -\frac{4{,}878{,}689{,}313}{10^{11}}\right]
\end{aligned}$$

which is an entirely negative interval; this confirms that the box $[\frac{1}{2}, \frac{5}{8}] \times [0, \frac{1}{8}]$ in Figure 4.1 is correctly depicted red.

4.3.1 The Correct Classification of Negative and Positive Boxes

As seen above, there are three types of boxes output by interval arithmetic: negative boxes, positive boxes and zero boxes (i.e. those that depict a region where the function evaluates to an interval that straddles zero). As will be shown later, not all zero boxes contain segments of the curve, i.e. not all boxes classified as zero boxes are genuine zero boxes. The prediction that the box contains at least a curve segment is reasonably accurate only for low-degree polynomials, but problems become manifest when the curve expression is of high degree.

However, we can prove that when a box is labelled as negative or positive it is indeed correctly classified. The proof will be carried out in the one-dimensional case, but can be easily generalised to any number of dimensions.

Given an implicit polynomial equation $f : \mathbb{R} \to \mathbb{R}$, $f(x) = 0$ and a 'box' $[\underline{x}, \overline{x}]$, we will first prove that if the box is labelled as positive then all the points in the box have a positive function value. In other words,

$$f([\underline{x}, \overline{x}]) > 0 \quad \overset{?}{\Longrightarrow} \quad f(x) > 0, \ \forall x \in [\underline{x}, \overline{x}].$$

Since $f(x, y)$ is chosen as a polynomial function, its expression is an algebraic combination of entities involving additions, subtractions, multiplications and exponentiations. Hence its corresponding interval expression will involve similar combinations. All that remains to be proved is that any arithmetic combination that yields a positive interval will yield a positive quantity when the calculation is performed with numbers instead of intervals.

Addition:

$$[\underline{x}, \overline{x}] + [\underline{y}, \overline{y}] = [\underline{x} + \underline{y}, \overline{x} + \overline{y}] > 0 \quad \stackrel{?}{\Longrightarrow} \quad x + y > 0, \; \forall x \in [\underline{x}, \overline{x}], y \in [\underline{y}, \overline{y}]$$

Proof.

$$x + y \geq \underline{x} + \underline{y} > 0$$

$\square$

Subtraction:

$$[\underline{x}, \overline{x}] - [\underline{y}, \overline{y}] = [\underline{x} - \overline{y}, \overline{x} - \underline{y}] > 0 \quad \stackrel{?}{\Longrightarrow} \quad x - y > 0, \; \forall x \in [\underline{x}, \overline{x}], y \in [\underline{y}, \overline{y}]$$

Proof.

$$x - y \geq \underline{x} - \overline{y} > 0$$

$\square$

Multiplication:

$$[\underline{x}, \overline{x}] \times [\underline{y}, \overline{y}] > 0 \quad \stackrel{?}{\Longrightarrow} \quad xy > 0, \; \forall x \in [\underline{x}, \overline{x}], y \in [\underline{y}, \overline{y}]$$

Proof.

$$[\underline{x}, \overline{x}] \times [\underline{y}, \overline{y}] = [\min\{\underline{x}\underline{y}, \underline{x}\overline{y}, \overline{x}\underline{y}, \overline{x}\overline{y}\}, \max\{\underline{x}\underline{y}, \underline{x}\overline{y}, \overline{x}\underline{y}, \overline{x}\overline{y}\}]$$
$$xy \geq \min\{\underline{x}\underline{y}, \underline{x}\overline{y}, \overline{x}\underline{y}, \overline{x}\overline{y}\} > 0$$

$\square$

Exponentiation:

For any natural number n, let us first consider the exponentation operator for odd powers:

$$[\underline{x}^{2n+1}, \overline{x}^{2n+1}] > 0 \quad \stackrel{?}{\Longrightarrow} \quad x^{2n+1} > 0, \; \forall x \in [\underline{x}, \overline{x}]$$

Proof.

$$x^{2n+1} \geq \underline{x}^{2n+1} > 0$$

$\square$

Now, let us do the same for even powers:

$$[\underline{x}, \overline{x}]^{2n} > 0 \quad \stackrel{?}{\Longrightarrow} \quad x^{2n} > 0, \; \forall x \in [\underline{x}, \overline{x}]$$

Proof. If $0 \leq \underline{x} < \overline{x}$ then $x^{2n} \geq \underline{x}^{2n} > 0$. If $\underline{x} < \overline{x} \leq 0$ then $x^{2n} \geq \overline{x}^{2n} > 0$. The case $\underline{x} < 0 \leq \overline{x}$ cannot be achieved because this would mean that $[\underline{x}, \overline{x}]^{2n} = [0, M^{2n}]$ (where $M = \max\{|\underline{x}|, |\overline{x}|\}$), which would contradict the strict inequality $[\underline{x}, \overline{x}]^{2n} > 0$. □

So, for any function f involving a combination of the arithmetic operations above, we have proved that

$$f([\underline{x}, \overline{x}]) > 0 \quad \Longrightarrow \quad f(x) > 0, \; \forall x \in [\underline{x}, \overline{x}]$$

There is another half to this proof, stating an analogous result for negative boxes.

$$f([\underline{x}, \overline{x}]) < 0 \quad \Longrightarrow \quad f(x) < 0, \; \forall x \in [\underline{x}, \overline{x}]$$

This result is based on the symmetry of the IA rules. Its proof is analogous to the one of the first part.

This theory can be easily extended to include rational functions, as interval division is expressed in terms of multiplication. Another important generalisation can be done to include more than one dimension. In fact, the one-dimensional case has been used merely for clarity of the argument, but since multidimensional IA rules are expressed componentwise, there is no reason why the result should not hold in any number of dimensions.

4.3.2 The Inaccurate Classification of Zero Boxes

Let us now examine the case where the resulting interval of the substitution straddles zero. At first sight this may seem to correspond to a situation where the box contains some curve segment or surface patch, independently of whether it belongs to the frontier of a solid or not. This section will illustrate a one-dimensional counterexample. We will show it is possible for the interval to straddle zero despite the box being an positive box indeed. Again, the phenomenon described can be easily observed and generalised to any number of dimensions.

Consider the following four real polynomial functions $f, g, h, k : [0, 1] \to \mathbb{R}$ given by

$$f(x) = 4x^2 - 12x + 9 \qquad \qquad (power\ form)$$

$$g(x) = (4x - 12)x + 9 \qquad \qquad (Horner\ form)$$

$$h(x) = 9(x-1)^2 - 6x(x-1) + x^2 \qquad (Bernstein\ form)$$

$$k(x) = (2x - 3)^2 \qquad \qquad (factored\ form)$$

Although they appear in different forms,[1] their definitions are chosen such that $f(x) = g(x) = h(x) = k(x)$. Despite the fact that they take only positive values over the interval $[0, 1]$, in some cases the membership test outputs intervals straddling zero, though of course they all contain the image of the function.

The functions f, g, h and k take the same values everywhere and have equivalent implicit expressions, so they must have the same image in the range—namely the interval $[1, 9]$. Depending on the form of the polynomial expression, the interval arithmetic method may give predictions for the image which are wider intervals *including* it. This phenomenon is known as *interval swell* or *interval over-estimation* and is responsible for the appearance of false zero boxes. Let us illustrate this with the previous four real-valued functions by replacing x by $[0, 1]$ in their expressions:

$$f([0, 1]) = 4([0, 1])^2 - 12[0, 1] + 9$$
$$= [-3, 13]$$
$$\supset [1, 9] = \text{Image } f$$

$$g([0, 1]) = (4\,[0, 1] - 12)\,[0, 1] + 9$$
$$= [-3, 9]$$
$$\supset [1, 9] = \text{Image } g$$

$$h([0, 1]) = 9([0, 1] - 1)^2 - 6[0, 1]([0, 1] - 1) + [0, 1]^2$$
$$= [0, 16]$$
$$\supset [1, 9] = \text{Image } h$$

$$k([0, 1]) = (2\,[0, 1] - 3)^2$$
$$= [1, 9]$$
$$= [1, 9] = \text{Image } k$$

After applying interval arithmetic to the functions f, g, h and k, we observe that only the prediction given by $k(x)$ gives an exact answer: the prediction in this case equals the exact image $[1, 9]$ of the function. The other examples illustrate the typical situation where the resulting interval straddles zero but the corresponding box is a false zero box because the box itself lies entirely in the positive half-space.

The boxes that interval arithmetic does label as negative or positive are always properly identified. However, not all zero boxes are correctly identified. But this is only a cautious box classification as the interval arithmetic technique cannot determine correctly the type of *all* the boxes in a given region of interest. In this scenario, the box classification is said to be *conservative*.

---

[1] For the definition of *Horner's scheme* (also known as nested multiplication), the reader is referred to the original article [194] as well as any good textbook on algebra or geometric algorithms [132]. A good *splines* textbook [132] will contain a definition and usage of the Bernstein polynomial basis.

Conservativeness is the main weakness of IA. Often the intervals produced are much wider than the true range of the computed quantities. This problem is particularly severe in long computational chains where the intervals computed at one stage are input into the next stage of the computation. The more variable occurrences there are in the algebraic expression, the wider the prediction and the larger the interval swell. However, this is not a general rule, because there are other aspects (such as the presence of even exponents and the order of the arithmetic operations) which may influence the final result. Several conservativeness examples and a suggested approach to this problem can be found in [40], [62] and [401].

4.4 The Influence of the Polynomial Form on IA

There is a wide variety of ways of writing and rewriting a polynomial. In the previous section, we have briefly approached four polynomial forms: power form, Horner form, Bernstein form and factored form. The reader is referred to de Boor [98] for other polynomial forms. Unfortunately, there is no known method to determine what is the best form to express a given polynomial function in order to get the sharpest possible bounds. This is so because the optimal way of representing and storing a polynomial is crucially determined by the kind of operations the user might want to perform on it afterwards. Studies and comparisons are given in Martin et al. [258, 259].

This section shows that the Bernstein form is the most stable numerically by comparing it to the power form as input to interval arithmetic. As suggested in the previous section for the equivalent functions $f(x)$, $g(x)$, $h(x)$ and $k(x)$, the resulting intervals obtained by replacing x by $[0, 1]$ may differ from one to another. This means that applying interval arithmetic to two equivalent functions has as a result two distinct space partitionings (Figure 4.2).

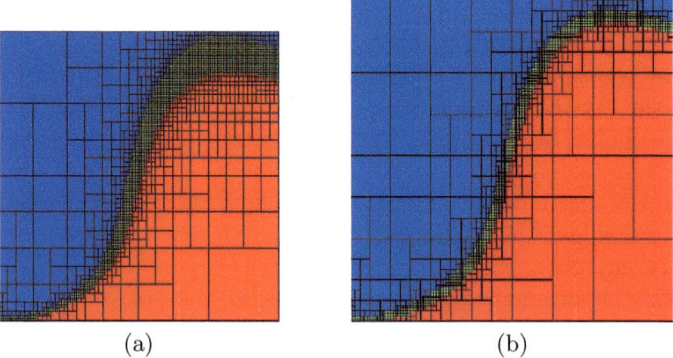

Fig. 4.2. The influence of the polynomial form on interval arithmetic applied to locate a curve: (a) the power-form polynomial $p(x, y)$ defined in Section 4.3 and (b) its equivalent Bernstein-form polynomial.

4.4 The Influence of the Polynomial Form on IA

As illustrated in Figure 4.2, it is visible the differences between the power form and Bernstein form of a polynomial, namely:

- *Number of sub-boxes.* Their corresponding 2-d tree space partitionings have a different number of sub-boxes. The power form polynomial on the left-hand side leads to a bigger number of space subdivisions than the Bernstein form polynomial on the right-hand side.
- *Box classification.* The zero boxes (in green) provide a better approximation to the curve when the function is in Bernstein form, so that there are fewer false zero boxes.

4.4.1 Power and Bernstein Form Polynomials

For brevity, we review univariate and multivariate polynomials in this section.

Univariate

A power form polynomial of degree $n \in \mathbb{N}$ in the variable x is defined by:

$$f(x) = \sum_{i=0}^{n} a_i x^i, \qquad (4.8)$$

where $a_i \in \mathbb{R}$. The equation $f(x) = 0$ is the *implicit equation* corresponding to the polynomial $f(x)$.

We have shown in the previous section that the form of the implicit expression supplied as input to interval arithmetic is crucial for the accuracy of the box classification. Since any input expression can be written in a number of equivalent forms, it makes sense to choose a transformation which will generate a more numerically stable polynomial form. If a base other than the power base is used in order to express the same polynomial, the interval arithmetic classification method will, in general, produce different results. The results which follow below encourage the use of the Bernstein base especially in the case of high-degree polynomials.

As seen in Section 3.1.3, the univariate Bernstein basis functions of degree n on the interval $[\underline{x}, \overline{x}]$ (see also Lorentz [248]) are defined by:

$$B_i^n(x) = \binom{n}{i} \frac{(x - \underline{x})^i (\overline{x} - x)^{n-i}}{(\overline{x} - \underline{x})^n}, \quad \forall x \in [\underline{x}, \overline{x}], \quad i = 0, 1, \ldots, n. \qquad (4.9)$$

For a given $n \in \mathbb{N}$, these $n + 1$ univariate degree-n Bernstein polynomials $(B_i^n)_{i=0,n}$ forms a basis for the ring of degree-n polynomials. This means that any univariate power form polynomial can be represented on the interval $[\underline{x}, \overline{x}]$ using its equivalent Bernstein form as follows:

$$f(x) = \underbrace{\sum_{i=0}^{n} a_i x^i}_{\text{power form } p(x)} = \underbrace{\sum_{i=0}^{n} b_i^n B_i^n(x)}_{\text{Bernstein form } B(x)}$$

where b_i^n are the Bernstein coefficients corresponding to the degree-n base. The two univariate representations $p(x)$ and $B(x)$ are equivalent on the interval $[\underline{x}, \overline{x}]$. For example, on the unit interval $[0, 1]$, determining $B(x)$ from $p(x)$ requires the computation of the univariate Bernstein coefficients in terms of the power coefficients:

$$b_i^n = \sum_{j=0}^{i} \frac{\binom{i}{j}}{\binom{n}{j}} a_j \qquad (4.10)$$

As referred in Section 3.1.3, Formula (4.10) can be used to design an algorithm of conversion between the power form and the Bernstein form of an univariate polynomial [133, 134].

Multivariate

The generalisation of Bernstein bases to multivariate polynomials is not immediate. The power form of a polynomial in d variables is written in terms of $x_1, \ldots, x_d$ like this:

$$f(x_1, \ldots, x_d) = \sum_{0 \leq k_1 + \cdots + k_d \leq n} a_{(k_1, \ldots, k_d)} x_1^{k_1} \cdots x_d^{k_d}$$

where the coefficients $a_{(k_1, \ldots, k_d)} \in \mathbb{R}$. Again, the equation $f(x_1, \ldots, x_d) = 0$ is the *implicit equation* corresponding to the *implicit polynomial* $f(x_1, \ldots, x_d)$. By convention, the degree of each term is $k_1 + \cdots + k_d$, and the degree of the polynomial is the maximum of all degrees of its terms.

The multivariate Bernstein form is defined recursively as a polynomial whose main variable is x_d and whose coefficients are multivariate Bernstein-form polynomials in $x_1, \ldots, x_{d-1}$.

Formula (4.10) can be generalised to more variables. Conversion between the power and the Bernstein representation is possible regardless of the number of variables (see Geisow [158] and Garloff [155, 425]). In [40, 41] Berchtold et al. give formulae and algorithms for the computation of the Bernstein form of bi- and trivariate polynomials, as needed for locating implicit curves in 2D and surfaces and solids in 3D, respectively. The following example makes use of this particular conversion method.

Example 4.1. Let us look again at Figure 4.2. The power-form polynomial appears on the left-hand side and is given by the polynomial defined in Section 4.3 and written now, for convenience, with $\mathbb{R}$-style coefficients, though under the understanding that the calculations are exact:

$$p(x, y) = 0.9446\, x\, y - 7.0044\, x^3\, y^2 + 7.6455\, x^4\, y^3 + 0.5640\, y^4 - x^3$$

The corresponding Bernstein form in $[0, 1] \times [0, 1]$ appears on the righ-hand side of Figure 4.2 and is as follows:

4.4 The Influence of the Polynomial Form on IA

$$B(x,y) = \left(-x^3(1-x) - x^4\right)(1-y)^4 +$$

$$4\left(0.2361x(1-x)^3 + 0.7084x^2(1-x)^2 - 0.2915x^3(1-x) - 0.7638x^4\right) \cdot$$

$$y(1-y)^3 +$$

$$6\left(0.4723x(1-x)^3 + 1.4169x^2(1-x)^2 - 0.7505x^3(1-x) - 1.6951x^4\right) \cdot$$

$$y^2(1-y)^2 +$$

$$4\left(0.7084x(1-x)^3 + 2.1253x^2(1-x)^2 - 2.3768x^3(1-x) - 1.8823x^4\right) \cdot$$

$$y^3(1-y) +$$

$$\left(0.564(1-x)^4 + 3.2006x(1-x)^3 + 6.2178x^2(1-x)^2 - 2.9146x^3(1-x) + 1.1497x^4\right)y^4$$

The power representation in Figure 4.2(a) is less effective in areas which are further away from the origin, whereas the Bernstein representation in Figure 4.2(b) starts classifying correctly boxes which are roughly at a constant distance away from the function.

For low-degree polynomials the advantage of using the Bernstein form is not immediately obvious. However, in rectangular areas which are further away from the origin, high-degree polynomials in the power form usually operate with large powers of large numbers. Any small errors in the coordinates can cause a significant change in the value of the polynomial. Thus, the Bernstein base is more numerically stable than the power base, which means that minor perturbations introduced in the coefficients tend not to affect the value of the polynomial.

Floating point errors can also be a reason for interval swell. Very small numbers on the "wrong" side of the origin are decisive in the classification procedure. Numerical stability helps correct this problem, though the Bernstein representation is not entirely error-free.

4.4.2 Canonical Forms of Degrees One and Two Polynomials

The standard form polynomial for three-dimensional quadrics is, as for any degree-two polynomial, written in the following manner:

$$A + 2Bx + 2Cy + 2Dz + 2Exy + 2Fxz + 2Gyz + Hx^2 + Iy^2 + Jz^2 = 0 \quad (4.11)$$

This is also known as the general *expanded* equation of a quadric. Quadric surfaces are always a special category of surfaces in geometric modelling because of various nice geometric properties they possess (see, for example, Sarraga [348]). Their importance comes from the fact that they are able to describe the geometry of most engineering mechanical parts designed by current CAD/CAM systems. This explains why CSG geometric kernels were designed and implemented from quadrics. For further details, the reader is referred to the sVLIs set-theoretic kernel geometric modeller [61].

Quadrics are more commonly known by their respective *canonical form* equations, where terms are grouped together in a symmetrical manner. By *canonical form* we mean the best-known implicit form in which quadrics are normally defined and studied (as shown in Figure 4.3):

$$\pm \frac{x^2}{a^2} \pm \frac{y^2}{b^2} \pm \frac{z^2}{c^2} = 1 \qquad (4.12)$$

that is, the normalised equation for a 3D quadric centred at the origin $(0,0,0)$. According to the sign of the coefficients of the expanded form (4.11) or the canonical form (4.12), the quadrics can be of different types. It can be easily proved (see, for example, Bronstein and Semendjajew [67]) that there are only a finite number of types of quadric surfaces.

Furthermore, empirical tests carried out by the geometric modelling research group at Bath suggest it is probably the case that IA yields perfect classifications of *all* the sub-boxes of a region, provided they are tested against the equation of a *plane* or a *quadric surface* in the canonical form. Otherwise, the classifications are only conservative.

The multiplication of an interval by a constant and the addition and subtraction of two intervals are all exact operations. Hence, when an interval is substituted into a linear equation of the type $Ax + By + Cz + D = 0$ the arithmetic is expected to be well-behaved. The immediate geometrical consequence is that it is always possible to determine precisely whether a plane in space cuts a given box. The 'perfect' results are due not only to the linearity of the polynomial form but also to the fact that each variable occurs in the expression of the polynomial exactly once and independently from other variables. Thus no interference occurs between the different sources of noise. The coefficients A, B, C and D in the linear form $Ax + By + Cz + D = 0$ are assumed to be obtained after all the reductions possible have been performed. Otherwise the swelling phenomenon reappears.

As an illustration, consider the polynomial $p(x) = 2x - x$. When studied over the unit interval, it yields a swollen result, despite its linearity:

$$2\,[0,1] - [0,1] = [0,2] - [0,1] = [-1,2] \;\supset\; [0,1]$$

With the exception of the plane equation and quadrics in their canonical form, these "perfect" results *cannot* be obtained for equations of degree two or higher. The functions $f(x)$, $g(x)$ and $h(x)$ given in Section 4.3.2 have already illustrated a counterexample. That is, they all had degree-two equations but the intervals which resulted after applying interval arithmetic were not the

4.4 The Influence of the Polynomial Form on IA 103

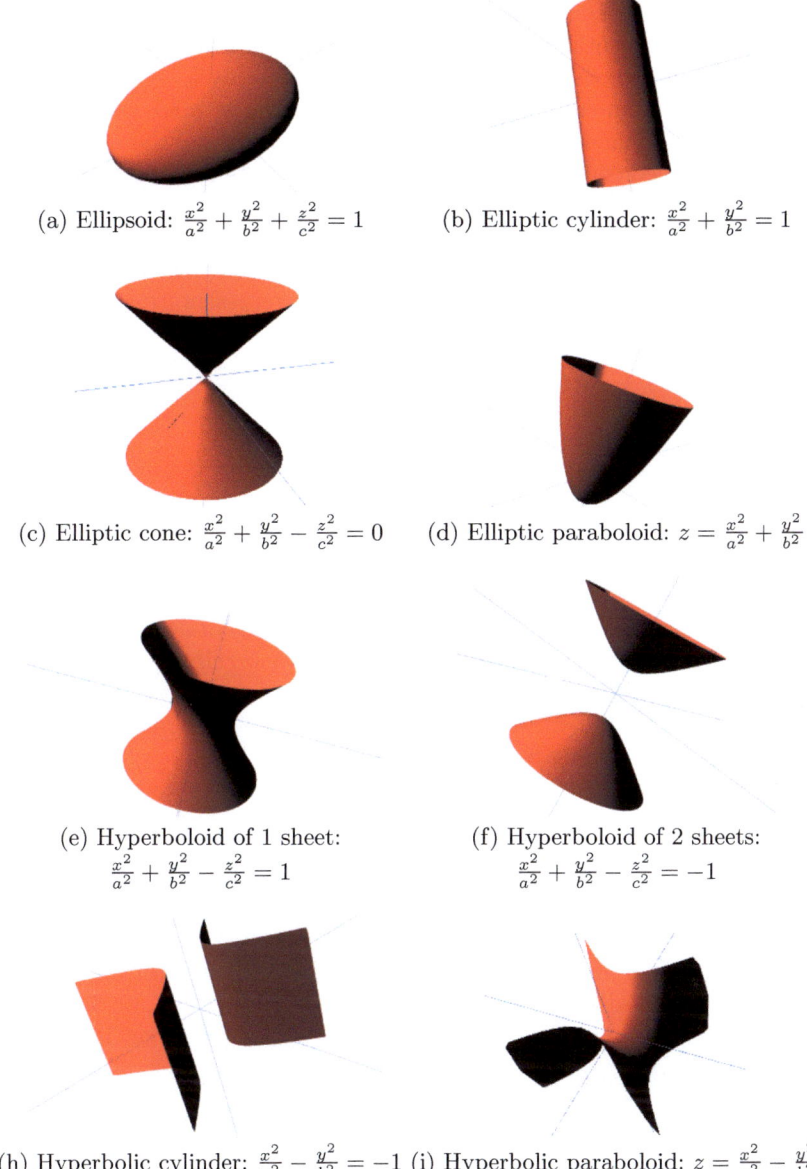

(a) Ellipsoid: $\frac{x^2}{a^2} + \frac{y^2}{b^2} + \frac{z^2}{c^2} = 1$

(b) Elliptic cylinder: $\frac{x^2}{a^2} + \frac{y^2}{b^2} = 1$

(c) Elliptic cone: $\frac{x^2}{a^2} + \frac{y^2}{b^2} - \frac{z^2}{c^2} = 0$

(d) Elliptic paraboloid: $z = \frac{x^2}{a^2} + \frac{y^2}{b^2}$

(e) Hyperboloid of 1 sheet:
$\frac{x^2}{a^2} + \frac{y^2}{b^2} - \frac{z^2}{c^2} = 1$

(f) Hyperboloid of 2 sheets:
$\frac{x^2}{a^2} + \frac{y^2}{b^2} - \frac{z^2}{c^2} = -1$

(h) Hyperbolic cylinder: $\frac{x^2}{a^2} - \frac{y^2}{b^2} = -1$

(i) Hyperbolic paraboloid: $z = \frac{x^2}{a^2} - \frac{y^2}{b^2}$

Fig. 4.3. Canonical forms for quadrics.

results of the exact calculations. When comparing the general expanded form with the canonical form of the same quadric, it is customarily the case that the former is the expansion of the latter and has degree-one terms as well as square terms. Most of the canonical forms of the quadrics have only degree-two terms, which constitutes an advantage for the application of interval arithmetic. This is due to the exponentiation rule stated in Equation (4.7), which causes the tightest positive intervals to be generated as results.

In general, the interval arithmetic technique can be used successfully for the location of the familiar quadrics, in that all the boxes of the spatial subdivision are classified correctly. The canonical form of the conic section surfaces, each of the variables occurs exactly once, independently from the others and with an exponent of one or two; thus it is expected that the resulting interval will coincide with the exact range. The technique starts suffering from conservativeness in the case of surfaces of an arbitrary representation, or of higher degree.

4.4.3 Nonpolynomial Implicits

One reason for extensively using polynomials is that the most important curves, surfaces and solids in geometric modelling can be expressed by means of polynomials. Perhaps the only significant exception is the helix. The helix is useful to represent as it is widely used in practice for such things as screw threads, but its formulation requires transcendental functions. Another reason why polynomials have been preferred is that algebraic theories provide extensive studies of polynomials. The findings concerning general algebraic functions cannot always be extended to transcendental functions.

As expected, conservativeness remains a problem for transcendental implicits. Whilst performing correctly for quite a large number of negative and positive boxes, the interval arithmetic technique still outputs some regions of space as zero boxes, although in reality they are purely negative or purely positive. Similarly to the polynomial case, the result is usable but not satisfactory.

Example 4.2. The expression $\sin(x)$ can legitimately be assumed to take values in the range $[-1, 1]$, but this may be quite a gross estimate. In the particular case where $x \in [\frac{1}{2}, \frac{7}{3}]$ the function's image is only $[\sin(\frac{1}{2}), 1] \subseteq [0.47, 1]$. When the $\sin(x)$ function is incorporated in further calculations, an initial range approximation as gross as $[-1, 1]$ will propagate the interval swell throughout the computation chain, affecting the final result.

An alternative evaluation method for periodic trigonometric functions (like $\sin(x)$ and $\cos(x)$) would be to calculate the range as a result of a circumstantial study of the domain. If the length of the domain interval is larger than the function period (2π in the case of $\sin(x)$ or $\cos(x)$), then the function takes values over the whole of the range $[-1, 1]$. If not, then a detailed study of the relative positions of the ends of the interval and multiples of the values 0, $\frac{\pi}{2}$, π, $\frac{3\pi}{2}$ and 2π will help establish the exact range. This is the case with the tangent and cotangent functions as well, with the further complication that these are not defined for certain values of their argument.

Other transcendental functions, like the logarithmic and exponential functions are slightly better behaved. Because they are monotone, an exact range can be obtained by evaluating the function at both ends of the interval. Problems may occur, however, when the function is not defined for the whole domain interval (e.g. $\log(x)$ is not defined for negative numbers or zero).

4.5 Affine Arithmetic Operations

As seen above, the conservativeness of algebraic methods that rely on interval arithmetic depends on the polynomial form used to represent implicit curves, surfaces and solids. We have also seen that the conservativeness is reduced when the input is provided in the Bernstein form. Furthermore, in the particular case of planes or quadrics represented by canonical form polynomials, the conservativeness vanishes.

As described in Section 4.3, the box classification method relies on substituting the interval coordinates of a box for the variables of an implicit function expression, performing interval arithmetic calculations, and studying the relative positions of the resulting interval and zero. It might be thought that the interval swell during the interval arithmetic evaluation depends merely on the number of occurrences of a variable in the implicit expression. There are other aspects (such as the presence of even exponents or the order of the arithmetic operations) which contradict this assumption. It is known that the Bernstein form of a polynomial has many more variable occurrences than the power form; despite this, the former behaves better with IA than the latter.

Still, whenever interval calculations are performed, no account is taken of the fact that each occurrence of any variable, such as x, always represents the same quantity. That is to say that each variable introduces the same error in all the terms of the polynomial. The method, called *affine arithmetic*, described in the rest of this chapter makes use of this observation and correlates the sources of error in the interval classification (see also Martin et al. [258] or Shou et al. [367]). And, more importantly, it does *not* depend on the polynomial form used to represent an implicit object. Thus, affine arithmetic can be viewed as a more sophisticated version of interval arithmetic.

4.5.1 The Affine Form Number

Affine arithmetic was proposed by Comba, Stolfi and others [90] in the early 1990s with a view to tackle the conservativeness problem caused by standard interval arithmetic. Like interval arithmetic, affine arithmetic can be used to manipulate imprecise values and to evaluate functions over intervals. While, like interval arithmetic, it provides guaranteed bounds for computed results, affine arithmetic also takes into account the dependencies between the sources of error. In this way it is able to produce much tighter and more accurate intervals than interval arithmetic, especially in long chains of computations.

In affine arithmetic an uncertain quantity x is represented by an affine form $\hat{x}$ that is a first-degree polynomial of a set of noise symbols ε_i.

$$\hat{x} = x_0 + x_1\varepsilon_1 + \cdots + x_m\varepsilon_m = x_0 + \sum_{i=1}^{m} x_i\varepsilon_i$$

Here the value of each noise symbol ε_i is unknown but defined to lie in the interval $[-1, 1]$. The corresponding coefficient x_i is a real number that determines the magnitude of the impact of the product $x_i\varepsilon_i$. Each product $x_i\varepsilon_i$ stands for an independent source of error or uncertainty which contributes to the total uncertainty in the quantity x. The number m may be chosen as large as necessary in order to represent all the sources of error. These may well be input data uncertainty, formula truncation errors, arithmetic rounding errors, and so on.

This piece of reasoning is not restricted to the univariate case. On the contrary, given a polynomial expression in any number of variables, the dependencies between them can be easily expressed by using the same noise symbol ε_i wherever necessary. If the same noise symbol ε_i appears in two or more affine forms (e.g. in both $\hat{x}$ and $\hat{y}$) it indicates the interdependencies and correlations that exist between the underlying quantities x and y. For example, in the bivariate case, computing with the affine forms is a matter of replacing x and y by $\hat{x}$ and $\hat{y}$ in $f(x,y)$, respectively, and each operation in $f(x,y)$ with the corresponding affine operation on $\hat{x}$ and $\hat{y}$. Of course, each affine operation must take into account the relationships between the noise symbols in x and y.

The rules for arithmetic operations on affine forms are explained below. The important thing to notice about the way affine arithmetic works is that algebraic expressions take into account the fact that the same variable may appear in them more than once. Thus using affine arithmetic, similar terms get cancelled when they appear in an expression (e.g. $2\hat{x} + \hat{y} - \hat{x} = \hat{x} + \hat{y}$). This is *not* the case with interval arithmetic.

4.5.2 Conversions between Affine Forms and Intervals

Conversions between affine forms and intervals are defined in various papers by Comba and Stolfi [90], Figueiredo [100] and Figueiredo and Stolfi [102].

Given an interval $[\underline{x}, \overline{x}]$ representing a quantity x, its affine form can be written as

$$\hat{x} = x_0 + x_1\varepsilon_x, \quad \text{where } x_0 = \frac{\underline{x} + \overline{x}}{2}, \quad x_1 = \frac{\overline{x} - \underline{x}}{2}. \tag{4.13}$$

Conversely, given an affine form $\hat{x} = x_0 + x_1\varepsilon_1 + \cdots + x_m\varepsilon_m$, the range of possible values of its corresponding interval is

$$[\underline{x}, \overline{x}] = [x_0 - \xi, x_0 + \xi], \quad \text{where } \xi = \sum_{i=1}^{m} |x_i|. \tag{4.14}$$

4.5.3 The Affine Operations

The affine arithmetic rules are fully defined in Comba and Stolfi [90]. Those that are relevant to the location of curves and surfaces are addition and multiplication, both of a scalar to an affine form, and of (two or more) affine forms to each other. Given the affine forms $\hat{x}$ and $\hat{y}$, and the real number $\alpha \in \mathbb{R}$ the simple arithmetic operations are carried out thus:

Addition:

$$\alpha + \hat{x} = (\alpha + x_0) + x_1\varepsilon_1 + \cdots + x_m\varepsilon_m \tag{4.15}$$
$$\hat{x} + \hat{y} = (x_0 + y_0) + (x_1 + y_1)\varepsilon_1 + \cdots + (x_m + y_m)\varepsilon_m \tag{4.16}$$

Subtraction:

$$\alpha - \hat{x} = (\alpha - x_0) + x_1\varepsilon_1 + \cdots + x_m\varepsilon_m \tag{4.17}$$
$$\hat{x} - \hat{y} = (x_0 - y_0) + (x_1 - y_1)\varepsilon_1 + \cdots + (x_m - y_m)\varepsilon_m \tag{4.18}$$

Multiplication:

$$\alpha\hat{x} = (\alpha x_0) + (\alpha x_1)\varepsilon_1 + \cdots + (\alpha x_m)\varepsilon_m \tag{4.19}$$
$$\hat{x}\hat{y} = (x_0 + x_1\varepsilon_1 + \cdots + x_m\varepsilon_m)(y_0 + y_1\varepsilon_1 + \cdots + y_m\varepsilon_m) \tag{4.20}$$
$$= \left(x_0 + \sum_{i=1}^{m} x_i\varepsilon_i\right)\left(y_0 + \sum_{i=1}^{m} y_i\varepsilon_i\right)$$
$$= \underbrace{x_0 y_0 + \sum_{i=1}^{m}(x_0 y_i + x_i y_0)\varepsilon_i}_{\mathcal{L}(\varepsilon_1,\ldots,\varepsilon_m)} + \underbrace{\left(\sum_{i=1}^{m} x_i\varepsilon_i\right)\left(\sum_{i=1}^{m} y_i\varepsilon_i\right)}_{\mathcal{Q}(\varepsilon_1,\ldots,\varepsilon_m)} \tag{4.21}$$

Now, $\mathcal{L}(\varepsilon_1,\ldots,\varepsilon_m)$ is an affine form in which the noise symbols ε_i occur only with degree 1, whereas $\mathcal{Q}(\varepsilon_1,\ldots,\varepsilon_m)$ is quadratic in the noise symbols. The quadratic term can be handled so that it becomes linear itself, at the expense of introducing a new noise symbol $\varepsilon_k \in [-1,1]$, with coefficient $\mu\nu$, where $\mu = \sum_{i=1}^{m}|x_i|$ and $\nu = \sum_{i=1}^{m}|y_i|$. So $\hat{x}\hat{y}$ can be expressed as an affine combination of first-degree polynomials in the noise symbols:

$$\hat{x}\hat{y} = x_0 y_0 + \sum_{i=1}^{m}(x_0 y_i + x_i y_0)\varepsilon_i + \mu\nu\varepsilon_k$$
$$= x_0 y_0 + (x_0 y_1 + x_1 y_0)\varepsilon_1 + \cdots + (x_0 y_m + x_m y_0)\varepsilon_m + \mu\nu\varepsilon_k$$

The index k can be chosen as $m+1$.

Division:

Division can be defined via inversion and multiplication in the same style as shown in Formula (4.5) for intervals. This is rarely used in calculations, as there is little scope for simplifying the polynomial expansions obtained.

Exponentiation:

$$\hat{x}^a = (x_0 + x_1\varepsilon_x)^a = x_0^a + \sum_{i=1}^{a} \binom{a}{i} x_0^{a-i} x_1^i \varepsilon_x^i, \quad a \in \mathbb{Z}. \qquad (4.22)$$

Unlike interval arithmetic, the affine exponentiation is a particular case of the affine multiplication because

$$\hat{x}^2 = \hat{x}.\hat{x}$$

and, consequently, there is no interval swell caused by exponentiation.

It is immediately apparent from the rules above that the affine arithmetic operations are commutative, associative and distributive. This was not the case with interval arithmetic, whose misbehaviour with the distributivity law caused the interval swell.

Practical experience with polynomials other than those of lowest degree, shows that simply using the rules of affine arithmetic directly gives relatively little advantage over ordinary interval arithmetic when localising polynomials (e.g. curves and surfaces), which are basically defined by additions, subtractions, multiplications and exponentiations. This is due to rapid introduction of many new error symbols. Much better results can be obtained by taking more care, in particular in handling exponentiations.

4.5.4 Affine Arithmetic Evaluation Algorithms

Various affine arithmetic schemes have been proposed for use in geometric modelling. One of the earlier ones (see Zhang and Martin [426] or Voiculescu et al. [402]) proposes to simplify exponentiations in a way that separates odd exponent terms from even exponent terms, and express any (univariate) polynomial with a degree-one polynomial of three terms and just two noise symbols:

$$\hat{x}^a = x_0^a + x_{odd}\varepsilon_{xodd} + x_{even}\varepsilon_{xeven}.$$

Whilst this yields results very efficiently and leads to reasonably narrow result intervals, it unfortunately does so at the expense of the loss of conservativeness. This comes from trying to share the noise symbols ε_{xodd} and ε_{xeven} between the computations of two distinct powers [373].

A more complete yet more expensive scheme is proposed in a related paper [258] where Martin et al. give a matrix-form evaluation of the affine interval polynomials that leads to a conservative interval result.

4.6 Affine Arithmetic-driven Space Partitionings

When applying affine arithmetic to algebraic surface location, the polynomial representing the implicit surface needs to be evaluated on the intervals over which its variables range. In particular, in order to locate a planar curve a polynomial $f(x,y)$ needs to be evaluated over the ranges in x and y representing a box. These are $[\underline{x}, \overline{x}]$ and $[\underline{y}, \overline{y}]$ or their affine equivalents $\hat{x}$ and $\hat{y}$ respectively.

Because the affine arithmetic form can be converted back into an interval, it can easily be used as an alternative to producing box classifications for power- or Bernstein-form polynomials using direct interval arithmetic rules.

To compare the relative merits of interval arithmetic and carefully evaluated affine arithmetic for curve drawing, we now present a practical example.

Example 4.3. Let us consider the following bivariate polynomial function $p(x,y)$ in the power form:

$$p(x,y) = \frac{945}{1000} x y - \frac{94{,}3214}{100{,}000} x^2 y^3 + \frac{74{,}554}{10{,}000} x^3 y^2 + y^4 - x^3$$

and then in its Bernstein form in the unit box $[0,1] \times [0,1]$:

$$B(x,y) = - x^3 (1-y)^4 +$$

$$4\left(\frac{23{,}625}{100{,}000} x (1-x)^2 + \frac{4725}{10{,}000} x^2 (1-x) - \frac{76{,}375}{100{,}000} x^3\right) y (1-y)^3 +$$

$$6\left(\frac{4725}{10{,}000} x (1-x)^2 + \frac{945}{1000} x^2 (1-x) + \frac{715{,}066{,}667}{1{,}000{,}000{,}000} x^3\right) y^2 (1-y)^2 +$$

$$4\left(\frac{70{,}875}{100{,}000} x(1-x)^2 - \frac{9{,}405{,}350{,}004}{10{,}000{,}000{,}000} x^2(1-x) + \frac{1{,}078{,}415}{1{,}000{,}000} x^3\right) y^3(1-y) +$$

$$\left((1-x)^3 + \frac{3945}{1000} x (1-x)^2 - \frac{4{,}542{,}140{,}001}{1{,}000{,}000{,}000} x^2 (1-x) - \frac{103{,}174}{100{,}000} x^3\right) y^4$$

The left-hand side of Figure 4.4 represents the interval arithmetic classification of the Bernstein form (i.e. the best polynomial form for IA). The right-hand side illustrates the result of applying affine arithmetic (AA) to the power-form polynomial $p(x, y)$. Both have been drawn using a minimum box size of $\frac{1}{2^7} \times \frac{1}{2^7}$. As apparent from Figure 4.4, AA definitely classifies a larger area, and in bigger chunks at a time, than either case of IA. The Table 4.1 gives the respective box percentages for $p(x,y)$ at a resolution $\Delta = \frac{1}{2^{10}} \times \frac{1}{2^{10}}$.

The complexity of each algorithm depends on the type of arithmetic used (i.e. standard interval arithmetic or affine arithmetic), as well as on the form of the input. Tables 4.2 and 4.3 summarise the running times and the number of subdivisions in each case. (Note that the times are interesting to compare, but not relevant in absolute terms, as the implementation depends on the interval package and hardware used.)

4 Interval Arithmetic

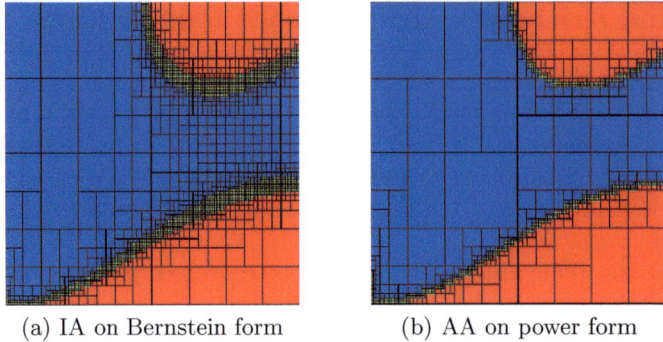

(a) IA on Bernstein form   (b) AA on power form

Fig. 4.4. Interval- and affine arithmetic box classification for $p(x,y)$ and $B(x,y)$ in the unit box $[0,1] \times [0,1]$.

Table 4.1. Box percentages for $p(x,y)$ at a resolution $\Delta = \frac{1}{2^{10}} \times \frac{1}{2^{10}}$.

	Negative boxes $[-,-]$	Zero boxes $[-,+]$	Positive boxes $[+,+]$
IA on power form	0.3171	0.0231	0.6597
IA on Bernstein form	0.3241	0.0088	0.6670
AA on power form	0.3266	0.0037	0.6695

Table 4.2. Running times and number of subdivisions for $p(x,y)$ at $\Delta = \frac{1}{2^{10}} \times \frac{1}{2^{10}}$.

	time (sec)	subdivisions
IA on power form	2338.121	39834
IA on Bernstein form	2783.140	15568
AA on power form	194.339	6447

Table 4.3. Running times and number of subdivisions for $p(x,y)$ at $\Delta = \frac{1}{2^7} \times \frac{1}{2^7}$.

	time (sec)	subdivisions
IA on power form	20.94	1854
IA on Bernstein form	51.52	947
AA on power form	10.07	433

The results in Table 4.2 and Table 4.3 have been obtained also using different minimum box sizes, $\frac{1}{2^{10}} \times \frac{1}{2^{10}}$ and $\frac{1}{2^7} \times \frac{1}{2^7}$, respectively.

For the example given above the affine arithmetic method produces results more quickly (and accurately) than either interval arithmetic method. The former involves slightly more calculations per box, but classifies big boxes in a very efficient manner. When interval arithmetic is applied there are fewer calculations per box than for affine arithmetic. Still, the Bernstein polynomial form is so much more complicated that the program runs much slower.

Regarding the number of subdivisions, interval arithmetic needs much finer subdivision of boxes for the power form than for the Bernstein form and ends up with a less accurate result. Affine arithmetic needs comparatively fewer subdivisions to reach a very accurate result.

In principle, rather than the interval arithmetic, one could also study the Bernstein form using the affine arithmetic approach. However, as it has been shown that affine arithmetic operations are associative, commutative and distributive, it is expected that different polynomial representations would produce the same results. This is because the various ways of expressing a polynomial function using different bases does nothing other than rearranging the terms. This rearrangement does not affect the arithmetic of the polynomial, and hence does not affect the result of applying affine arithmetic to an equivalent polynomial form. Therefore, when studying affine arithmetic, it is only the power basis that needs to be considered. The proof of this final statement has been published in [259].

As a final remark in this section, it is worth noting that when interval arithmetic produces a correct estimate of the range of values, then affine arithmetic is expected to produce an exact range too. For example, in the case of function $k(x) = (2x - 3)^2$ studied in Section 4.3.2, interval arithmetic gives the correct range $[1, 9]$, and so does affine arithmetic.

4.7 Floating Point Errors

Recursive subdivision using interval arithmetic relies fundamentally on the arithmetic operations carried out on the end values of the intervals being accurate. This is why the examples given so far have involved polynomials with rational coefficients and subdivisions of boxes stored as rational intervals. Implementations in languages without a *rational number* data type will compromise the precision of the calculations by storing the numbers as *floating point* values.

The current section illustrates the extent to which floating point errors propagate through the evaluation process, often making the classification process impracticable. Let us recall the polynomial p, defined in Section 4.3, in its rational and floating point power forms:

$$p_{rat}(x,y) = \frac{9446}{10{,}000}\,x\,y - \frac{700{,}443{,}214}{100{,}000{,}000}\,x^3\,y^2 + \frac{764{,}554}{100{,}000}\,x^4\,y^3 + \frac{564}{1000}\,y^4 - x^3$$

$$p_{flt}(x,y) = 0.9446\,y\,x - 7.00443214\,y^2\,x^3 + 7.64554\,x^4\,y^3 + 0.564\,y^4 - x^3$$

This was originally defined in the unit box, and had the zero set illustrated in Figure 4.1.

112   4 Interval Arithmetic

We now aim to translate p so that it has the same zero set in a general box, say $[9.62, 10.62] \times [7.31, 8.31]$. This can be achieved in several ways, all involving the substitution of x by x-minus-some-quantity, and y by y-minus-some-quantity, in either p_{rat} or p_{flt}:

- substitute $x := x - \frac{962}{100}$ and $y := y - \frac{731}{100}$ in p_{rat}, yielding p_1;
- substitute $x := x - \frac{962}{100}$ and $y := y - \frac{731}{100}$ in p_{flt}, yielding p_2;
- substitute $x := x - 9.62$ and $y := y - 7.31$ in p_{rat}, yielding p_3;
- substitute $x := x - 9.62$ and $y := y - 7.31$ in p_{flt}, yielding p_4.

Floating point errors already start occurring at the stage where brackets are multiplied out. In the particular case of $p(x, y)$, when using a precision of 10 significant digits, $p_3 = p_4$. The three zero sets (corresponding to p_1, p_2 and p_3 respectively) in the box $[9.62, 10.62] \times [7.31, 8.31]$ are plotted in Figure 4.5, in the order *cyan, magenta, yellow*.

The affine arithmetic method necessarily complies to one of the four schemes above. Our study uses two schemes in parallel: all the way through the subdivision process described above. Any subdivision decisions are taken using p_1 and a "totally rational" scheme. At the same time, the subboxes are also converted to their floating point equivalents and subjected to the sign test against the floating point polynomial p_4. Thus it is certain that subdivision is carried out correctly. The respective ranges (given by the two different approximations) can be compared.

When the signs of the two ranges agree (in that they both indicate a negative or a positive box), the same conventions for colours as before has been used—that is, red for negative and blue for positive. However, when the rational arithmetic predicts a negative box and the floating point arithmetic calculation disagrees, the box is coloured *magenta*. Similarly, when the rational arithmetic predicts a positive box but the floating point arithmetic calculation disagrees, the box is coloured *cyan*. Zero boxes are still coloured green. The result is illustrated in Figure 4.6(a).

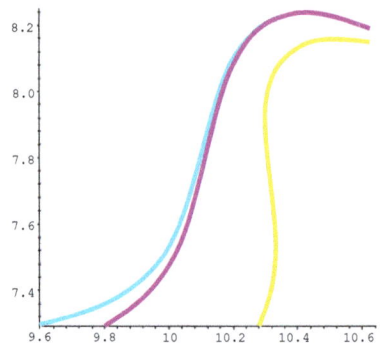

Fig. 4.5. Zero sets of p_1 (cyan), p_2 (magenta) and p_3 (yellow).

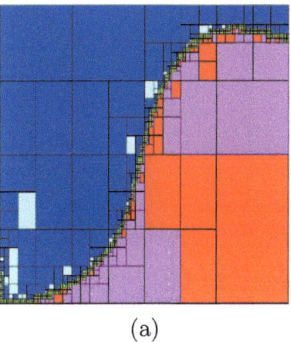
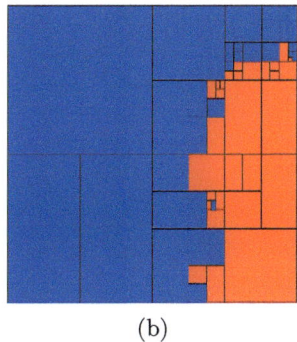

(a)　　　　　　　　　　　　(b)

Fig. 4.6. (a) Affine arithmetic classification of floating-point polynomial using rational evaluations in the box $[9.62, 10.62] \times [7.31, 8.31]$; (b) Affine arithmetic (mis)classification using only floating-point evaluations in the box $[9.62, 10.62] \times [7.31, 8.31]$.

The frequent occurrence of magenta and cyan boxes indicates to what extent floating point errors can influence the affine arithmetic calculations. Had there been only floating point evaluations, the classification would have been totally irrelevant, as decisions for further subdivision would have been taken in completely the wrong places. Indeed, when running such a test it simply returns an inconsistent collection of negative and positive boxes, which only vaguely evokes the shape of the initial curve (Figure 4.6(b)).

This is a typical illustration of the propagation of floating point errors. Let us now consider a single magenta box and examine the way in which the four possibilities there are for approximating either the coefficients or the box can influence the final range given as a result.

Take the rational box:

$$\left[\frac{8471}{800}, \frac{33{,}909}{3200}\right] \times \left[\frac{1637}{200}, \frac{13{,}121}{1600}\right] =$$
$$[10.58875000, 10.59656250] \times [8.185000000, 8.200625000]$$

This is one of the boxes coloured magenta in Figure 4.6(a). Its corresponding affine forms are:

$$\hat{x} = \frac{67{,}793}{6400} + \frac{1}{256}\varepsilon_x = 10.59265625 + 0.003906250000\,\varepsilon_x$$
$$\hat{y} = \frac{26{,}217}{3200} + \frac{1}{128}\varepsilon_y = 8.192812500 + 0.007812500000\,\varepsilon_y$$

Let us evaluate the results returned by affine arithmetic when classifying these affine forms and/or their floating point equivalents against the various forms of p. The results differ according to the amount of floating point approximation carried out:

affine_eval$(p_1(x,y), \hat{x}_{rat}, \hat{y}_{rat})$ =
$$\left[\frac{-8{,}180{,}237{,}644{,}390{,}479{,}080{,}447}{56{,}294{,}995{,}342{,}131{,}200{,}000{,}000}, \frac{-2{,}383{,}608{,}804{,}974{,}363}{140{,}737{,}488{,}355{,}328{,}000}\right]$$
$$= [-0.1453102109, -0.01693655921]$$

affine_eval$(p_1(x,y), \hat{x}_{\text{float}}, \hat{y}_{\text{float}}) = [0.09670367511, 0.2243380429]$

affine_eval$(p_4(x,y), \hat{x}_{rat}, \hat{y}_{rat}) = [-0.1038870246, 0.02492879264]$

affine_eval$(p_4(x,y), \hat{x}_{\text{float}}, \hat{y}_{\text{float}}) = [0.09641367500, 0.2246280630]$

To summarise, the intervals generated as answers vary in their signs and positions relative to zero. The results are not conservative anymore; on the contrary, some of them have completely misclassified the box type, as shown in Table 4.4:

Table 4.4. Box classification for p_1 and p_4.

$p(x,y)$	$\hat{x} \times \hat{y}$	interval type	box classification
rat	rat	$[-,-]$	negative
rat	float	$[+,+]$	positive
float	rat	$[-,+]$	zero
float	float	$[+,+]$	positive

Of course, "mixed" forms of the polynomial (such as p_2) could have been used in the experiments as well, generating a potentially wider variety of answers. Nevertheless the study outlined above illustrates the point being made in this section, which is that floating point errors are not negligible.

All the floating point calculations in this section have been carried out using a precision of 10 significant digits. Increasing the precision of the calculations may eliminate the problem for particular cases. Indeed, in the case of $p(x,y)$ a precision of 40 significant digits seems to be enough for a box classification comparable to the one where rational arithmetic had been used. However this is not a general solution, as the result depends thoroughly on the precision with which the polynomial coefficients and the edges of the box are being calculated in the first place.

4.8 Final Remarks

There are, of course, a variety of ways in which the polynomial can be input, such as storing it in some canonical form, using a planar basis [61], or using an implicitisation of some Bernstein form [41], a Taylor expansion [368], etcetera. Overall, we conclude that the conservativeness problem which occurs in surface location can be reduced in at least two major ways: either the input is given in Bernstein form instead of power form and interval arithmetic is used, or the calculations are carried out on the power form, but a careful strategy based on affine arithmetic is used instead of interval arithmetic.

When the Bernstein form is used the improvement is significant: boxes can be located much more accurately in a given region of interest. The shape of the surface is outlined in enough detail for it to be located.

When affine arithmetic is used as shown above, our results demonstrate that curves can be located even more closely. This is because the intervals produced during polynomial evaluation are tighter.

Affine arithmetic calculations are more complicated than interval arithmetic ones. This is why the method is more error-prone when using floating point calculations. This is also why, in some cases, we have found it to be perhaps twice as slow as simple interval arithmetic, although this is strongly dependent on the implementation.

However, affine arithmetic has a speed advantage in some cases when interval arithmetic performs particularly badly. This advantage arises in the subdivision method because fewer boxes need to be considered, even though the amount of computation for any single box is greater.

All in all, it is fully expected that the benefits shown in curve drawing are also applicable to other uses of solutions to implicit equations, such as surface intersection, surface location, etc. Although the examples shown here have used polynomials, similar approaches could also be used if non polynomial functions are needed for modelling. Different suitable basis functions and affine evaluation methods will need to be found for such cases.

There is also a need to express the operations defined here in a more compact form (perhaps using matrices). This would facilitate generalisations and would help study the operations and properties at a higher level of abstraction.

5
Root-Finding Methods

Broadly speaking, the study of *numerical methods* is known as "numerical analysis," but also as "scientific computing," which includes several sub-areas such as sampling theory, matrix equations, numerical solution of differential equations, and optimisation. Numerical analysis does not seek exact answers, because exact answers rarely can be obtained in practice. Instead, much of numerical analysis aims at determining approximate solutions and at the same time keeping reasonable bounds on errors. In fact, computations with floating-point numbers are performed on a computer through approximations, instead of exact values, of real numbers, so that it is inevitable that some errors will creep in. Besides, there are frequently many different approaches to solve a particular numerical problem, being some methods faster, more accurate or requiring less memory than others.

The ever-increasing advances in computer science and technology have enabled us to apply numerical methods to simulate physical phenomena in science and engineering, but nowadays they are also found and applied to interesting scientific computations in life sciences and even arts. For example, ordinary differential equations are used in the study of the movement of heavenly bodies (planets, stars and galaxies); optimisation appears in portfolio management; numerical linear algebra plays an important role in quantitative psychology; stochastic differential equations and Markov chains are employed in simulating living cells for medicine and biology; and, the chaotic behaviour of numerical methods associated to colour theory in computer graphics can be used to generate art on computer.

Nevertheless, in computer graphics, numerical techniques have mainly found applications in the design of parametric curves and surfaces (i.e. CAGD); they also appear in ray tracing of parametric and implicit surfaces. In this book, numerical methods are essentially used to *approximate* the roots of real functions in two and three variables as a way of sampling implicit curves in $\mathbb{R}^2$ and surfaces in $\mathbb{R}^3$. Recall that a root solver involves two main steps: *root isolation* and *root approximation* (also called *root-finding*). Relevant root isolation techniques were approached in the last two chapters. This chapter deals with the so-called numerical *approximation methods* or *root-finding methods*.

5.1 Errors of Numerical Approximations

There are various potential sources of errors in numerical computation. Two of these errors are universal because they occur in any numerical computation: *round-off* and *truncation* errors. Inaccuracies of numerical computations due to the errors lead to a deviation of a numerical solution from the exact solution, independently of the latter is known *a priori* or not. To better understand the effects of finite precision of a numerical solution, let us consider the definition of *relative error* as follows:

$$e = \frac{|x - \rho|}{|\rho|} \tag{5.1}$$

where ρ and x denote the exact solution and its approximate value, respectively. The numerator $|x - \rho|$ of this fraction denotes the *absolute error*.

5.1.1 Truncation Errors

As known, floating-point numbers are represented in a computer with a finite number of digits of precision. The simplest hardware implementation is to keep the first n digits after the period, and then to chop off all other digits. A *truncation error* occurs when a decimal number is cut off beyond the maximum number of digits allowed by the computer accuracy, also called *machine precision*.

Machine precision is the smallest number $\epsilon = 2^{-N}$ that a computer recognises as nonzero. On a 32-bit computer, single precision is 2^{-23} (approximately 10^{-7}) while double precision is 2^{-52} (approximately 10^{-16}). Algorithm 4 computes not only the machine precision ϵ, but also the largest number N of bits such that the difference between 1 and $1 + 2^{-N}$ is nonzero.

It is worthy of noting that truncation errors are present even in a scenario of infinite-precision arithmetic because the computer accuracy and termination criteria associated to algorithms lead to the truncation of the infinite

---

Algorithm 4 The Machine Precision

1: **procedure** MACHINEPRECISION(ϵ, N)
2:     $\epsilon \leftarrow 1.0$
3:     $N \leftarrow 0$
4:     **while** $\epsilon + 1 > 1$ **do**
5:         $\epsilon \leftarrow \epsilon/2$
6:         $N \leftarrow N + 1$
7:     **end while**
8:     $\epsilon \leftarrow 2.0\,\epsilon$
9: **end procedure**

Taylor series that approximate mathematical functions (e.g. transcendental functions) to a finite number of terms (see [83] for further details).

5.1.2 Round-off Errors

A more accurate alternative to truncation is to round the nth digit to the nearest integer. This cutting off of digits leads to round-off errors. For example, the irrational number $\pi = 3.14159265358979...$ has infinitely many digits after the period, and let us round its 6th digit so that $\pi = 3.141593$. Everyone agrees that 3.14 is a reasonable approximation for π, so the resulting absolute error is $|x - \rho| = |3.14 - 3.141593| = 0.001593$ and the relative error is $e = 0.0507$ percent.

Thus, the round-off error of a floating-point number also depends on how many digits are left out. A major problem in numerical analysis is how to keep the accuracy of numerical computations despite the accumulation and propagation of round-off errors in computer arithmetic. That is, round-off errors are a consequence of using finite precision floating-point numbers on computers.

Numerical errors produced by computers affect the quality of computations, which are particularly important for sampling implicit curves or surfaces with self-intersections and other singularities. For example, a self-intersection of a curve may be detected by a convergent sequence of points, each determined by some numerical method. But, this requires that a stopping or termination criterion has been defined very carefully in order to get a trade-off between accuracy and time performance; otherwise, the result may be unpredictable (e.g. divergence caused by inaccurate computations).

5.2 Iteration Formulas

In 1824, the Norwegian mathematician Niels Abel proved the impossibility of a quintic formula by radicals. Later on, the French mathematician Évariste Galois extended Abel's result that it is impossible to obtain a general analytic formula to determine the roots of fifth-order or higher polynomials. In other words, unlike quadratic equations, higher nonlinear equations cannot be solved through a general analytic formula.

This fact led to the development of root-finding numerical methods. There are many numerical formulas and methods to determine a root of a nonlinear equation, namely: the Newton-Raphson method, bisection method, secant method, and false position (or *regula falsi*) method (see Press et al. [329] for a classical treatment of numerical methods).

In numerical analysis, the generic iteration formula is as follows

$$x_{i+1} = F_i(x_i, x_{i-1}, \ldots, x_{i-n+1}). \tag{5.2}$$

and is called the *n-point iteration function*. Most implicit surface (curve) polygonisers use 2-point iteration functions. For example, the *bisection method* and the *false position method* are two examples of 2-point numerical methods. Recall that the acclaimed marching cubes polygoniser [247] uses a 2-point numerical method for sampling implicit surfaces. Sampling a surface consists in computing the intersection points between the surface and each edge of every single cubic cell enclosed in an axis-aligned ambient bounding box. Each intersection point is determined by applying a numerical 2-point method over the edge, i.e. its endpoints work as initial guesses. These 2-point polygonisers are based on the intermediate value theorem (IVT):

Theorem 5.1. (Intermediate Value Theorem) *Let f be a continuous real function on the interval $[x_{i-1}, x_i]$. If $f(x_{i-1}).f(x_i) < 0$, then there exists $x_{i+1} \in [x_{i-1}, x_i]$ such that $f(x_{i+1}) = 0$.*

That is, polygonisers based on IVT are sign-based polygonisers because the next estimate x_{i+1} is determined from two previous estimates x_i and x_{i-1} on which the function f has values with different signs. Consequently, as stated by the IVT, there must be at least one root (unless a singularity is present) in the interval $[x_{i-1}, x_i]$. In these circumstances, a root is said to be bracketed in the interval defined by those two points x_i and x_{i-1}.

However, these 2-point iteration functions are not able to detect sign-invariant branches and sign-invariant components of implicit curves and surfaces. These sign-invariant subsets of curves and surfaces enjoy the property that their functions do not change sign in the neighbourhood of each of their points. For example, the spherical surface defined by the level set $f(x, y, z) = (x^2 + y^2 + z^2 - 9)^2 = 0$ cannot be sampled by any signed 2-point iteration function because f is positive everywhere, except on the surface points where it is zero.

We could ask ourselves why not to use a 1-point iteration function such as, for example, the Newton-Raphson iteration formula, which is sign-independent. However, if the initial estimate is not sufficiently close to the root, the method may not converge. Besides, it is necessary to guarantee that all roots have already been isolated properly.

5.3 Newton-Raphson Method

As suggested above, each numerical method has a specific iteration function. There are many ways to construct iteration functions. These functions are often formulated from the problem itself. For example, solving the equation $x - \sin x = 0$ can be intuitively done by the iterative formula

$$x_{i+1} = \sin x_i, \quad i = 0, 1, 2, \ldots,$$

for which the iteration function is given by $F_i(x) = \sin x_i$.

5.3 Newton-Raphson Method

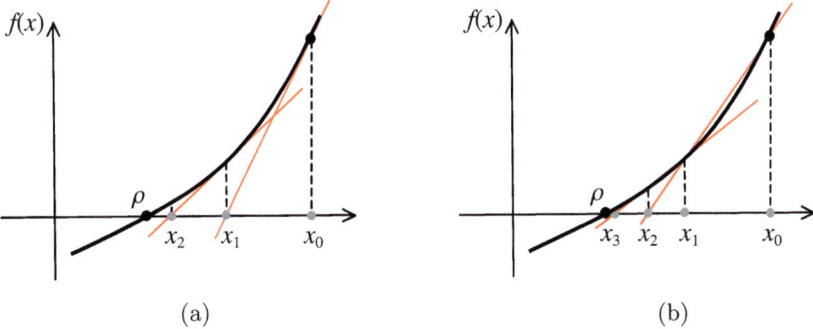

Fig. 5.1. (a) Newton's method uses tangents; (b) secant method uses secants.

5.3.1 The Univariate Case

However, the idea of the Newton-Raphson method is a bit different. Given an initial guess x_0 reasonably close to a zero ρ, one approximates the function by its tangent line at x_0, computing then the x-intercept x_1 of this tangent line. Typically, this x-intercept better approximates such a zero that the original guess, as illustrated in Figure 5.1(a). This process can be repeated until we obtain a sufficiently close estimate to function zero at x, or until a predefined maximum number of iterations have passed.

It is clear that we are assuming that $f(x)$ is differentiable in the neighbourhood of any zero. So, let us start with the point $(x_0, f(x_0))$ in Figure 5.1(a). We easily see that the gradient of the tangent to the function at this point is

$$f'(x_0) = \frac{y - f(x_0)}{x - x_0}$$

where (x, y) is a point on the tangent. The x-intercept of the tangent is the point $(x_1, 0)$, i.e.

$$f'(x_0) = \frac{0 - f(x_0)}{x_1 - x_0}$$

or

$$x_1 = x_0 - \frac{f(x_0)}{f'(x_0)}$$

Repeating this process for the next estimates, we come to the Newton iteration formula

$$x_{i+1} = x_i - \frac{f(x_i)}{f'(x_i)} \tag{5.3}$$

The previous geometric construction of the Newton-Raphson iteration formula agrees with its standard construction from the Taylor series expansion. In fact, if x is a zero of $f : \mathbb{R} \to \mathbb{R}$, and f is sufficiently differentiable in a neighbourhood $N(x)$ of x, then the Taylor series expansion of f about $x_0 \in N(x)$ is given by

$$f(x) = 0 = f(x_0) + (x - x_0)f'(x_0) + \frac{(x - x_0)^2}{2!}f''(x_0) + \ldots$$

By neglecting the 2- and higher order terms, we quickly come to (5.3) by iterating over the indices. That is, the standard Newton-Raphson iteration formula results from the linearisation of f. Obviously, taking into account that an analytic expression exists for quadratic polynomials, we might use a quadratic approximation to f at x_0 by ignoring the 3- and higher power terms. In geometric terms, this is equivalent to use a parabola, instead of a tangent, to approximate f at x_0.

As easily seen from (5.3), Newton's method is an 1-point iterative numerical method viewing that the next estimate x_{i+1} is determined from a single estimate x_i (see also Figure 5.1)(a). Besides, this method uses both the values of the function f and its derivative f'.

The root finding algorithm for the 1-dimensional Newton method is then as appears described in Algorithm 5. This algorithm stops when two consecutive guesses are sufficiently close to each other, i.e. within a small tolerance $\tau > 0$. Note that ϵ is not the machine accuracy, but just and hereinafter called the approximation accuracy, i.e. the absolute-valued difference between two consecutive estimates in the process of convergence to the root.

Convergence

Newton's method converges quadratically to a single root ρ provided that the initial guess is close to it. In mathematical terms, this is equivalent to say that there exists a constant C such that

$$|\rho - x_{n+1}| \leq C|\rho - x_n|^2, \quad n \geq 0. \tag{5.4}$$

Let $\epsilon_n = \rho - x_n$ be the error at the step n. Then, from (5.3), it follows that

$$\epsilon_{n+1} = \rho - x_{n+1} = \rho - x_n + \frac{f(x_n)}{f'(x_n)}$$

Algorithm 5 The Univariate Newton Method

1: **procedure** NEWTON($f, x_0, \epsilon, \tau, x_i$)
2:     $i \leftarrow 0$
3:     **while** $\epsilon > \tau$ **do**   ▷ stopping condition
4:        Evaluate $f(x_i)$ and $f'(x_i)$
5:        $x_{i+1} \leftarrow x_i - \frac{f(x_i)}{f'(x_i)}$ ▷ iteration formula
6:        $\epsilon \leftarrow |x_{i+1} - x_i|$ ▷ approximation accuracy
7:        $i \leftarrow i + 1$
8:     **end while**
9: **end procedure**

that is,
$$\epsilon_{n+1} = \frac{\epsilon_n f'(x_n) + f(x_n)}{f'(x_n)} \quad (5.5)$$

On the other hand, from the Taylor expansion, we have
$$f(\rho) = 0 = f(x_n) + (\rho - x_n)f'(x_n) + \frac{(\rho - x_n)^2}{2}f''(\xi_n)$$

with ξ between x_n and $x_n + \epsilon_n = \rho$, that is
$$0 = f(x_n) + \epsilon_n f'(x_n) + \frac{\epsilon_n^2}{2}f''(\xi_n)$$

or
$$\epsilon_n f'(x_n) + f(x_n) = -\frac{\epsilon_n^2}{2}f''(\xi_n) \quad (5.6)$$

Replacing (5.6) in (5.5) we obtain
$$\epsilon_{n+1} = -\frac{1}{2}\frac{f''(\xi_n)}{f'(x_n)}\epsilon_n^2$$

So, if the method converges, then for x_n and ξ_n near to ρ we get
$$|\epsilon_{n+1}| \approx \frac{1}{2}\frac{|f''(\rho)|}{|f'(\rho)|}|\epsilon_n|^2$$

or
$$|\epsilon_{n+1}| \approx C|\epsilon_n|^2$$

with $C = \frac{1}{2}|\frac{f''(\rho)}{f'(\rho)}|$, which proves that the Newton method has quadratic convergence. This means that the number of exact significant digits in the approximate root doubles from one iteration to the next. But, this is only true if the initial estimate is close enough to the root. Also, this is only true for single roots, not for multiple roots. In fact, the order of convergence at a double root is only linear [374].

5.3.2 The Vector-valued Multivariate Case

The iteration formula (5.3) can be generalised to higher dimensions. Let $f : \mathbb{R}^n \to \mathbb{R}^n$ a real vector-valued function of several real variables, i.e. $f(\mathbf{p})$ is defined by n real-valued function components $f_1, , f_2, \ldots, f_n$ of n real variables $x_1, x_2, \ldots, x_n$; equivalently, $f(\mathbf{x}) = (f_1(\mathbf{x}), f_2(\mathbf{x}), \ldots, f_n(\mathbf{x}))$ with $\mathbf{x} = (x_1, x_2, \ldots, x_n)$.

The multidimensional Newton-Raphson formula can be derived similarly as in the 1-dimensional case. Starting again with the Taylor series of the jth function component centred at the current estimate $\mathbf{x}_i$, we have

Algorithm 6 The Vector-Valued Multivariate Newton Method

1: **procedure** VECTORVALUEDMULTIVARIATENEWTON($f, \mathbf{x}_0, \epsilon, \tau, x_i$)
2:    $i \leftarrow 0$
3:    **while** $\epsilon > \tau$ **do**   ▷ stopping condition
4:       Evaluate $f(\mathbf{x}_i)$ and $Jf(\mathbf{x}_i)$
5:       Solve $Jf(\mathbf{x}_i).\mathbf{c}_i = f(\mathbf{x}_i)$ for $\mathbf{c}_i$ ▷ yields correction percentage for $\mathbf{x}_i$
6:       $\mathbf{x}_{i+1} \leftarrow \mathbf{x}_i - \mathbf{c}_i$ ▷ next estimate after correction
7:       $\epsilon \leftarrow |\mathbf{x}_{i+1} - \mathbf{x}_i|$ ▷ approximation accuracy
8:       $i \leftarrow i + 1$
9:    **end while**
10: **end procedure**

$$f_j(\mathbf{x}) = 0 = f_j(\mathbf{x}_i) + (\mathbf{x} - \mathbf{x}_i)\sum_{k=1}^{n}\frac{\partial f_j(\mathbf{x}_i)}{\partial x_k} + \ldots$$

In matrix notation, after neglecting the quadratic and higher-power terms, this is equivalent to

$$\mathbf{x}_{i+1} = \mathbf{x}_i - \frac{f(\mathbf{x}_i)}{Jf(\mathbf{x}_i)}, \quad i = 0, 1, 2, \ldots \tag{5.7}$$

where $Jf(\mathbf{x}_i)$ is the Jacobian of the vector-valued function $f = (f_1, f_2, \ldots, f_n)$, i.e. the multidimensional counterpart of the derivative of f. (See Stoer and Bulirsch [374], and Ortega and Rheinboldt [314] for an insight into multidimensional numerical methods.) The formula (5.7) is commonly used in numerical analysis to solve nonlinear equation systems.

Although (5.3) and (5.7) are analogous, there is an important difference between the multidimensional Newton formula (5.7) and its 1-dimensional counterpart. Looking at the multidimensional formula, we readily come across that we need to compute the Jacobian matrix inverse. In practice, we do not need to do so explicitly. In fact, solving $Jf(\mathbf{x}_i).\mathbf{c}_i = f(\mathbf{x}_i)$, with $\mathbf{c}_i = \mathbf{x}_{i+1} - \mathbf{x}_i$ saves about a factor of three in computing time over computing the inverse. This improvement appears in Algorithm 6 that describes the vector-valued multivariate Newton method. Apart these subtleties, this version of Newton's method is identical to the univariate case.

5.3.3 The Multivariate Case

Algorithm 6 is commonly used to solve systems of n equations with n variables. In computer graphics, we use Algorithm 7 instead for sampling implicit curves and surfaces. This is so because an implicit curve in $\mathbb{R}^2$ (respectively, a surface in $\mathbb{R}^3$) is defined by a single function in two (respectively, three) real variables. This means that, instead of using the Jacobian, we use the gradient of f as follows:

$$\mathbf{x}_{i+1} = \mathbf{x}_i - \frac{f(\mathbf{x}_i)}{\nabla f(\mathbf{x}_i)}, \quad i = 0, 1, 2, \ldots \tag{5.8}$$

5.3 Newton-Raphson Method

Algorithm 7 The Multivariate Newton Method
1: **procedure** MULTIVARIATENEWTON($f, \mathbf{x}_0, \epsilon, \tau, x_i$)
2:     $i \leftarrow 0$
3:     **while** $\epsilon > \tau$ **do**   ▷ stopping condition
4:        Evaluate $f(\mathbf{x}_i), \nabla f(\mathbf{x}_i)$ and $||\nabla f(\mathbf{x}_i)||$
5:        $\mathbf{x}_{i+1} \leftarrow \mathbf{x}_i - \frac{\nabla f(\mathbf{x}_i)}{||\nabla f(\mathbf{x}_i)||^2} f(\mathbf{x}_i)$ ▷ iteration formula
6:        $\epsilon \leftarrow |\mathbf{x}_{i+1} - \mathbf{x}_i|$ ▷ approximation accuracy
7:        $i \leftarrow i + 1$
8:     **end while**
9: **end procedure**

However, this is not as simple as for (5.7) because the multiplicative inverse of a vector (i.e. the gradient) is not defined for vector spaces, also called linear spaces. Vector spaces are the core objects studied in linear algebra. Informally speaking, a vector space is a set of vectors that may be scaled and added. The geometric product of vectors is not defined; consequently, the multiplicative inverse of a vector is not defined either.

Fortunately, such a geometric product is defined in a geometric algebra, also called multilinear algebra, described technically as a Clifford algebra over a real vector space. Intuitively, a multilinear algebra is more general than a linear algebra because the essential objects of study are multivector spaces. A multivector is an object defined as the addition of a scalar and a vector, in exactly the same way that the addition of real and imaginary numbers yields an object known as complex number.

In multilinear algebra, the geometric product $\mathbf{uv}$ of two vectors $\mathbf{u}$ and $\mathbf{v}$ is defined as follows:

$$\mathbf{uv} = \mathbf{u} \cdot \mathbf{v} + \mathbf{u} \times \mathbf{v} \tag{5.9}$$

where $\mathbf{u} \cdot \mathbf{v}$ and $\mathbf{u} \times \mathbf{v}$ are their scalar and cross products, respectively.

The multiplicative inverse of a vector $\mathbf{u}$, denoted by $1/\mathbf{u}$ or $\mathbf{u}^{-1}$, is the vector which yields 1 when multiplied by $\mathbf{u}$, that is

$$\mathbf{uu}^{-1} = 1 \quad \text{or} \quad \mathbf{u}^{-1} = \frac{1}{\mathbf{u}} = \frac{1}{\mathbf{u}} \frac{\mathbf{u}}{\mathbf{u}} = \frac{\mathbf{u}}{\mathbf{uu}}$$

hence, by the definition of geometric product, we obtain

$$\mathbf{u}^{-1} = \frac{\mathbf{u}}{\mathbf{u} \cdot \mathbf{u}} \tag{5.10}$$

That is, the multiplicative inverse vector $\mathbf{u}^{-1} = \frac{\mathbf{u}}{||\mathbf{u}||^2}$ is the normalised vector of $\mathbf{u}$ divided by its norm, so it is parallel to but smaller than the normalised vector of $\mathbf{u}$. So, (5.8) can be rewritten as

$$\mathbf{x}_{i+1} = \mathbf{x}_i - \frac{\nabla f(\mathbf{x}_i)}{||\nabla f(\mathbf{x}_i)||^2} f(\mathbf{x}_i), \quad i = 0, 1, 2, \ldots \tag{5.11}$$

Newton's method has not been often used for sampling implicit curves and surfaces in computer graphics because of its unreliable convergence. Even so, the formula (5.11) appears in a couple of polygonisers such as, for example, that one due to Hartmann [179].

5.4 Newton-like Methods

Newton's method has quadratic convergence. Unfortunately, it may fail to converge under the following two circumstances:

- PROBLEM I—*The initial guess is far from the root.* The convergence of the Newton method is only guaranteed if the starting estimate is "sufficiently close" to the root [374]. Otherwise, the method risk to converge slowly or even diverge (Figure 5.2(a)).
- PROBLEM II—*The derivative is very small or vanishes.* The Newton iteration formula requires that the derivative f' does not vanish, i.e. $f'(x) \neq 0$. This means that method blows up at local extrema (i.e. local minima and maxima) and inflection points (Figure 5.2(b)).

These two difficulties have led several researchers to modify the Newton method in a variety of ways. Altogether, these modified Newton methods are called *Newton-like methods*. As seen in Chapter 3, Problem I can be solved at the root isolation stage by computing closer guesses. In fact, Newton-like methods have concentrated almost all efforts in solving Problem II concerning the derivative annihilation.

In [416], Wu proposes a family of continuation Newton-like methods for finding a root of a univariate function, for which the derivative is allowed to vanish on some points. In [215], Kou et al. extend Wu's results to the vector-valued multivariate functions such that the Jacobian is allowed to be

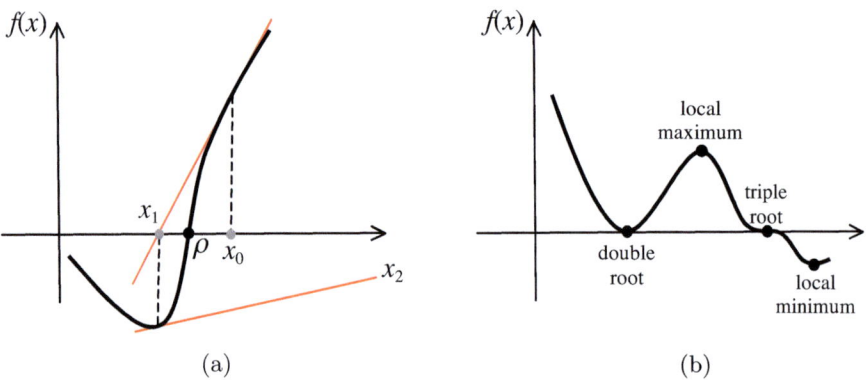

Fig. 5.2. (a) Problem I: divergence; (b) Problem II: null derivative.

numerically singular on some points. The iterative methods of this family also have quadratic convergence.

Recently, similar methods with cubic convergence have been studied and proposed by Kou et al. [216] to solve the same problem. For that, they used the modifications of Newton described in [147] and [406], and the discretisation of a variant of Halley's method [173, 314, 350] to get two new modified Newton methods with cubic convergence which allow for points with null derivative. See also Sharma [358] for other recent work on Newton-like methods.

5.5 The Secant Method

In addition to those two main problems, Newton's method suffers from a third problem related to time performance.

- *The convergence slows down on multiple roots.* As illustrated in Figure 5.2(b), $f'(x)$ also vanishes on multiple roots. In the presence of multiple roots, Newton's method slows down so that it no longer converges quadratically. This is the first "slowing-down" problem.
- *Computation of the derivative expression.* Unless we know the derivative expression in advance, we have to determine it using algebraic and symbolic techniques, what may pose a significant burden on performance of the method. This is the second "slowing-down" problem.
- *Derivative evaluation.* The evaluation of the derivative for every new estimate often is more time-consuming than the function evaluation itself. This is the third "slowing-down" problem.

The secant method is a derivative-free method. It is an attempt to overcome the problems posed by the use of the derivative by Newton's method. In this sense, the secant method can be considered as a Newton-like method. The secant method avoids the problems posed by the derivative by using the discrete derivative, also called difference quotient, which approximates the derivative as follows:

$$f'(x_i) \approx \frac{f(x_i) - f(x_{i-1})}{x_i - x_{i-1}}. \tag{5.12}$$

Then, substituting the expression (5.12) of $f'(x_i)$ into (5.3), we get the iteration formula of the secant method:

$$x_{i+1} = x_i - f(x_i) \frac{x_i - x_{i-1}}{f(x_i) - f(x_{i-1})}, \quad i \geq 1, \tag{5.13}$$

which forces the method to start with two initial estimates, say x_0 and x_1, instead the single initial estimate of Newton's method.

The secant method also has a geometric interpretation. Replacing the derivative by the difference quotient in (5.3) results in using a secant line instead a tangent line to approximate a root of a function. In fact, the next estimate x_{i+1} is the x-intercept of the secant line to the graph of f at the points $(x_{i-1}, f(x_{i-1}))$ and $(x_n, f(x_n))$. This secant line is given by

$$y - f(x_i) = \frac{f(x_i) - f(x_{i-1})}{x_i - x_{i-1}}(x - x_i).$$

But we know that $y = 0$ at the x-intercept of the secant line, so we have

$$0 - f(x_i) = \frac{f(x_i) - f(x_{i-1})}{x_i - x_{i-1}}(x - x_i)$$

or, solving for x,

$$x = x_i - f(x_i)\frac{x_i - x_{i-1}}{f(x_i) - f(x_{i-1})} \tag{5.14}$$

which is equivalent to the secant iteration formula (5.13).

Note that the secant method is a 2-point method because the next estimate is determined from the two previous ones. However, it is not a bracketing method because the root is not necessarily between the last two estimates, so the IVT does not apply. The successive estimates converge to the root in a similar way to Newton's method. This is shown in Figure 5.1. However, when we force both estimates to bracket the root for every single iteration, the secant method turns into the so-called false position method. This explains why they use the same iteration formula. Algorithm 8 describes the secant method in its multivariate version for brevity.

5.5.1 Convergence

Let $\epsilon_i = \rho - x_i$ be the error at the step i. Then, from (5.13), it follows that

$$\epsilon_{i+1} = \rho - x_{i+1} = \rho - [x_i - f(x_i) + \frac{x_i - x_{i-1}}{f(x_i) - f(x_{i-1})}]$$

Algorithm 8 The Multivariate Secant Method

1: **procedure** MULTIVARIATESECANTMETHOD($f, \mathbf{x}_0, \mathbf{x}_1, \epsilon, \tau, x_i$)
2:    $i \leftarrow 1$
3:    **while** $\epsilon > \tau$ **do**   ▷ stopping condition
4:       Evaluate $f(\mathbf{x}_i), f(\mathbf{x}_{i-1})$
5:       $\mathbf{x}_{i+1} \leftarrow \mathbf{x}_i - f(\mathbf{x}_i)\frac{\mathbf{x}_i - \mathbf{x}_{i-1}}{f(\mathbf{x}_i) - f(\mathbf{x}_{i-1})}$ ▷ iteration formula
6:       Evaluate $f(\mathbf{x}_{i+1})$
7:       **if** $f(\mathbf{x}_{i+1}) = 0$ **then**
8:          $\epsilon \leftarrow 0$
9:       **else**
10:         $\epsilon \leftarrow |\mathbf{x}_{i+1} - \mathbf{x}_i|$ ▷ approximation accuracy
11:       **end if**
12:       $i \leftarrow i + 1$
13:    **end while**
14: **end procedure**

that is,

$$\epsilon_{i+1} = \frac{\rho f(x_i) - \rho f(x_{i-1}) + x_i f(x_{i-1}) - f(x_i) x_{i-1}}{f(x_i) - f(x_{i-1})}$$
$$= \frac{\epsilon_{i-1} f(x_i) - \epsilon_i f(x_{i-1})}{f(x_i) - f(x_{i-1})}$$

after factoring and replacing $\rho - x_i$ (respectively, $\rho - x_{i-1}$) by ϵ_i (respectively, ϵ_{i-1}). Equivalently, we have

$$\epsilon_{i+1} = \frac{x_i - x_{i-1}}{f(x_i) - f(x_{i-1})} \frac{\epsilon_{i-1} f(x_i) - \epsilon_i f(x_{i-1})}{x_i - x_{i-1}} \qquad (5.15)$$
$$= \frac{x_i - x_{i-1}}{f(x_i) - f(x_{i-1})} \frac{\frac{f(x_i)}{\epsilon_i} - \frac{f(x_{i-1})}{\epsilon_{i-1}}}{x_i - x_{i-1}} \epsilon_i \epsilon_{i-1}.$$

On the other hand, from the Taylor expansion at x_i, we have

$$f(x_i) = f(\rho - \epsilon_i) = f(\rho) + \epsilon_i f'(\rho) + \frac{1}{2}\epsilon_i^2 f''(\rho) + \mathcal{O}(\epsilon_i^3)$$

where $f(\rho) = 0$; thus, dividing by ϵ_i we have

$$\frac{f(x_i)}{\epsilon_i} = f'(\rho) + \frac{1}{2}\epsilon_i f''(\rho) + \mathcal{O}(\epsilon_i^2) \qquad (5.16)$$

and analogously

$$\frac{f(x_{i-1})}{\epsilon_{i-1}} = f'(\rho) + \frac{1}{2}\epsilon_{i-1} f''(\rho) + \mathcal{O}(\epsilon_{i-1}^2) \qquad (5.17)$$

Subtracting (5.17) from (5.16) we obtain

$$\frac{f(x_i)}{\epsilon_i} - \frac{f(x_{i-1})}{\epsilon_{i-1}} = \frac{1}{2}(\epsilon_i - \epsilon_{i-1})f''(\rho) + \mathcal{O}(\epsilon_i^2) - \mathcal{O}(\epsilon_i^2),$$

or

$$\frac{f(x_i)}{\epsilon_i} - \frac{f(x_{i-1})}{\epsilon_{i-1}} \approx \frac{1}{2}(\epsilon_i - \epsilon_{i-1})f''(\rho)$$

or still

$$\frac{f(x_i)}{\epsilon_i} - \frac{f(x_{i-1})}{\epsilon_{i-1}} \approx \frac{1}{2}(x_i - x_{i-1})f''(\rho). \qquad (5.18)$$

Now, by combining (5.15) and (5.18) we get

$$\epsilon_{i+1} \approx \frac{x_i - x_{i-1}}{f(x_i) - f(x_{i-1})}\left(-\frac{1}{2}\right)f''(\rho)\epsilon_i \epsilon_{i-1}$$
$$\approx \frac{1}{f'(\rho)}\left(-\frac{1}{2}\right)f''(\rho)\epsilon_i \epsilon_{i-1}$$

provided that $\frac{1}{f'(\rho)} \approx \frac{x_i - x_{i-1}}{f(x_i) - f(x_{i-1})}$ for x_i, x_{i-1} near to ρ. Consequently,

$$|\epsilon_{i+1}| \approx \frac{1}{2} \frac{|f''(\rho)|}{|f'(\rho)|} |\epsilon_i \epsilon_{i-1}|$$

or

$$|\epsilon_{i+1}| \approx K |\epsilon_i \epsilon_{i-1}| \quad (5.19)$$

Intuitively, this shows us that the rate of convergence of the secant method is superlinear, though not quite quadratic. To make sure about that, let us recall the definition of rate of convergence

$$\lim_{i \to \infty} \frac{|\epsilon_{i+1}|}{|\epsilon_i|^\alpha} = C$$

or, similarly,

$$\lim_{i \to \infty} \frac{|\epsilon_i|}{|\epsilon_{i-1}|^\alpha} = C$$

so that, from these two expressions, we have

$$\frac{|\epsilon_{i+1}|}{|\epsilon_i|^\alpha} \approx \frac{|\epsilon_i|}{|\epsilon_{i-1}|^\alpha}$$

or

$$|\epsilon_{i+1}| \approx \frac{|\epsilon_i|^{\alpha+1}}{|\epsilon_{i-1}|^\alpha} \quad (5.20)$$

so, inserting (5.20) into (5.19) we get

$$\frac{|\epsilon_i|^{\alpha+1}}{|\epsilon_{i-1}|^\alpha} \approx K |\epsilon_i| |\epsilon_{i-1}|,$$

that is,

$$|\epsilon_i| \approx K^{\frac{1}{\alpha}} |\epsilon_{i-1}|^{\frac{\alpha+1}{\alpha}}$$

or, equivalently,

$$|\epsilon_{i+1}| \approx K^{\frac{1}{\alpha}} |\epsilon_i|^{\frac{\alpha+1}{\alpha}} \quad (5.21)$$

but, by definition, we know that

$$|\epsilon_{i+1}| \approx C |\epsilon_i|^\alpha. \quad (5.22)$$

Therefore, from (5.21) and (5.22), we conclude that

$$\frac{\alpha + 1}{\alpha} = \alpha.$$

But, this is equivalent to

$$\alpha^2 - \alpha - 1 = 0$$

that is, a quadratic equation whose solutions are given by $\alpha = \frac{1 \pm \sqrt{5}}{2}$.

Taking the positive solution, we can say that the rate of convergence is then $\alpha = \frac{1+\sqrt{5}}{2} \approx 1.618$, i.e. the *golden ratio*. Thus, the convergence is superlinear. Similar to Newton method, this result only holds if f is twice continuously differentiable and the root is simple (i.e. it is not a multiple root). Analogously, if the initial guesses are not sufficiently near to the root, then there is no guarantee that the method converges.

It is worth noting that, despite its slower rate of convergence, the secant method converges faster than Newton's method in practice. This is so because the secant method only requires the evaluation of the function f for each iteration, while Newton's method requires the evaluation of both f and its derivative.

5.6 Interpolation Numerical Methods

Interpolation numerical methods, also called *bracketing methods*, are 2-point numerical methods. Let us then describe two of these methods: the *bisection method* and the *false position method*.

5.6.1 Bisection Method

As any other bracketed method, the bisection method combines the recursive subdivision of the initial interval with the IVT to ensure that the sequence of intervals converge to the root. In the process of interval subdivision and convergence to the root, we choose the subinterval that contains such a root after applying the IVT. The search continues on such a subinterval recursively until the root is found within a subinterval of minimum length.

The Univariate Case

Roughly speaking, the procedure behind the bisection method consists of three major steps: (i) *interval subdivision* by its midpoint; (ii) function *evaluation* at the midpoint; (iii) *selection* of the subinterval which satisfies the IVT. This procedure continues recursively until the function approximately vanishes at the midpoint. The midpoint of the search interval $[x_{i-1}, x_i]$ is given by

$$x_{i+1} \leftarrow \frac{x_{i-1} + x_i}{2} \tag{5.23}$$

The next subinterval is the one that satisfies the IVT. That is, either $[x_{i-1}, x_{i+1}]$ if $f(x_{i-1}).f(x_{i+1}) < 0$ or $[x_{i+1}, x_i]$ if $f(x_{i+1}).f(x_i) < 0$. Therefore, without loss of generality, the bisection method produces a sequence of shrinking subintervals $[x_{i-1}, x_i]$, $i \geq 1$, satisfying $f(a_i).f(b_i) < 0$. The corresponding algorithm is described in Algorithm 9, though it appears here in its multivariate version for brevity.

Algorithm 9 The Multivariate Bisection Method

1: **procedure** MULTIVARIATEBISECTION($f, [\mathbf{x}_0, \mathbf{x}_1], \epsilon, \tau$)
2:     $i \leftarrow 1$
3:     **while** $\epsilon > \tau$ **do**   ▷ stopping condition: smallest interval
4:         Evaluate $f(\mathbf{x}_i), f(\mathbf{x}_{i-1})$
5:         $\mathbf{x}_{i+1} \leftarrow \frac{\mathbf{x}_{i-1}+\mathbf{x}_i}{2}$
6:         Evaluate $f(\mathbf{x}_{i+1})$
7:         **if** $|f(\mathbf{x}_{i+1})| < \tau$ **then**   ▷ stopping condition: root found
8:             $r \leftarrow \mathbf{x}_{i+1}$
9:         **else**
10:             **if** $f(\mathbf{x}_{i-1}).f(\mathbf{x}_{i+1}) < 0$ **then**   ▷ bracketing through IVT
11:                 $\mathbf{x}_i \leftarrow \mathbf{x}_{i+1}$
12:             **else**
13:                 $\mathbf{x}_{i-1} \leftarrow \mathbf{x}_{i+1}$
14:             **end if**
15:         **end if**
16:         $\epsilon \leftarrow |\mathbf{x}_i - \mathbf{x}_{i-1}|$ ▷ approximation accuracy
17:         $i \leftarrow i+1$
18:     **end while**
19:     **if** $\epsilon < \tau$ **then**
20:         $r \leftarrow \frac{\mathbf{x}_{i-1}+\mathbf{x}_i}{2}$
21:     **end if**
22: **end procedure**

The Multivariate Case

The formula (5.23) easily generalises to the multivariate case as follows:

$$\mathbf{x}_{i+1} \leftarrow \frac{\mathbf{x}_{i-1}+\mathbf{x}_i}{2} \tag{5.24}$$

The recursive computation of the midpoint terminates when the zero is found within a small tolerance $\tau > 0$ in the latest interval, i.e. $f(\mathbf{x}_{i+1}) < \tau$ at the latest midpoint $\mathbf{x}_{i+1}$, as expressed in Algorithm 9. The value of $\epsilon = |\mathbf{x}_i - \mathbf{x}_{i-1}|$ yields a measure of the error of in the approximation to the root. In the following, we prove that the IVT-driven subdivision of the interval guarantees that the bisection method always converge.

Convergence

Let $e_i = \mathbf{x}_i - \rho$ be the absolute error at the step i, with ρ denoting the exact root we are looking for. Taking into consideration that

$$|\mathbf{x}_i - \rho| \leq |\epsilon_i|,$$

with $\epsilon_i = \mathbf{x}_i - \mathbf{x}_{i-1}$, it follows from the iteration formula (5.24) that

$$|e_i| = \frac{|\epsilon_0|}{2^i} = \frac{x_1 - x_0}{2^i}, \quad i \geq 0.$$

Consequently, we have $|e_i| \leq \frac{x_1 - x_0}{2^i}$, $i \geq 0$, which implies $\lim_{i \to \infty} |e_i| = 0$. Therefore, the bisection method is *globally convergent*. That is, it is guaranteed to converge. However, in comparison to other methods, some of which are discussed in this chapter, bisection tends to converge slowly.

5.6.2 False Position Method

The *false position method*, also called *regula falsi method*, can be described as the *bracketed* secant method. The iteration formula is exactly the same as for secant method, but before using it we have to guarantee that the root remains bracketed through the IVT. Therefore, the false position method combines the features of both bisection and secant methods.

The Univariate Case

Like the bisection method, the false position method is a bracketed 2-point method. It starts with an interval $[x_0, x_1]$, then it checks whether $f(x_0)$ and $f(x_1)$ are of opposite signs or not by using the IVT. If so, then a root exists in the interval surely, and the method proceeds by recursively subdividing the interval into two sub-intervals, discarding those which do not satisfy the IVT. The result is a sequence of shrinking intervals $[x_{i-1}, x_i]$ ($i = 1, \ldots n$) converging to a root of f.

The $(i+1)$-th estimate that subdivides the interval $[x_{i-1}, x_i]$ into two is given by the iteration formula (5.13), here rewritten for convenience:

$$x_{i+1} = x_i - f(x_i) \frac{x_i - x_{i-1}}{f(x_i) - f(x_{i-1})}, \quad i \geq 1. \tag{5.25}$$

As illustrated in Figure 5.3, x_{i+1} is the root of the secant line passing through $(x_{i-1}, f(x_{i-1}))$ and $(x_i, f(x_i))$. Now, we use the IVT in order to guarantee that the guess x_{i+1} remains bracketed. If $f(x_{i-1}).f(x_{i+1}) < 0$, then we set $x_i = x_{i+1}$; otherwise, we set $x_{i-1} = x_{i+1}$. This process continues until the root of the secant line approximates the function root inside a small tolerance $\tau > 0$.

Note that we can also get the iteration formula (5.25) by geometric means. For that, we use the point-slope equation of the secant line through $(x_{i-1}, f(x_{i-1}))$ and $(x_i, f(x_i))$, which is defined as follows:

$$y - f(x_i) = \frac{f(x_i) - f(x_{i-1})}{x_i - x_{i-1}} (x - x_i).$$

Solving this equation after substituting the secant root $(x_{i+1}, 0)$, we just obtain the formula (5.25) above.

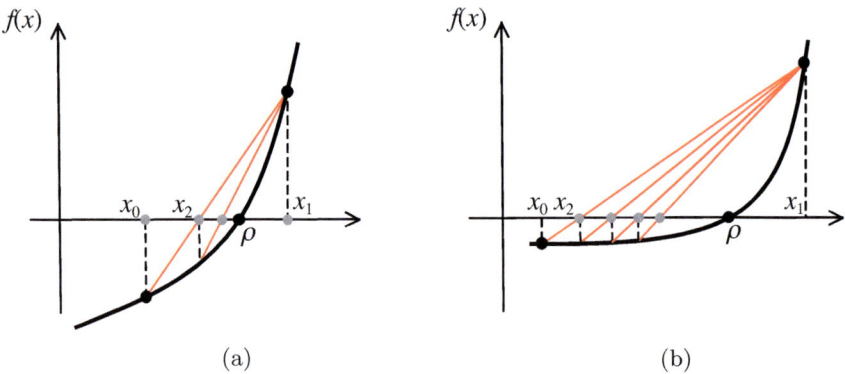

Fig. 5.3. (a) The false position (FP) method; (b) The "slowing-down" problem of FP method.

The Multivariate Case

Similar to bisection method, the false position method also generalises to higher dimensions easily. Instead of a simple variable x, we use a multi-variable $\mathbf{x} = (x_1, x_2, \ldots, x_n)$ in the iteration formula (5.25), which yields

$$\mathbf{x}_{i+1} = \mathbf{x}_i - f(\mathbf{x}_i)\frac{\mathbf{x}_i - \mathbf{x}_{i-1}}{f(\mathbf{x}_i) - f(\mathbf{x}_{i-1})}. \tag{5.26}$$

This multivariate iteration formula allows for sampling curves and surfaces defined by implicit functions in two and three variables in $\mathbb{R}^2$ and $\mathbb{R}^3$, respectively. In this case, a curve or surface is sampled against a general straight line (not necessarily the x-axis) in $\mathbb{R}^2$ or $\mathbb{R}^3$, respectively.

The pseudocode of the multivariate false position method appears in Algorithm 10. This algorithm produces a sequence of shrinking bracketed subintervals $[\mathbf{x}_{i-1}, \mathbf{x}_i]$, $i \geq 1$, satisfying $f(\mathbf{x}_{i-1}).f(\mathbf{x}_i) < 0$. But, unlike the bisection method, it does not terminate when the length of the current interval is less than or equal to a small tolerance $\tau > 0$. It stops when the absolute value of f for the guess $\mathbf{x}_{i+1}$ is approximately zero, i.e. it is within a small tolerance τ. This is so because, unlike the bisection method, the length of the brackets does not tend to zero. In fact, only one of the endpoints converges to the root, the other remains fixed (Figure 5.3).

Convergence

Despite the false position method has the same iteration formula as the secant method, it only converges linearly. This happens in this manner because only one of the interval endpoints converges to the root, the other remains unchanged (Figure 5.3).

Algorithm 10 The Multivariate False Position Method

1: **procedure** MULTIVARIATEFALSEPOSITION($f, [\mathbf{x}_0, \mathbf{x}_1], \tau, r$)
2:    $i \leftarrow 1$
3:    **while** $i < i_{MAX}$ **do**   ▷ stopping condition: maximum number of iterations
4:       Evaluate $f(\mathbf{x}_{i-1})$
5:       Evaluate $f(\mathbf{x}_i)$
6:       $\mathbf{x}_{i+1} \leftarrow \mathbf{x}_i - f(\mathbf{x}_i) \frac{\mathbf{x}_i - \mathbf{x}_{i-1}}{f(\mathbf{x}_i) - f(\mathbf{x}_{i-1})}$
7:       Evaluate $f(\mathbf{x}_{i+1})$
8:       **if** $|f(\mathbf{x}_{i+1})| < \tau$ **then**   ▷ stopping condition: root found
9:          $r \leftarrow \mathbf{x}_{i+1}$
10:          **return** true
11:       **else**
12:          **if** $f(\mathbf{x}_{i-1}).f(\mathbf{x}_{i+1}) < 0$ **then**   ▷ bracketing through IVT
13:             $\mathbf{x}_i \leftarrow \mathbf{x}_{i+1}$
14:          **else**
15:             $\mathbf{x}_{i-1} \leftarrow \mathbf{x}_{i+1}$
16:          **end if**
17:       **end if**
18:       $i \leftarrow i + 1$
19:    **end while**
20:    **return** false
21: **end procedure**

Let us then consider that the false position method produces a sequence $\mathbf{x}_0, \mathbf{x}_1, \ldots, \mathbf{x}_i, \ldots$ of approximations or estimates to a root ρ, that is $\lim_{i \to \infty} \mathbf{x}_i = \rho$. Let $\epsilon_i = \mathbf{x}_i - \rho$ be the error in the ith iterate. The speed of convergence is determined by subtracting and dividing the iteration formula (5.26) by $\mathbf{x}$ and $\epsilon_i = \mathbf{x}_i - \rho$, respectively, to get

$$\frac{\epsilon_{i+1}}{\epsilon_i} = 1 - \frac{f(\mathbf{x}_i)}{f(\mathbf{x}_i) - f(\mathbf{x}_{i-1})} \cdot \frac{(\mathbf{x}_i - \alpha) - (\mathbf{x}_{i-1} - \alpha)}{\mathbf{x}_i - \alpha}$$

$$= 1 + \left(\frac{\mathbf{x}_{i-1} - \alpha}{\mathbf{x}_i - \alpha} - 1\right) \cdot \frac{f(\mathbf{x}_i)}{f(\mathbf{x}_i) - f(\mathbf{x}_{i-1})}$$

$$= 1 + \left(\frac{\epsilon_{i-1}}{\epsilon_i} - 1\right) \cdot \frac{f(\mathbf{x}_i)}{f(\mathbf{x}_i) - f(\mathbf{x}_{i-1})}$$

Since $\frac{f(\mathbf{x}_i)}{f(\mathbf{x}_i) - f(\mathbf{x}_{i-1})} > 0$ for bracketing intervals, $\lim_{n \to \infty} \frac{\epsilon_{i+1}}{\epsilon_i} = \lim_{i \to \infty} \frac{\epsilon_i}{\epsilon_{i-1}}$ and $\lim_{i \to \infty} \mathbf{x}_{i-1} = \lim_{i \to \infty} \mathbf{x}_i = \rho$, we get

$$\lim_{i \to \infty} \frac{\epsilon_{i+1}}{\epsilon_i} = 1 + \left(\frac{1}{\lim_{i \to \infty} \frac{\epsilon_{i+1}}{\epsilon_i}} - 1\right) \cdot \frac{1}{2}$$

that is $\lim_{i \to \infty} \frac{\epsilon_{1+1}}{\epsilon_1} = 1$, as expected.

5.6.3 The Modified False Position Method

Taking into account that brackets do not converge to zero, it may happen that the speed of convergence is too slow that the algorithm easily gets the maximum number of iterations without finding the root. For example, the univariate function $f(x) = \tan(x)^{\tan(x)} - 10^3$ has a root in the interval $[1.3, 1.4]$, but the false position method is not capable of finding it if the maximum number of iterations $i_{MAX} = 200$. This situation is similar to that one shown in Figure 5.3(b).

Intuitively, this problem arises when the absolute function values at the endpoints differ significantly. Therefore, the idea behind the modified method is down-weighting one of the endpoint function values to force the next estimate $\mathbf{x}_{i+1}$ to approximate the root more rapidly. In other words, we reduce the weight of the "bad" or higher function value to a half or any other value found appropriate.

Rewriting the false position formula (5.26) as follows

$$\mathbf{x}_{i+1} = \frac{f(\mathbf{x}_i)\mathbf{x}_{i-1} - f(\mathbf{x}_{i-1})\mathbf{x}_i}{f(\mathbf{x}_i) - f(\mathbf{x}_{i-1})} \tag{5.27}$$

we readily come to a weighted iteration formula. So, using a factor of 2, we can fix the problem by changing the weight of a function value at an endpoint such as

$$\mathbf{x}_{i+1} = \frac{\frac{1}{2}f(\mathbf{x}_i)\mathbf{x}_{i-1} - f(\mathbf{x}_{i-1})\mathbf{x}_i}{\frac{1}{2}f(\mathbf{x}_i) - f(\mathbf{x}_{i-1})} \tag{5.28}$$

or

$$\mathbf{x}_{i+1} = \frac{f(\mathbf{x}_i)\mathbf{x}_{i-1} - \frac{1}{2}f(\mathbf{x}_{i-1})\mathbf{x}_i}{f(\mathbf{x}_i) - \frac{1}{2}f(\mathbf{x}_{i-1})}. \tag{5.29}$$

The factor 2 in cutting down the function value at one of the endpoints guarantees superlinear convergence. Other rescaling factors are possible to work out in order to speed up the convergence to the root.

5.7 Interval Numerical Methods

An interval numerical method combines a numerical method with interval arithmetic. This way, we end up by getting an iterative method that can be used both to isolate and to approximate zeros of a real function, in a way similar to the Bernstein solvers dealt with in Chapter 3. For brevity, we only show here how this can be done for Newton's method.

5.7.1 Interval Newton Method

Before proceeding any further, let us say that the foundations of the mathematical theory behind the interval Newton method appears described in Alefeld and Herzberger [4].

The Univariate Case

Basically, the *interval Newton method* computes an enclosure of a zero of a real function $f(x)$ defined on the interval $\mathbf{X} = [\underline{x}, \overline{x}]$, that is $f : \mathbf{X} = [\underline{x}, \overline{x}] \to \mathbb{R}$. Let us assume that its derivative $f'(x)$ is continuous and does not vanish in $\mathbf{X}$, i.e. $0 \notin f'(\mathbf{X})$, and that $f(\underline{x}).f(\overline{x}) < 0$.

Starting with the 0th inclusion $\mathbf{X}_0$, if $\mathbf{X}_i$ is the ith inclusion of the root, the smaller $(i+1)$-th inclusion $\mathbf{X}_{i+1}$ may be computed by

$$\mathbf{X}_{i+1} = \mathbf{N}(x, \mathbf{X}_i) \cap \mathbf{X}_i, \quad x \in \mathbf{X}_i. \tag{5.30}$$

where $\mathbf{N}(x, \mathbf{X}_i)$ is the interval Newton operator defined over the interval $\mathbf{X}_i$ as follows

$$\mathbf{N}(x, \mathbf{X}_i) = x - \frac{f(x)}{f'(\mathbf{X}_i)}, \quad x \in \mathbf{X}_i. \tag{5.31}$$

Usually, x is the midpoint of $\mathbf{X}_i$, but any other value within the interval $\mathbf{X}_i$ is eligible.

Therefore, the interval Newton method produces a sequence of shrinking intervals which converge to the zero, so that it can be considered as a 2-point numerical method. Let us see an example:

Example 5.2. Let $f(x) = x^2 - 4$ a real function defined over the interval $\mathbf{X}_0 = [-3, 3]$, and $x = 0$ the midpoint of $\mathbf{X}_0$ chosen to subdivide it. The range of f' is then $f'(\mathbf{X}_0) = [-6, 6]$. The first iteration yields

$$\mathbf{N}(0, \mathbf{X}_0) = 0 - \frac{-4}{[-6, 6]}$$
$$= 0 - \left([-\infty, -\tfrac{2}{3}] \cup [\tfrac{2}{3}, \infty] \right)$$
$$= \left[-\infty, -\tfrac{2}{3} \right] \cup \left[\tfrac{2}{3}, \infty \right]$$

Therefore,

$$\mathbf{X}_1 = \mathbf{N}(0, \mathbf{X}_0) \cap \mathbf{X}_0 = \left[-3, -\tfrac{2}{3} \right] \cup \left[\tfrac{2}{3}, 3 \right]$$

what yields the subintervals $\mathbf{X}_1^0 = [-3, -\tfrac{2}{3}]$ and $\mathbf{X}_1^1 = [\tfrac{2}{3}, 3]$. Now, the first sub-interval is put on a stack for posterior processing, while the second is again subdivided by interval Newton method.

In geometric terms, the leading idea of the interval Newton method [175, 210, 252, 339] is the following:

- First, to enclose the graph of f in a cone given by $\mathbf{N}(x, \mathbf{X}_i)$. This cone is defined by the extremal tangents of $f'(\mathbf{X}_i)$, i.e. the enclosure or interval function for f'.
- Second, to enclose the zeros in the intersection of such a cone and the x-axis (cf. Equation (5.30)).

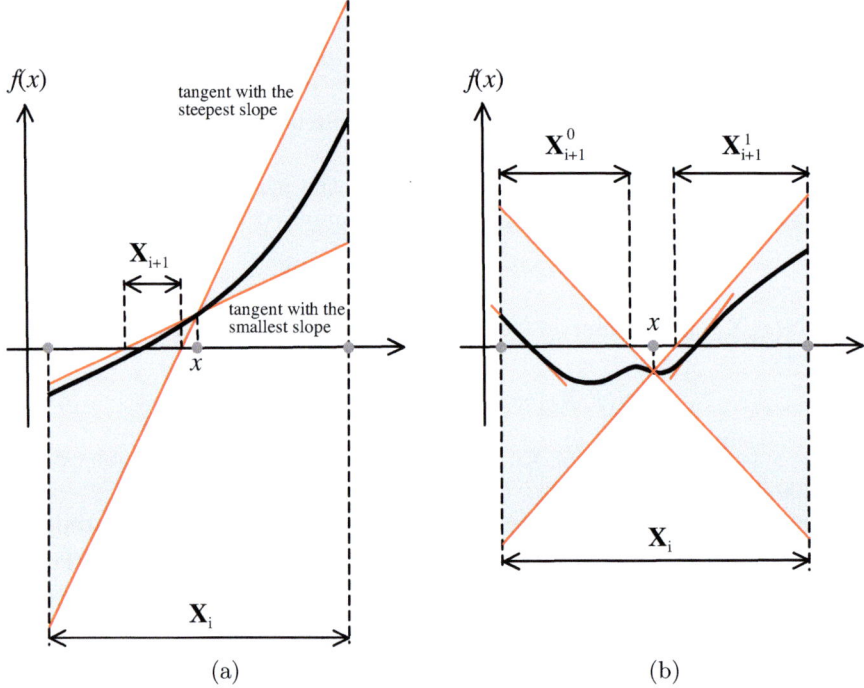

Fig. 5.4. Interval Newton method: (a) one subinterval; (b) two subintervals.

This is illustrated in Figure 5.4(a) for a single zero in an interval $\mathbf{X}_i$. But, if $\mathbf{X}_i$ contains more zeros, the derivative f' vanishes somewhere on it, as shown in Figure 5.4(b). In this case, to surround the problem of $f'(\mathbf{X}_i)$ containing 0, an extended interval division is performed to compute the cone, after which two subintervals $\mathbf{X}^0_{i+1}$ and $\mathbf{X}^1_{i+1}$ are produced by intersecting the cone and the interval $\mathbf{X}_i$, as illustrated in Figure 5.4(b).

Note that no root belonging to the initial interval is missed out by using the interval Newton algorithm. Moreover, every zero appears isolated in one of the intervals of the final list of intervals. Eventually, this list may include intervals without any root, but in this case they can be discarded if the interval image of f does not contain 0.

Univariate Interval Newton Algorithm

The univariate interval Newton algorithm appears described in Algorithm 11. The first stopping condition is global in the sense the algorithm stops when there is no interval to process further. The second stopping condition is satisfied when $l(\mathbf{X}) \leq \tau$, i.e. the length of the interval $\mathbf{X}$ which contains a root is less or equal to a given tolerance τ.

Algorithm 11 The Univariate Interval Newton Method
───
1: **procedure** UNIVARIATEINTERVALNEWTONMETHOD($f, \mathbf{X}, \tau, L$)
2:     $I \leftarrow \mathbf{X}$ ▷ initialise auxiliary list of intervals
3:     **while** $I \neq \varnothing$ **do**   ▷ first stopping condition
4:         Remove an interval $\mathbf{X}$ from I
5:         **while** $l(\mathbf{X}) > \tau$ **do**   ▷ second stopping condition: the length of $\mathbf{X}$
6:             Compute $\mathbf{N}(x, \mathbf{X}) \cap \mathbf{X}$ for some $x \in \mathbf{X}$
7:             **if** $\mathbf{N}(x, \mathbf{X}) \cap \mathbf{X}$ consists of a single interval **then**
8:                 $\mathbf{X} \leftarrow \mathbf{N}(x, \mathbf{X}) \cap \mathbf{X}$
9:             **else**
10:                 Put the first interval into I
11:                 Set the second interval as $\mathbf{X}$
12:             **end if**
13:         **end while**
14:         $L \leftarrow \mathbf{X}$ ▷ new tight interval with root found
15:     **end while**
16: **end procedure**
───

5.7.2 The Multivariate Case

The multivariate interval Newton method is analogous to the univariate case. Thus, by analogy to the iteration Formula (5.8), the multivariate interval Newton operator is as follows:

$$\mathbf{N}(\mathbf{x}, \mathbf{X}_i) = \mathbf{x} - \frac{f(\mathbf{x})}{\nabla f(\mathbf{X}_i)}, \quad \mathbf{x} \in \mathbf{X}_i. \tag{5.32}$$

This operator provides a robust tool for sampling implicit curves and surfaces in 2D and 3D, respectively, in particular in those algorithms using axis-aligned space partitioning such as quadtrees and octrees. Surprisingly, it seems that there is no polygonisation algorithm in the computer graphics literature using this or any other interval numerical method. Instead, intervals and numerical methods have been used separately for isolation and approximation, respectively. For further details on the mathematics of multivariate interval Newton methods, the reader is referred to Hansen [174].

5.8 Final Remarks

In this chapter we have presented several classical root-finding numerical algorithms in the context of sampling implicit curves and surfaces. Hence, we have focused on multivariate numerical methods, and this makes a difference in relation to the classical numerical analysis textbooks.

We have seen that there are essentially two broad classes of numerical methods:

- *1-Point Methods.* Starting from a single guess, we try to move it closer to the root.
- *2-Point or Interval Methods.* Starting from a bracket that contains the root, we attempt to shrink such an interval until the desired accuracy is reached.

We can say that most interval techniques are reliable, but slow, while 1-point techniques tend to be faster, but do not guarantee convergence. With all their advantages and shortcomings, we can also say that numerical methods are still an active research area in mathematics and computing. In particular, interval numerical methods seem to be so promising in sampling implicit curves and surfaces in respect to both speed and reliability (or quality) of numerical computations.

Part III

Reconstruction and Polygonisation

Overview

Part III develops from two main ideas. The first is how to *polygonise* an implicit surface from its multivariate real function. The second is more recent and has to do with the reconstruction of an implicit surface from an unorganised scattered dataset or cloud of points acquired by a range scanner. The reconstruction problem involves a process of first finding the function that approximates or interpolates the acquired points, after which the polygonisation follows.

The first two chapters of Part III detail the most important polygonisation algorithms, while the third describes reconstruction algorithms for implicit surfaces, and occasionally curves. This allows us to have a clean view of this rich collection of algorithms, and to get a better understanding of the nature of implicit objects.

6
Continuation Methods

Continuation methods are based on piecewise linear approximation of a variety (e.g. curve or surface) by means of numerical solution of an initial value problem [7]. In other words, they compute solution varieties of nonlinear systems usually expressed in terms of an equation

$$f(\mathbf{p}) = \mathbf{0} \qquad (6.1)$$

with $f : \mathbb{R}^{n+d} \to \mathbb{R}^n$ a real function. The solution of this equation is called *zero set* (i.e. a particular level set).

As studied in Chapter 1, a zero set is a variety that consists of regular pieces called manifolds, which are joined at singular solutions (which are also solution manifolds, but of a system with lower d). The regular pieces are manifold curves when $d = 1$, manifold surfaces when $d = 2$, and d-manifolds in general. These systems arise frequently in engineering and scientific problems, because these problems are often formulated in terms of the computation of a function that satisfies some set of equations, for example, the Navier-Stokes equations, Maxwell's equations, or Newton's law.

6.1 Introduction

The essential idea behind a continuation method is very simple: first, compute a piece of the solution manifold near one solution, then select another solution from this set and repeat the process. As long as the new piece covers some new part of the solution manifold the computation progresses. So the basic issues are:

1. How to compute the solution manifold near some point $\mathbf{p}_i$ at which $f(\mathbf{p}_i) = 0$.
2. How to select a new point.
3. How to avoid recomputing the same part of the manifold.

There are two ways to perform the first task, which lead us to two types of continuation methods: *simplicial continuation methods* and *predictor-corrector methods*.

6.2 Piecewise Linear Continuation

In the mid-1960s, Lemke and Howson introduced piecewise linear methods in mathematics for calculating solutions to complementarity problems [227, 228], as needed in economics. In the 1970s and 1980s, the research on piecewise linear methods moved on to computing fixed points in mathematics. In the 1990's they started to be used to approximate implicitly defined manifolds and varieties in computer graphics.

Piecewise linear continuation (or PL continuation), also called *simplicial continuation*, operates on a triangulation (simplicial complex) of a given domain $\Omega \subset \mathbb{R}^{n+d}$ to approximate and to sample a a manifold (e.g. a curve or a surface). This is illustrated in Figure 6.1 for a curve that lies in $\Omega \subset \mathbb{R}^2$. The triangulation (a) splits Ω into equilateral triangles, where the red ones are those triangles which the curve passes through. The triangulation (b) partitions Ω into isosceles triangles, where the green ones are those which better approximate the curve; they intersect the curve indeed.

6.2.1 Preliminary Concepts

Simplicial continuation methods are used to trace a piecewise linear (PL) approximation of a zero set

$$S = \{\mathbf{x} \in \mathbb{R}^{n+d} : f(\mathbf{x}) = 0\} \qquad (6.2)$$

given by the map $f : \mathbb{R}^{n+d} \to \mathbb{R}^n$. For implicit curves, the PL approximation is restricted to $n = 1$ and $d = 1$; for implicit surfaces, $n = 1$ and $d = 2$. The idea is to construct a connected set of simplices that approximate S

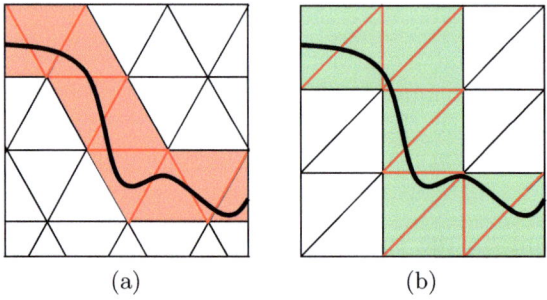

Fig. 6.1. Two simplicial approximations of a curve in $\Omega \subset \mathbb{R}^2$.

by stepping through transverse simplices (or simplexes) of a particular triangulation in $\mathbb{R}^N$ ($N = n + d$). For any positive integer N, and for any set $\{\mathbf{p}_0, \ldots, \mathbf{p}_N\}$ of points in some linear space which are affinely independent (or, equivalently, $\{\mathbf{p}_1 - \mathbf{p}_0, \ldots, \mathbf{p}_N - \mathbf{p}_0\}$ are linearly independent), the convex hull $[\mathbf{p}_0, \ldots, \mathbf{p}_N]$ is called the *d-simplex* with vertices $\mathbf{p}_0, \ldots, \mathbf{p}_N$. As known, the possible N-simplices in $\mathbb{R}^3$ are: vertices ($N = 0$), edges ($N = 1$), triangles ($N = 2$), and tetrahedra ($N = 3$). Also, for each subset of $K + 1$ vertices $\{\mathbf{q}_0, \ldots, \mathbf{q}_K\} \subset \{\mathbf{p}_0, \ldots, \mathbf{p}_N\}$, the K-simplex $[\mathbf{q}_0, \ldots, \mathbf{q}_K]$ is called a *K-face* of $[\mathbf{p}_0, \ldots, \mathbf{p}_N]$. In particular, 0-faces are *vertices*, 1-faces are *edges*, 2-faces are *triangles*, and $(K-1)$-faces are *facets*. Simplices are the "building bricks" that allow us to construct different sorts of triangulations in $\mathbb{R}^N$.

Definition 6.1. *Let $\mathcal{T}$ be a non-empty collection of d-simplices in $\mathbb{R}^N$. We call $\mathcal{T}$ a **triangulation** of $\mathbb{R}^N$ if the following properties are satisfied:*

(1) $\bigcup_{\sigma \in \mathcal{T}} \sigma = \mathbb{R}^N$;
(2) *the intersection $\sigma_1 \bigcap \sigma_2$ of two simplices $\sigma_1, \sigma_2 \in \mathcal{T}$ is empty or a **common** facet of both simplices;*
(3) *the collection $\mathcal{T}$ is locally finite, i.e. any compact subset of $\mathbb{R}^d$ meets only a finite number of simplices of $\sigma \in \mathcal{T}$.*

This definition applies not only to triangulations of $\mathbb{R}^N$ but also to its subspaces, as needed in computer graphics.

6.2.2 Types of Triangulations

As Dobkin et al. noted in [114], we would like to have triangulations with the following properties:

(1) It should be easy to find the simplex that shares a facet with a given simplex.
(2) It should be possible to label the vertices of all the simplexes at the same time with indexes $0, \ldots, N$, such that each of the $N + 1$ vertices of an N-simplex has a different label.
(3) It should be desirable for all the simplexes to have almost the same size.
(4) It should be desirable for all the simplexes to have roughly the same dimensions in all directions.

There is a dimension-independent class of triangulations that fit these requirements, and are called *Coxeter triangulations* [92]. The Coxeter triangulations are *monohedral triangulations generated by reflections*. Monohedral means that all N-simplexes are congruent, whereas *generated by reflections* means that all N-simplexes can all be obtained from a fixed one by successive reflections in its facets [114].

In Figure 6.2 we can see three different Coxeter triangulations of a domain $\Omega \subset \mathbb{R}^N$ ($N = 2$), called a *bounding box*. So, after finding a starting or seeding transverse N-simplex, we proceed to the next transverse N-simplex

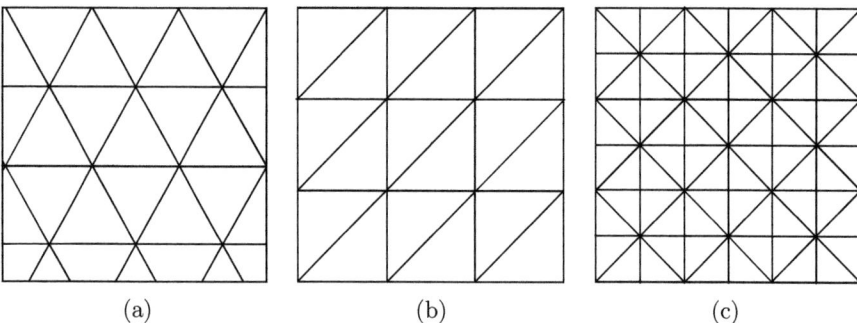

Fig. 6.2. Distinct types of Coxeter triangulations in $\mathbb{R}^2$: (a) equilateral triangulation; (b) Freudenthal's triangulation; (c) Todd's triangulation J_1.

in sequence. This is done by reflection in the common $(N-1)$-simplex of those adjacent N-simplices. Recall that two simplices $\sigma_1, \sigma_2 \in \mathcal{T}$ are called **adjacent** if they meet in a common facet [8].

6.2.3 Construction of Triangulations

Coxeter's triangulations are generated in a computer by moving from one simplex to an adjacent one through a common facet, a process known as **pivoting** [8]. Pivoting is essential for the dynamics of PL methods.

As suggested above, different pivoting rules generate different triangulations, but the same rule applied to different triangles also generate distinct triangulations. For example, in Figure 6.2, the triangulations (a) and (b) were generated by applying the same pivoting rule to distinct triangles, while the triangulation (c) was generated using a different rule.

Freudenthal's Triangulation

The fact that Coxeter's triangulations are monohedral means that a simplex must have dihedral angles that are each a submultiple of 2π. That is, for each pair of facets of the simplex, there is an integer $i > 1$ such that the dihedral angle between the two facets is $\frac{2\pi}{i}$. To understand how this is done, consider the angle $\frac{2\pi}{6}$ between two facet edges of a triangle in Figure 6.2(a), and then alternately reflect the triangle in each of the two facet edges. After six reflections, we must get back the original triangle; otherwise, this triangle does not triangulate by reflection. Therefore, we get six equilateral triangles around a common vertex, as illustrated in Figure 6.3.

The triangulation depicted in Figure 6.2(a) can be also obtained by pivoting across the midpoint of each of the two facet edges. This rule is known as Freudenthal's pivoting rule. The Freudenthal's triangulation is shown in Figure 6.2(b), which is generated by pivoting isosceles triangles. Recall that

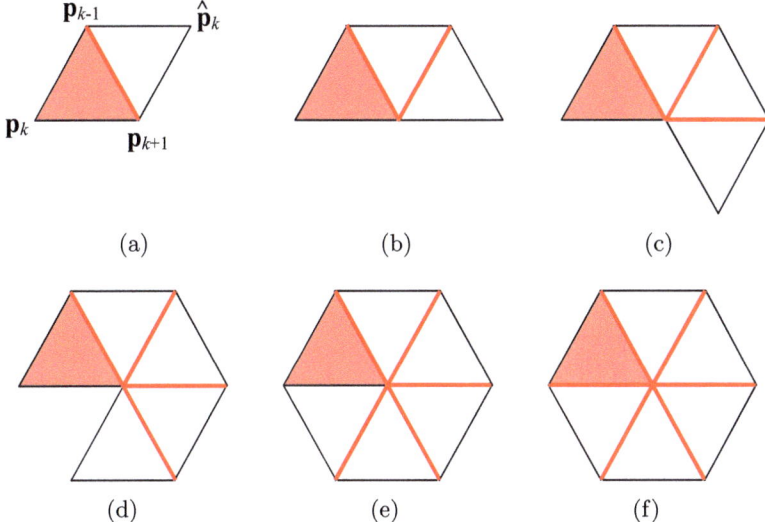

Fig. 6.3. Generation of equilateral triangles by reflection in $\mathbb{R}^2$.

the reflection of a vertex is across the midpoint of the reflection facet edge. As a result, we obtain six triangles around the common vertex, but the dihedral angle is not always the same; sometimes it is $2\pi/8$, sometimes it is $2\pi/4$.

Let us now formalise the Freudenthal rule. Let $\sigma = [\mathbf{p}_0, \ldots, \mathbf{p}_N]$ be an N-simplex. We will consider that the vertices of σ have inherited the following cyclic ordering. For each $k \in \{0, \ldots, N\}$ let us define $k-1$ and $k+1$ as the "left" and "right" neighbours of k in the cyclic ordering of $(0, 1, \ldots, N)$. Analogously, for vertices, $\mathbf{p}_{k-1}$ ($k \neq 0$) and $\mathbf{p}_{k+1}$ ($k \neq N$) are defined to be the "left" and "right" neighbours of the vertex $\mathbf{p}_k$. It is clear that $\mathbf{p}_N$ is the left neighbour of $\mathbf{p}_0$ and, conversely, $\mathbf{p}_0$ is the right neighbour of $\mathbf{p}_N$.

So, the vertex obtained as follows

$$\hat{\mathbf{p}}_k = \mathbf{p}_{k-1} - \mathbf{p}_k + \mathbf{p}_{k+1}$$

is called the *reflection* of $\mathbf{p}_k$ across the centre of the "neighbouring edge" $[\mathbf{p}_{k-1}, \mathbf{p}_{k+1}]$. Pivoting $\mathbf{p}_k$ of $\sigma = [\mathbf{p}_0, \ldots, \mathbf{p}_k, \ldots, \mathbf{p}_N]$ by reflection follows the rule $\sigma \to \hat{\sigma}$, where $\hat{\sigma} = [\mathbf{p}_0, \ldots, \hat{\mathbf{p}}_k, \ldots, \mathbf{p}_N]$. This is illustrated in Figure 6.4, where $\hat{\mathbf{p}}_k$ was obtained by reflection of $\mathbf{p}_k$ across the edge $[\mathbf{p}_{k-1}, \mathbf{p}_{k+1}]$. As in Figure 6.3, we end up getting an hexagon after six reflections, though the triangles are isosceles.

Thus, Freudenthal's triangulations are invariant under the pivot operation $\Phi_k([\mathbf{p}_0, \ldots, \mathbf{p}_k, \ldots, \mathbf{p}_N]) = [\mathbf{p}_0, \ldots, \hat{\mathbf{p}}_k, \ldots, \mathbf{p}_N]$, where

$$\hat{\mathbf{p}}_k = \begin{cases} \mathbf{p}_N - \mathbf{p}_k + \mathbf{p}_{k+1}, & k = 0 \\ \mathbf{p}_{k-1} - \mathbf{p}_k + \mathbf{p}_{k+1}, & 0 < k < N \\ \mathbf{p}_{k-1} - \mathbf{p}_k + \mathbf{p}_0, & k = N \end{cases} \quad (6.3)$$

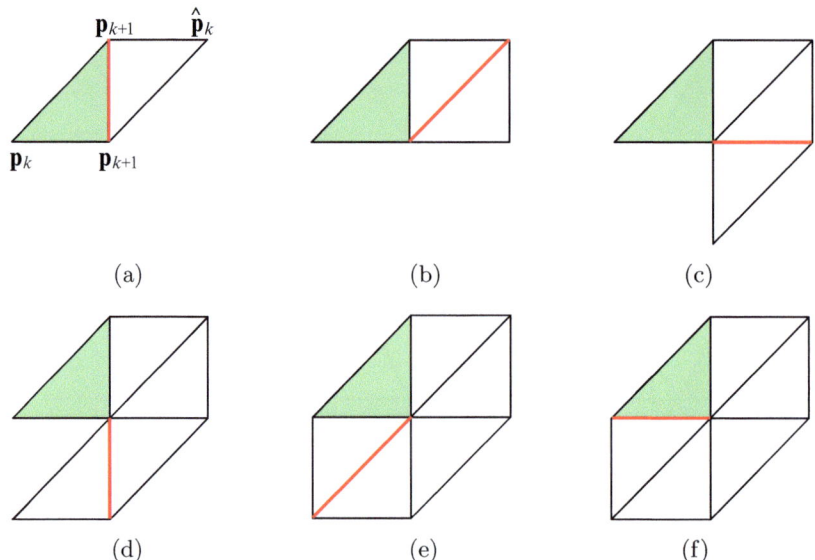

Fig. 6.4. Generation of isosceles triangles by reflection in $\mathbb{R}^2$.

This pivoting rule allows us to generate a Freudenthal's triangulation of the an axially aligned bounding box $\Omega \in \mathbb{R}^N$ (or even the entire $\mathbb{R}^N$).

Equivalently, the pivoting rule can be expressed in terms of interchange permutations [8]. As a particular Coxeter's triangulation, Freudenthal's triangulation has the advantage that any simplex can be concisely stored by means of a single integer vector Z and a permutation π, being most pivot steps achieved by interchanging two components of π. The reader is referred to [8, 9, 101, 400] for more details about Freudenthal's triangulations and their usage in the implementation of piecewise linear algorithms, in particular those concerned with implicit curves and surfaces in computer graphics.

Todd's Triangulation J_1

Todd introduced the "Union Jack" triangulations, namely J_1 triangulations [390]. Figure 6.5 shows the J_1 triangulation of a subspace of $\mathbb{R}^2$. The J_1 triangulation is invariant under the pivot operation $\Theta_k([\mathbf{p}_0, \ldots, \mathbf{p}_k, \ldots, \mathbf{p}_N]) = [\mathbf{p}_0, \ldots, \hat{\mathbf{p}}_k, \ldots, \mathbf{p}_N]$, where

$$\hat{\mathbf{p}}_k = \begin{cases} 2\mathbf{p}_{k+1} - \mathbf{p}_k, & k = 0 \\ \mathbf{p}_{k-1} - \mathbf{p}_k + \mathbf{p}_{k+1}, & 0 < k < N \\ 2\mathbf{p}_{k-1} - \mathbf{p}_k, & k = N \end{cases} \qquad (6.4)$$

This pivoting rule generates a triangulation as shown in Figure 6.5. Similar to the discussion carried out for pivoting in Freudenthal's triangulations, the pivoting rule for J_1 can be also expressed by interchange permutations [8].

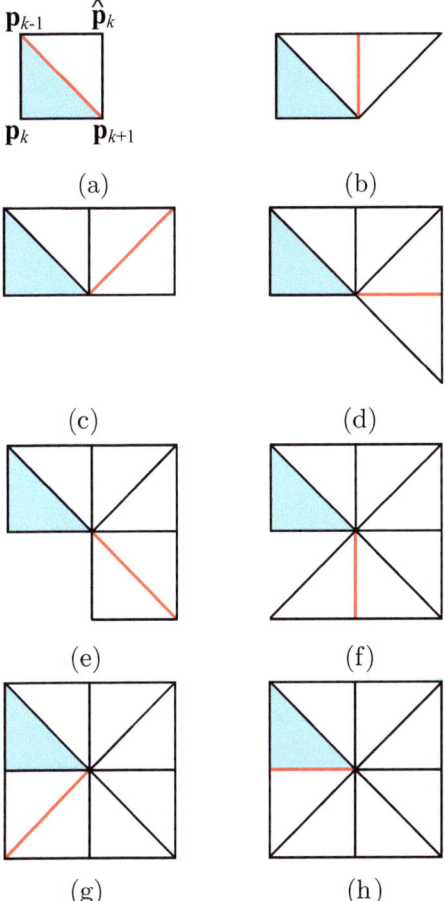

Fig. 6.5. Generation of an 8-gon of isosceles triangles by applying J_1 pivoting rule.

6.3 Integer-Labelling PL Algorithms

In addition to pivoting rules to generate triangulations, PL algorithms use labellings with the following purposes:

- *To keep track the PL approximation of a manifold.* That is, by labelling the simplexes that intersect a manifold, also called transverse simplexes, we are able to follow such a manifold. Labellings work as a way of distinguishing simplexes which intersect a zero set of a map $f : \mathbb{R}^{n+d} \to \mathbb{R}^n$ from those which do not.
- *To prevent the cycling phenomenon.* This is the classical problem of continuation algorithms. Unless the transverse simplexes are labelled or stored in a separate data structure, we have no way to know whether a new pivoted triangle has already been determined.

Simplicial continuation provides approximations of zero sets of maps by using *labellings*. A simplicial algorithm provides a piecewise linear zero set of f via an auxiliary map L_f called *labelling* induced by f. The values of L_f at vertices are then used in determining whether a given n-simplex is "completely labelled" or not. These "completely labelled simplices" are those that intersect the zero set (e.g. a curve or surface) of f in $\mathbb{R}^{n+d}$.

There are two major techniques to label the simplexes of a triangulation, namely:

- integer labelling
- vector labelling.

This section describes algorithms based on integer labelling. Those algorithms based on vector labelling will be dealt with in the next section.

Integer Labelling

The idea of labelling is to attach integer labels to vertices or nodes of a triangulation. The integer labelling scheme proposed by Allgower and Schmidt [11] is based on the following definition:

Definition 6.2. *For $\mathbf{p} \in \mathbb{R}^{n+d}$, the **labelling** of $\mathbf{p}$ is defined by $L_f(\mathbf{p}) = j$, where $j \in \{0, \ldots, n\}$ is the number of leading nonnegative components of $f(\mathbf{p}) \in \mathbb{R}^n$.*

Definition 6.3. *An n-simplex $[\mathbf{p}_0, \ldots, \mathbf{p}_n]$ is said to be **completely labelled** if $L_f\{\mathbf{p}_0, \ldots, \mathbf{p}_n\} = \{0, \ldots, n\}$.*

Let us look at Figure 6.6, where the parabola curve $y - x^2 + \frac{1}{2} = 0$ appears depicted across a Freudenthal triangulation of the bounding box $\Omega = [-2, 2] \times [-2, 2]$ in $\mathbb{R}^2$. Each two triangles result from splitting a square into two isosceles triangles, whose identical sides have length 1. Such parabolic zero set is thus described by a real map in two real variables $f : \Omega \subset \mathbb{R}^2 \to \mathbb{R}$, with $f(x, y) = y - x^2 + \frac{1}{2}$; hence $n = 1$ and $d = 1$. Thus, in this case, f has only one component function ($n = 1$), which means that there are only two possible labels for any vertex v of the triangulation: either 0 when $f(v) < 0$ or 1 when $f(v) \geq 0$.

Example 6.4. Looking again at Figure 6.6, we note that the vertices $\mathbf{p}_0 = (0, 0)$, $\mathbf{p}_1 = (1, 0)$, $\mathbf{p}_2 = (1, 1)$ define a 2-simplex or triangle in $\Omega = [-2, 2] \times [-2, 2]$. The labels of these three vertices are:

- $L_f(0, 0) = 1$ because $f(0, 0) = \frac{1}{2} > 0$;
- $L_f(1, 0) = 0$ because $f(1, 0) = -\frac{1}{2} < 0$
- $L_f(1, 1) = 1$ because $f(1, 1) = \frac{1}{2} > 0$

Thus, the 1-simplices or edges $[\mathbf{p}_0, \mathbf{p}_1]$ and $[\mathbf{p}_1, \mathbf{p}_2]$ are completely labelled, but not the edge $[\mathbf{p}_0, \mathbf{p}_2]$.

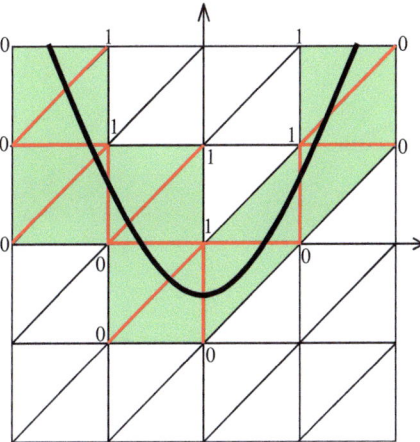

Fig. 6.6. Integer labelling of a triangulation in $\mathbb{R}^2$.

This labelling language hides a very simple idea. The fact that a given 1-simplex is completely labelled means that, by the intermediate value theorem, it intersects the zero set. Therefore, as argued by Allgower and Schmidt [11], completely labelled 1-simplices yield "nearly zero-points" of f. When the labels of the vertices of a 1-simplex are identical, or, equivalently, the values of f at those vertices have identical signs, we say that such a 1-simplex is not completely labelled.

Interestingly, as Allgower and Schmidt proved in [11], we have:

Proposition 6.5. *If the $(n+d)$-simplex σ contains a completely integer labelled n-face τ, then the number of completely labelled n-faces of σ is between $(d+1)$ and 2^d.*

Note that the number of completely labelled n-faces of σ does not depend on n; it only depends on d. For example, in $\mathbb{R}^2$, a triangle or 2-simplex ($n=1$, $d=1$) has exactly either 0 or 2 completely labelled edges or 1-simplices (Figure 6.6); in $\mathbb{R}^3$, a tetrahedron or 3-simplex ($n=1$, $d=2$) has exactly either 0, 3 or 4 completely labelled edges or 1-simplices (Figure 6.7).

An important question is then how to obtain the *approximate zero set* or *PL zero set* of f in the $(n+d)$-simplex σ. For that purpose, we use the barycentre of each of its completely labelled n-simplices; the approximate zero of f within σ is then the convex hull of these barycentres. Thus, the PL zero set for f in σ is a convex d-dimensional polytope having between $d+1$ and 2^d vertices [11]. For example, the PL zero set in a triangle lying in $\mathbb{R}^2$ is a 1-dimensional polytope (or line segment), while the PL zero set in a tetrahedron in $\mathbb{R}^3$ is a 2-dimensional polytope, i.e. a triangle or a quadrilateral (Figure 6.7). In short, these polytopes intersect all faces that contain a completely labelled n-face transversally. This leads to the following definition:

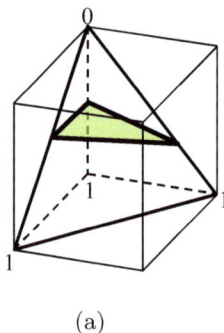 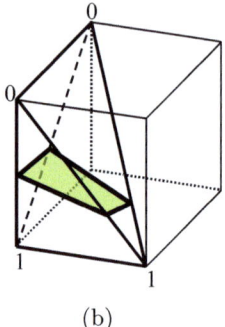

Fig. 6.7. Integer labelling of a triangulation in $\mathbb{R}^3$ ($n = 1, d = 2$): (a) three and (b) four completely labelled edges.

Definition 6.6. *An $(n+m)$-simplex ($m = 0, \ldots, d$) is said to be **transverse** if it contains a completely labelled n-face.*

Integer-labelling PL Algorithm

Allgower and Georg introduced a multidimensional algorithm for curves ($d = 1$) via integer labellings [7]. This algorithm follows the door-in-door-out principle [124]:

Proposition 6.7. *An $(n+1)$-simplex σ in $\mathbb{R}^{n+1}$ has either zero or exactly two completely labelled n-faces.*

This principle follows from Proposition 6.5. The first completely labelled face of σ is viewed as an "entrance" and the second as an "exit". Then, by using a pivoting process on σ, one determines a new $(n+1)$-simplex $\hat{\sigma}$. Now σ and $\hat{\sigma}$ have a common n-face—the pivot n-face—which is simultaneously the exit face of σ and the entrance face of $\hat{\sigma}$. For example, in Figure 6.6, the 1-simplex $[\mathbf{p}_1, \mathbf{p}_2]$ is the "exit" of the 2-simplex $[\mathbf{p}_0, \mathbf{p}_1, \mathbf{p}_2]$, where $\mathbf{p}_0 = (0,0)$, $\mathbf{p}_1 = (1,0)$ and $\mathbf{p}_2 = (1,1)$, as well as the "entrance" of the 2-simplex $[\mathbf{p}_1, \mathbf{p}_2, \mathbf{p}_3]$, with $\mathbf{p}_3 = (2,1)$. This algorithm produces an alternate sequence of completely labelled n-faces and $(n+1)$-simplices from which one obtains approximate zero points and a 1-dimensional approximation to the zero set. In other words, it generates a sequence of completely labelled n-faces in the triangulation, by entering an $(n+1)$-simplex through one n-face and leaving it through the other.

Allgower-Georg algorithm for implicit curves was later generalised to higher dimensional manifolds ($d > 1$) by Allgower and Schmidt [11]. In this case, the door-in-door-out (DIDO) principle above for curves has to be reformulated because —as stated by Proposition 6.5—the number of possible exit doors is now greater than one. For example, in Figure 6.7, the tetrahedron (a) has three doors, while the tetrahedron (b) has four doors; consequently, we

Algorithm 12 Integer-Labelling PL Algorithm for Manifolds

1: **procedure** ALLGOWER-GEORG-SCHMIDT($f,\Omega,\mathcal{T}$)
2:     $\mathcal{T} \leftarrow \varnothing$ ▷ set of pivoted, transverse $(n+d)$-simplices of $\mathcal{T}$
3:     $S \leftarrow \varnothing$ ▷ set of non-pivoted, transverse $(n+d)$-simplices of $\mathcal{T}$
4:     Find a transverse starting $(n+d)$-simplex $\sigma \in \mathcal{T}$
5:     $S \leftarrow S \cup \{\sigma\}$
6:     **while** $S \neq \varnothing$ **do**
7:         Get $\sigma \in S$
8:         Label vertices of σ
9:         Determine F ▷ set of non-pivoted, transverse facets of σ
10:         **while** $F \neq \varnothing$ **do**
11:             Choose a pivot facet $\tau \in F(\sigma)$
12:             Determine the $(n+d)$-simplex $\hat{\sigma}$ by pivoting σ across τ
13:             $S \leftarrow S \cup \{\hat{\sigma}\}$
14:             $S \leftarrow S \setminus \{\sigma\}$
15:             $\mathcal{T} \leftarrow \mathcal{T} \cup \{\sigma\}$
16:             $F \leftarrow F \setminus \{\tau\}$
17:         **end while**
18:     **end while**
19: **end procedure**

may assume that the first tetrahedron has one "entrance" door and two "exit" doors, while the second has one "entrance" door and three "exit" doors in the continuation process of tracking a surface. Thus, applying the DIDO principle as many times as the number of "exit" doors, we can easily program the door-in-door-out step by pivoting only those vertices having the same label.

Algorithm 12 is the Allgower-Georg-Schmidt algorithm using integer labelling. The integer labelling of each transverse $(n+d)$-simplex occurs in the beginning of the outer **while** statement (step 8), while the multiple pivoting of this $(n+d)$-simplex is done in the inner **while** statement (steps 11–12). The number of times a $(n+d)$-simplex is pivoted equals the number of "exit" doors (i.e. nonpivoted, transverse facets). Figure 6.8 illustrates the pivoting of a tetrahedron for a surface in $\mathbb{R}^3$, but, for simplicity, one uses only one "entrance" door and one "exit" door for tracking the surface.

In order to guarantee that the algorithm terminates after a finite number of steps, one assumes that:

- *Compactness.* Not the whole $\mathbb{R}^{n+d}$, but the compact domain $\Omega \subset \mathbb{R}^{n+d}$ (e.g. an axis-aligned bounding box of finite size) is triangulated by $\mathcal{T}$.
- *Finiteness.* The triangulation $\mathcal{T}$ contains a finite number of $(n+d-1)$-facets. This is reinforced by the fact that each transverse facet is found exactly twice: once when it is created and once more when it is "pivoted across" or when "bumped into" as expected for a pivot facet.
- *Cycling.* Labelling and pivoting constitute an important mechanism to avoid the cycling phenomenon, i.e. the recomputation of transverse or intersecting simplices.

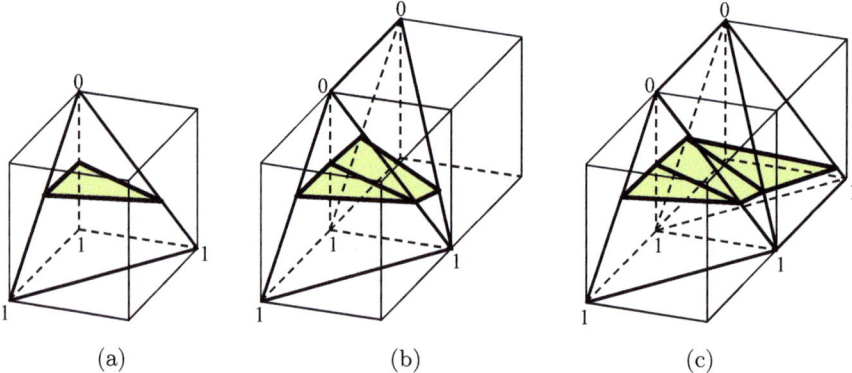

Fig. 6.8. The Allgower-Georg-Schmidt algorithm using integer labelling for approximating a surface in $\mathbb{R}^3$.

As it stands, Algorithm 12 only produces a sequence of transverse $(n+d)$-simplices by pivoting across transverse $(n+d-1)$-simplices or facets of the triangulation $\mathcal{T}$. To output a manifold PL zero set of a map $f : \mathbb{R}^{n+d} \to \mathbb{R}^n$, it is necessary two additional steps immediately after step 8 (labelling). The first would determine the set of completely labelled n-faces of the current $(n+d)$-simplex σ. The second would determine the approximate zero points (e.g. the barycentres) in those n-faces, whose convex hull is the polytope that approximates the zero set of f in σ.

6.4 Vector Labelling-based PL Algorithms

Integer labelling has the disadvantage that it leads to a very coarse approximation of the zero set of $f : \mathbb{R}^{n+d} \to \mathbb{R}^n$. This is due to the fact that the PL zero set is built upon the barycentres of transverse n-faces. In comparison to vector labelling, the advantage of integer labelling is that numerical linear algebra (i.e. matrix calculations) is not necessary to drive the pivoting process. But, as shown below, vector labelling provides a finer PL zero set of f than integer labelling.

Vector Labelling

Vector labelling is based on the barycentric coordinates of the vertices of the current $(n+d)$-simplex. Using barycentric coordinates leads to the computation of zero points by linear interpolation of the values of $f : \mathbb{R}^{n+d} \to \mathbb{R}^n$ on the transverse n–faces of the triangulation $\mathcal{T}$. We may leave $\mathcal{T}$ unspecified; the only fact to retain is that $\mathcal{T}$ is generated by a repeated use of some pivoting rule. For simplicity, we are also assuming that f never vanishes at the vertices of $\mathcal{T}$ and is never constant on a $(n+d)$–simplex.

6.4 Vector Labelling-based PL Algorithms

Let us first consider the contour plotting in $\mathbb{R}^2$ ($n = 1$, $d = 1$), i.e. tracing an implicit curve in $\mathbb{R}^2$. For convenience, let us assume that the axially-aligned bounding box $\Omega \subset \mathbb{R}^2$ is divided into squares, each of them is in turn split into two triangles by its right diagonal. This is Freudenthal's triangulation in $\mathbb{R}^2$, whose counterpart in $\mathbb{R}^3$ is known as Kuhn's triangulation (see Chapter 7 for more details). We also assume that the square sides have length one.

Now, we are able to compute the PL zero set on each transverse simplex using linear interpolation. This computation can be done using one of the following two alternatives:

- Computing the convex hull of the zero points on the edges of the simplex.
- Computing the equation of the hyperplane that contains the convex hull.

Using the first alternative, we can determine the contour that passes through a triangle by computing the points where the line intersects the edges of such triangle. In this case, any transverse edge with vertices $\mathbf{p}_0$ and $\mathbf{p}_1$ can be written as

$$\mathbf{p} = \mathbf{p}_0 + t(\mathbf{p}_1 - \mathbf{p}_0) \tag{6.5}$$

with $t \in [0, 1]$, and the linear interpolant over such an edge is given by

$$F(\mathbf{p}) = f(\mathbf{p}_0) + t[f(\mathbf{p}_1) - f(\mathbf{p}_0)] \tag{6.6}$$

Setting $F(\mathbf{p}) = 0$, we get $t = -\frac{f(\mathbf{p}_0)}{f(\mathbf{p}_1) - f(\mathbf{p}_0)}$ from Equation (6.6). Substituting the value of t in Equation (6.5), we obtain the zero point on the transverse edge. Note that this zero point on the edge $[\mathbf{p}_0, \mathbf{p}_1]$ only depends on the value of f at the vertices, so any triangle sharing this edge produces the same zero point; hence the contour is continuous. Thus, applying linear interpolation to each transverse edge of $\mathcal{T}$ ends up producing the entire contour or PL curve.

The second alternative uses the interpolant over the triangle as a whole, not over its edges. Analogously, the values of f at the corners of a triangle $[\mathbf{p}_0, \mathbf{p}_1, \mathbf{p}_2]$ define a unique piecewise linear interpolant $F(\mathbf{p})$ to $f(\mathbf{p})$ over each triangle, which can be written in terms of the equations

$$\mathbf{p} = \mathbf{p}_0 + (\mathbf{p}_1 - \mathbf{p}_0)s + (\mathbf{p}_2 - \mathbf{p}_0)t \tag{6.7}$$

and

$$F(\mathbf{p}) = f(\mathbf{p}_0) + [f(\mathbf{p}_1) - f(\mathbf{p}_0)]s + [f(\mathbf{p}_2) - f(\mathbf{p}_0)]t \tag{6.8}$$

where $s \geq 0$, $t \geq 0$, and $s + t = 1$. The piecewise linear interpolant F is continuous over the whole triangulation. The interpolant F is piecewise linear because its contour (or PL zero set) across an individual triangle is a line segment, whose line equation can be easily determined by solving the system of Equations (6.7) and (6.8), after setting $F(\mathbf{p}) = 0$.

Using linear interpolation amounts to use barycentric coordinates. In fact, Equation (6.7) can be written as

$$\mathbf{p} = \alpha_0 \mathbf{p}_0 + \alpha_1 \mathbf{p}_1 + \alpha_2 \mathbf{p}_2 \tag{6.9}$$

with $\alpha_0 = 1 - (s+t)$, $\alpha_1 = s$, $\alpha_2 = t$, $\alpha_0 + \alpha_1 + \alpha_2 = 1$ and $0 \le s, t \le 1$. The scalars α_0, α_1 and α_2 are called the *barycentric coordinates* of **p**. In general, every point **p** in a simplex can be expressed as a convex combination of its vertices, and more importantly this representation is unique [224].

Analogously, Equation (6.8) can be re-written as follows:

$$F(\mathbf{p}) = \alpha_0 f(\mathbf{p}_0) + \alpha_1 f(\mathbf{p}_1) + \alpha_2 f(\mathbf{p}_2) \qquad (6.10)$$

In general, every point $\mathbf{p} \in \mathbb{R}^{n+d}$ of an $(n+d)$–simplex $\sigma = [\mathbf{p}_0, \ldots, \mathbf{p}_{n+d}] \subset \mathbb{R}^{n+d}$ can be expressed in barycentric coordinates

$$\mathbf{p} = \alpha_0 \mathbf{p}_0 + \cdots + \alpha_{n+d} \mathbf{p}_{n+d} \qquad (6.11)$$

with $\alpha_i \ge 0$ $(i = 0, \ldots, n+d)$ and $\alpha_0 + \cdots + \alpha_{n+d} = 1$.

Similarly, we have

$$F(\mathbf{p}) = \alpha_0 f(\mathbf{p}_0) + \cdots + \alpha_{n+d} f(\mathbf{p}_{n+d}) \qquad (6.12)$$

or, equivalently,

$$L_f(\sigma) . \alpha = \begin{pmatrix} 1 \\ F(\mathbf{p}) \end{pmatrix} \qquad (6.13)$$

where $\alpha = (\alpha_0, \ldots, \alpha_{n+d})^T$ are the barycentric coordinates of a point $\mathbf{p} \in \mathbb{R}^{n+d}$ and

$$L_f(\sigma) = \begin{pmatrix} 1 & \cdots & 1 \\ f(\mathbf{p}_0) & \cdots & f(\mathbf{p}_{n+d}) \end{pmatrix} \qquad (6.14)$$

The matrix $L_f(\sigma)$ is known as *labelling matrix* of a $(n+d)$-simplex $\sigma = [\mathbf{p}_0, \ldots, \mathbf{p}_{n+d}] \subset \mathbb{R}^{n+d}$. It consists of $n+d$ labelling column vectors, each vector storing the value of f, which works as a label, at each vertex. In general, the standard *vector labelling* induced by $f : \mathbb{R}^{n+d} \to \mathbb{R}^n$ is then

$$l_f(\mathbf{p}) = \begin{pmatrix} 1 \\ f(\mathbf{p}) \end{pmatrix} \qquad (6.15)$$

where $\mathbf{p} \in \mathbb{R}^{n+d}$.

Following Allgower and Gnutzmann [9], we have:

Definition 6.8. *Let $\tau = [\mathbf{p}_0, \ldots, \mathbf{p}_n] \subset \mathbb{R}^{n+d}$ be a n-simplex and let $f : \mathbb{R}^{n+d} \to \mathbb{R}^n$. Then τ is said to be* **completely labelled** *with respect to the vector labelling l_f if the labelling matrix $L_f(\tau)$ has a lexicographically positive inverse.*

In other words, τ is completely labelled if and only if the following two conditions are satisfied [8]:

- $L_f(\tau)$ is nonsingular:
- $L_f(\tau)^{-1}$ is lexicographically positive, i.e. the first nonvanishing entry in any row of $L_f(\tau)^{-1}$ is positive.

6.4 Vector Labelling-based PL Algorithms

Intuitively, this means that f changes sign at vertices of τ; consequently, any $(n+d)$-simplex σ having τ as a n-face is said to be **transverse** to the zero set of f. The labelling matrix then plays an important role for numerically tracing the zero set of f.

As noted above, piecewise linear algorithms produce approximations of zero points of maps by means of induced auxiliary maps called labellings, vector labellings in this case. So, from Equation (6.13), the PL zero set across the simplex σ is the set of points whose barycentric coordinates satisfy

$$L_f(\sigma).\alpha = \begin{pmatrix} 1 \\ 0 \\ \vdots \\ 0 \end{pmatrix} \qquad (6.16)$$

with $\alpha_i \geq 0$ $(i = 0, \ldots, n+d)$, that is

$$\alpha = L_f(\sigma)^{-1} \begin{pmatrix} 1 \\ 0 \\ \vdots \\ 0 \end{pmatrix} \qquad (6.17)$$

This gives us the barycentric coordinates of at least a point $\mathbf{b}_0$ in the d-dimensional hyperplane that approximates the zero set inside a given $(n+d)$-simplex σ. The parametric equation in barycentric coordinates corresponding to the general Equation (6.16) of such hyperplane can be written as follows

$$\mathbf{b}(t_1, \ldots, t_d) = \mathbf{b}_0 + \sum_{i=0}^{d} t_i \mathbf{b}_i \qquad (6.18)$$

where $t_i \in \mathbb{R}$ is the real parameter on the line defined by the vector $\mathbf{b}_i - \mathbf{b}_0$, and $\{\mathbf{b}_i\}$ $(i = 1, \ldots, d)$ is a linearly independent set of points. This linear independence implies that the barycentric coordinates of a nonzero $(n+d)$-tuple $\mathbf{b}_i$ have sum zero. Thus, computing $\mathbf{b}_i$ in the zero set hyperplane inside σ reduces to determine a nontrivial solution of the homogeneous equation

$$L_f(\sigma).\mathbf{b}_i = \begin{pmatrix} 0 \\ 0 \\ \vdots \\ 0 \end{pmatrix} \qquad (6.19)$$

Finding $\mathbf{b}_i$ reduces to a standard linear algebra problem (e.g. using the reduced row-echelon form). The hyperplane passes through the facets of σ, the completely labelled facets, opposite to vertices for which $t_i = -\frac{\mathbf{b}_0}{\mathbf{b}_i}$ is negative. We are here assuming that all the vertices of the triangulation, including σ, have been assigned an index, as well as the current completely labelled facet is the "door we are currently entering"; the remaining facet is the exit or pivoting facet. Found the index i of the vertex $\mathbf{p}_i$ opposite to the exit facet, $\mathbf{p}_i$ is pivoted into a new vertex $\hat{\mathbf{p}}_i$, and the labelling matrix $L_f(\hat{\sigma})$ is obtained by replacing the ith label or column of the $L_f(\sigma)$ by

$$l_{f_i}(\hat{\mathbf{p}}_i) = \begin{pmatrix} 1 \\ f(\hat{\mathbf{p}}_i) \end{pmatrix} \tag{6.20}$$

So, the new labelling matrix can be algebraically obtained as follows:

$$L_f(\hat{\sigma}) = L_f(\sigma) + [l_{f_i}(\hat{\sigma}) - L_f(\sigma) \cdot \mathbf{e}_i] \cdot \mathbf{e}_i^T \tag{6.21}$$

where $\mathbf{e}_i$ is the ith unit basis vector. This leads to the implementation of the DIDO principle for vector labelling.

Example 6.9. Let $f : \Omega \subset \mathbb{R}^2 \to \mathbb{R}$ a real function in two real variables defined by $f(x,y) = -2x + y + \frac{1}{4}$ (see Figure 6.9). In this case, the zero set of f is the straight line $-2x + y + \frac{1}{4} = 0$ in $\mathbb{R}^2$, so it coincides with its PL zero set. Let us also consider that the domain $\Omega \subset \mathbb{R}^2$ is to be triangulated according to Freudenthal's pivoting rule, where the coordinates of the vertices are all integer. For brevity, we let us consider the 2-simplex or triangle $\sigma = [\mathbf{p}_0, \mathbf{p}_1, \mathbf{p}_2] \subset \mathbb{R}^2$, with $\mathbf{p}_0 = (0,0)$, $\mathbf{p}_1 = (1,0)$ and $\mathbf{p}_2 = (1,1)$. The labelling matrix is then

$$L_f(\sigma) = \begin{pmatrix} 1 & 1 & 1 \\ f(\mathbf{p}_0) & f(\mathbf{p}_1) & f(\mathbf{p}_2) \end{pmatrix} = \begin{pmatrix} 1 & 1 & 1 \\ \frac{1}{4} & -\frac{7}{4} & -\frac{3}{4} \end{pmatrix} \tag{6.22}$$

Strictly speaking, a rectangular $(m \times n)$–matrix does not have an inverse. But, in some cases such a matrix may have a left or right inverse. In this

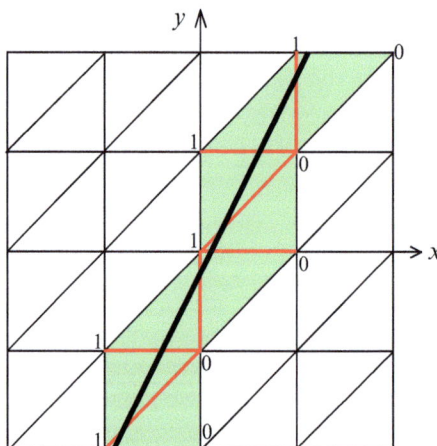

Fig. 6.9. Vector labelling of a triangulation in $\mathbb{R}^2$.

6.4 Vector Labelling-based PL Algorithms

example, the rank of $L(\sigma)$ is equal to $m = 2$, so $L_f(\sigma)$ has a right inverse $L_f(\sigma)^{-1}$ such that $L_f(\sigma) . L_f(\sigma)^{-1} = I$, where I is the identity matrix and $L_f(\sigma)^{-1} = L_f(\sigma)^T . [L_f(\sigma) . L_f(\sigma)^T]^{-1}$, that is

$$L_f(\sigma)^{-1} = \begin{pmatrix} 1 & \frac{1}{4} \\ 1 & -\frac{7}{4} \\ 1 & -\frac{3}{4} \end{pmatrix} \left[\begin{pmatrix} 1 & 1 & 1 \\ \frac{1}{4} & -\frac{7}{4} & -\frac{3}{4} \end{pmatrix} \begin{pmatrix} 1 & \frac{1}{4} \\ 1 & -\frac{7}{4} \\ 1 & -\frac{3}{4} \end{pmatrix} \right]^{-1} = \frac{1}{6} \begin{pmatrix} \frac{68}{16} & 3 \\ -\frac{1}{4} & -3 \\ 2 & 0 \end{pmatrix} \quad (6.23)$$

Thus, the barycentric coordinates of a point $\mathbf{b}_0$ in the 1-dimensional PL zero set (or hyperplane) of f are given by Equation (6.17)

$$\mathbf{b}_0 = \frac{1}{6} \begin{pmatrix} \frac{68}{16} & 3 \\ -\frac{1}{4} & -3 \\ 2 & 0 \end{pmatrix} \begin{pmatrix} 1 \\ 0 \end{pmatrix} = \begin{pmatrix} \frac{17}{24} \\ -\frac{1}{24} \\ \frac{1}{3} \end{pmatrix} \quad (6.24)$$

Now, by solving Equation (6.19), we get the second barycentric-valued point

$$\mathbf{b}_1 = \begin{pmatrix} -\frac{1}{2} \\ -\frac{1}{2} \\ 1 \end{pmatrix} \quad (6.25)$$

So, substituting $\mathbf{b}_0$ and $\mathbf{b}_1$ in Equation (6.18), we have

$$\begin{pmatrix} \frac{17}{24} \\ -\frac{1}{24} \\ \frac{1}{3} \end{pmatrix} + t \begin{pmatrix} -\frac{1}{2} \\ -\frac{1}{2} \\ 1 \end{pmatrix} = \begin{pmatrix} 0 \\ 0 \\ 0 \end{pmatrix} \quad (6.26)$$

or

$$\begin{cases} t = \frac{-\frac{17}{24}}{-\frac{1}{2}} = \frac{17}{12} \\ t = \frac{\frac{1}{24}}{-\frac{1}{2}} = -\frac{1}{12} \\ t = \frac{-\frac{1}{3}}{1} = -\frac{1}{3} \end{cases}$$

That is, $t < 0$ for the second and third coordinates, so the zero hyperplane intersects the facets $\tau_2 = [\mathbf{p}_0, \mathbf{p}_2]$ and $\tau_0 = [\mathbf{p}_0, \mathbf{p}_1]$, respectively, of σ. For $t = -\frac{1}{12}$, we obtain the barycentric coordinates of the solution point that results from the intersection between the PL zero set and the facet $\tau_2 = [\mathbf{p}_0, \mathbf{p}_2]$ as follows:

$$\begin{cases} \alpha_0 = \frac{17}{24} + (-\frac{1}{12})(-\frac{1}{2}) = \frac{3}{4} \\ \alpha_1 = 0 \\ \alpha_2 = \frac{1}{3} + (-\frac{1}{2})(1) = \frac{1}{4} \end{cases}$$

so that the corresponding point in Cartesian coordinates is then

$$\mathbf{p} = \frac{3}{4} \begin{pmatrix} 0 \\ 0 \end{pmatrix} + 0 \begin{pmatrix} 1 \\ 0 \end{pmatrix} + \frac{1}{4} \begin{pmatrix} 1 \\ 1 \end{pmatrix} = \begin{pmatrix} \frac{1}{4} \\ \frac{1}{4} \end{pmatrix}.$$

Likewise, for $t = -\frac{1}{3}$, we have solution point in $\tau_0 = [\mathbf{p}_0, \mathbf{p}_1]$ with the following barycentric coordinates:

$$\begin{cases} \alpha_0 = \frac{17}{24} + (-\frac{1}{3})(-\frac{1}{2}) = \frac{7}{8} \\ \alpha_1 = -\frac{1}{24} + (-\frac{1}{3})(-\frac{1}{2}) = \frac{1}{8} \\ \alpha_2 = 0 \end{cases}$$

hence the corresponding point in Cartesian coordinates

$$\mathbf{p} = \frac{7}{8}\begin{pmatrix}0\\0\end{pmatrix} + \frac{1}{8}\begin{pmatrix}1\\0\end{pmatrix} + 0\begin{pmatrix}1\\1\end{pmatrix} = \begin{pmatrix}\frac{1}{8}\\0\end{pmatrix}$$

Note that the index of the null barycentric coordinate tell us which is the pivoting vertex. For example, for $t = -\frac{1}{12}$, the pivoting vertex is $\mathbf{p}_1$ because $\alpha_1 = 0$, which is opposite to the transverse facet $\tau_2 = [\mathbf{p}_0, \mathbf{p}_2]$.

Vector-labelling PL Algorithm

We can now describe an algorithm, using vector labelling, that provides a PL approximation of a curve implicitly defined by the equation $f(\mathbf{p}) = 0$, where $f : \Omega \subset \mathbb{R}^{n+d} \to \mathbb{R}^n$. Such an algorithm (Algorithm 13) generates a sequence

$$\sigma_0 \supset \tau_0 \subset \sigma_1 \supset \tau_1 \cdots$$

Algorithm 13 Vector-labelling PL Algorithm for Curves

1: **procedure** ALLGOWER-GEORG-GNUTZMANN$(f,\Omega,\mathcal{T})$
2:     Find a transverse $(n+1)$-simplex $\sigma \in \mathcal{T}$ with c.l. n-face τ opposite to $\mathbf{p}_i$.
3:     Calculate labelling matrix $L_\sigma = \begin{pmatrix} 1 & \cdots & 1 \\ f(\mathbf{p}_0) & \cdots & f(\mathbf{p}_{n+1}) \end{pmatrix}$.
4:     **repeat**
5:         Solve $L_\sigma \alpha = \mathbf{e}_1$, with $\alpha_i = 0$.   ▷ first hyperplane point
6:         **if** $\alpha \not\geq \mathbf{0}$ **then**
7:             stop
8:         **end if**
9:         Solve $L_\sigma \beta = \mathbf{0}$.   ▷ find other hyperplane points
10:        Find index j of the next pivoting vertex.   ▷ door-in-door-out step
11:        Pivot $\mathbf{p}_j$ into $\hat{\mathbf{p}}_j$.   ▷ pivoting step
12:        $\mathbf{p}_j \leftarrow \hat{\mathbf{p}}_j$
13:        Update j-component of σ with the new $\mathbf{p}_j$.   ▷ adjacent $(n+1)$–simplex
14:        Calculate new label $l_j = \begin{pmatrix} 1 \\ f(\mathbf{p}_j) \end{pmatrix}$.
15:        $L_\sigma \leftarrow L_\sigma + (l_j - L_\sigma \mathbf{e}_j)\mathbf{e}_j^T$ ▷ update labelling matrix
16:        $i \leftarrow j$
17:     **until**
18: **end procedure**

of transverse $(n+1)$-simplices σ_i bounded by the two completely labelled n-faces τ_{i-1} and τ_i. This is performed by pivoting a vertex of an $(n+1)$-simplex σ_i across a completely labelled facet τ in order to find another adjacent $(n+1)$-simplex sharing the same facet τ (steps 11-13). Altogether, steps 11-13 form the *pivoting step*. Steps 5-9 allow us to determine PL zero set inside σ_i. These latter steps altogether are known as the *piecewise linear step*. The piecewise linear step is usually more expensive, in computational terms, than the pivoting step because it involves linear algebra operations (i.e. matrix operations). From the piecewise linear step we can determine the index of of the next vertex to be pivoted, i.e. the DIDO step.

For a more comprehensive discussion of piecewise linear algorithms for curves using vector labelling, the reader is referred to Allgower and Georg[8]. Dobkin et al. [114] proposed a similar algorithm for curves ($d = 1$). Interestingly, the first multidimensional algorithm (Algorithm 14) using vector labelling approximation was described by Allgower and Gnutzmann in [9] for

---

Algorithm 14 PL Algorithm for Manifolds

1: **procedure** ALLGOWER-GNUTZMANN($f,\Omega,\mathcal{T}$)
2:　　$\mathcal{T} \leftarrow \varnothing$　　▷ set of pivoted, transverse $(n+d)$-simplices of $\mathcal{T}$
3:　　$S \leftarrow \varnothing$　　▷ set of nonpivoted, transverse $(n+d)$-simplices of $\mathcal{T}$
4:　　Find a transverse starting $(n+d)$-simplex $\sigma \in \mathcal{T}$.
5:　　$S \leftarrow S \cup \{\sigma\}$
6:　　$V(\sigma) \leftarrow$ set of nonpivoted vertices of σ
7:　　**while** $S \neq \varnothing$ **do**
8:　　　　Get $\sigma \in S$.
9:　　　　**while** $V(\sigma) \neq \varnothing$ **do**
10:　　　　　Get $\mathbf{p} \in V(\sigma)$.
11:　　　　　Pivot $\mathbf{p}$ into $\hat{\mathbf{p}}$ to get an adjacent $(n+d)$-simplex $\hat{\sigma}$.
12:　　　　　**if** $\hat{\sigma} \cap \sigma$ is not transverse or $\hat{\sigma} \cap \Omega = \varnothing$ **then**
13:　　　　　　$V(\sigma) \leftarrow V(\sigma) \setminus \{\mathbf{p}\}$　　▷ delete $\mathbf{p}$ from $V(\sigma)$
14:　　　　　**else**
15:　　　　　　**if** $\hat{\sigma} \in \mathcal{T}$ or $\hat{\sigma} \in S$ **then**　　▷ $\hat{\sigma}$ is not new
16:　　　　　　　$V(\sigma) \leftarrow V(\sigma) \setminus \{\mathbf{p}\}$　　▷ delete $\mathbf{p}$ from $V(\sigma)$
17:　　　　　　　$V(\hat{\sigma}) \leftarrow V(\hat{\sigma}) \setminus \{\hat{\mathbf{p}}\}$　　▷ delete $\hat{\mathbf{p}}$ from $V(\hat{\sigma})$
18:　　　　　　**else**
19:　　　　　　　$S \leftarrow S \cup \{\hat{\sigma}\}$
20:　　　　　　　$V(\hat{\sigma}) \leftarrow$ set of nonpivoted vertices of $\hat{\sigma}$
21:　　　　　　　$V(\sigma) \leftarrow V(\sigma) \setminus \{\mathbf{p}\}$　　▷ delete $\mathbf{p}$ from $V(\sigma)$
22:　　　　　　　$V(\hat{\sigma}) \leftarrow V(\hat{\sigma}) \setminus \{\hat{\mathbf{p}}\}$　　▷ delete $\hat{\mathbf{p}}$ from $V(\hat{\sigma})$
23:　　　　　　**end if**
24:　　　　　**end if**
25:　　　**end while**
26:　　　$S \leftarrow S \setminus \{\sigma\}$
27:　　　$\mathcal{T} \leftarrow \mathcal{T} \cup \{\sigma\}$
28:　**end while**
29: **end procedure**

implicit surfaces ($d = 2$). Note that the PL algorithms described so far only apply to curves and surfaces that are manifolds. This is so because it is not possible to approximate a self-intersecting curve or surface by a hyperplane inside a given simplex of the triangulation. In this case, PL methods approach non-manifold curves and surfaces (varieties, in general) using small perturbations of the zero value of the map f as a way to rid off possible singularities [9]. Thus, a general piecewise linear algorithm for approximating manifolds consists in pivoting through simplices which subdivide the domain of the map f. Algorithm 14 describes such an algorithm.

6.5 PC Continuation

Predictor-corrector (PC) methods constitute the second class of continuation methods. They also output a piecewise linear approximation of the zero set defined by an arbitrary smooth function $f : \mathbb{R}^{n+d} \to \mathbb{R}^n$. However, this PL approximation is obtained using different devices. Instead of using a fixed triangulation of the ambient space, predictor-corrector algorithms directly triangulate the variety (e.g. curve or surface) on the fly in a progressive manner. This means that the next vertex of the polyline that approximates a curve is determined from the current vertex; analogously, the next vertex of a new triangle that approximates a surface is determined from two consecutive vertices of the boundary of the current growing mesh.

Every PC algorithm comprises two major stages: the *growing* stage and the *filling* stage. The growing stage consists of two steps, the predictor and corrector steps. The predictor step estimates a point in the tangent hyperplane to the variety at the current vertex; the corrector step settles the predicted point onto the surface producing a new vertex on the surface. The correction is usually done using a Newton corrector, but a 2-point numerical corrector (e.g. bisection method) may also be used.

The filling stage is only needed for closed curves and surfaces. For example, expanding the mesh on a closed surface requires to avoid that the mesh overlaps; otherwise, the meshing of the surface will never stop. This stopping condition on the triangulation creates cracks or gaps in the mesh that need to be filled with new triangles in order to close the surface.

6.6 PC Algorithm for Manifold Curves

Algorithm 15 outputs a 1-dimensional piecewise linear approximation for implicit curves in $\mathbb{R}^2$ ($n = 1$, $d = 1$).

Algorithm 15 consists of the following steps. Step 2, as well as step 5, uses Newton's method (or some Newton-like method) to iterate a point near the curve onto the curve. Steps 3–4 are illustrated in Figure 6.10 for the computation of two curve points, $\mathbf{x}_{i+1}$ and $\mathbf{x}_{i+2}$.

6.6 PC Algorithm for Manifold Curves

Algorithm 15 Derivative-based Predictor–Corrector Algorithm for Curves

1: **procedure** PREDICTORCORRECTORFORCURVES(f,Ω,δ)
2:     Determine one point $\mathbf{x}_i$ on the curve.
3:     Determine a tangent vector $\mathbf{t}_i$ to the curve at $\mathbf{x}_i$.
4:     Step out a small amount δ along $\mathbf{t}_i$ to get a predicted point $\mathbf{p}_0$.
5:     Map $\mathbf{p}_0$ onto the curve in order to obtain the next curve point $\mathbf{x}_{i+1}$.
6:     $\mathbf{x}_i \leftarrow \mathbf{x}_{i+1}$
7:     Go to step 3.
8: **end procedure**

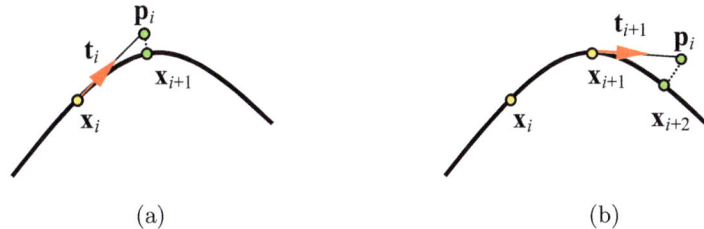

Fig. 6.10. The Rheinboldt algorithm for curves in $\mathbb{R}^2$.

Step 3 is carried out by solving the linear system $Jf(\mathbf{x}_i).\mathbf{t}_i = 0$ where $Jf(\mathbf{x}_i)$ is the $(n+d) \times n$ Jacobian matrix of f evaluated at a point $\mathbf{x}_i$ on the curve, and where $\mathbf{t}_i$ is the unit tangent vector to $f = 0$ at $\mathbf{x}_i$.

Step 4 (*predictor step*) computes a predicted point $\mathbf{p}_i = \mathbf{x}_i + \delta.\mathbf{t}_i$, where δ is a given step size.

Step 5 (*corrector step*), this predicted point $\mathbf{p}_i = \mathbf{p}_i^0$ is the starting point of a Newton-like procedure of a sequence $\mathbf{p}_i^0, \mathbf{p}_i^1, \ldots, \mathbf{p}_i^k \approx \mathbf{x}_{i+1}$ of points converging to a curve point $\mathbf{x}_{i+1}$ since $\mathbf{p}_i$ is sufficiently near the curve.

In the end, we end up having a sequence $\mathbf{x}_0, \mathbf{x}_1, \mathbf{x}_2, \ldots \in \mathbb{R}^2$ of points on the curve such that $\mathbf{x}_{i+1}$ is obtained from $\mathbf{x}_i$ using a predictor step and its subsequent corrector step. This shows us that a curve can be traced in relatively few steps for coarse approximations (i.e. with a step size not very small).

Despite its simplicity, this class of algorithms has some deficiencies because they may fail under some circumstances. Let us enumerate two of them:

1. *Drifting away from the curve.* One may fail to keep on moving along the curve if by some misfortune a predictor step comes to close to some unwanted point of $f^{-1}(0)$, as illustrated in Figure 6.11(a).
2. *Cycling.* This a variant of the previous situation, where the unwanted point has been already determined a few steps earlier, so that the algorithm risks cycling forever (Figure 6.11(b)).

This "nasty behaviour" in the predictor step can be resolved using more sophisticated machinery such as Runge-Kutta or Adam's methods [329] or,

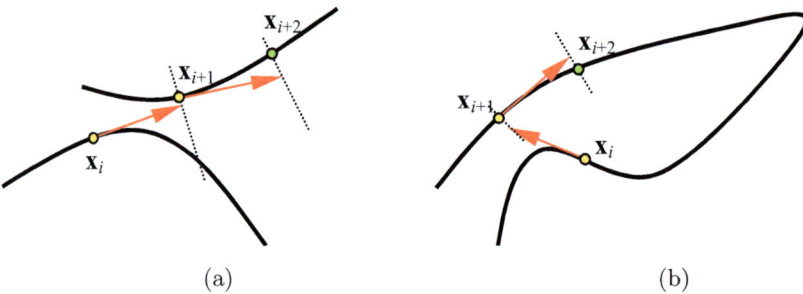

Fig. 6.11. (a) Drifting and (b) cycling phenomena for curves in $\mathbb{R}^2$.

alternatively, to halve the step length δ. These precautions help to ensure that one stays in the domain of quadratic convergence about the curve of Newton's method. This is important because Newton's method may not converge if the curve oscillates too much in the interval $[\mathbf{x}_i, \mathbf{x}_{i+1}]$.

Another point of concern is whether the curve possesses special points such as:

- *Turning points*, where one of the partial derivatives to the curve vanishes.
- *Singular points*, where $f = 0$ and $\text{grad}(f) = (0,0)$, i.e. where both partial derivatives vanish.

For example, the circle defined as the zero set $f(x,y) = x^2 + y^2 - 4 = 0$ has four turning points at $(0,-2)$, $(0,2)$, $(-2,0)$ and $(2,0)$. The first two result from $\frac{\partial f}{\partial x} = 2x = 0$, whereas the other two are found by means of $\frac{\partial f}{\partial y} = 2y = 0$. These derivatives vanish simultaneously at $(0,0)$, but it is not a singular point (or singularity) because it does not belong to the circle. A singularity on an curve is where it is not smooth. For example, the curve $f(x,y) = y^2 - x^3 = 0$ has a cusp at $(0,0)$ because the both partial derivatives vanish at it and because it is on the curve.

The fact that partial derivatives vanish at a singularity makes any algorithm based upon Newton's numerical method to break down because such derivatives appear in the denominator of the iteration formula (cf. Equation (5.11)), which is here re-written for our convenience:

$$\mathbf{p}_{k+1} = \mathbf{p}_k - \frac{\nabla f(\mathbf{p}_k)}{||\nabla f(\mathbf{p}_k)||^2} f(\mathbf{p}_k) \qquad (6.27)$$

There are two main approaches to overcome this breakdown problem:

- *Derivative-free methods*. Using a derivative-free numerical method as, for example, the false position method is a good numerical device for sampling implicit curves and surfaces. But, since it is based on the sign variation of function at two distinct points, it fails sampling sign-invariant components of curves and surfaces. For example, the zero set

$(x^2 + y^2 + z^2 - 25)^2 = 0$ is a sphere, inside and outside which the corresponding function $f(x,y,z) = (x^2+y^2+z^2-25)^2$ always evaluates positive. This means that 2-points numerical methods using the intermediate value theorem cannot be used to sample zero sets with sign-invariant components. To overcome this problem, and assuming that Newton's method breaks down at singularities, we have to use a sign-invariant 2-points numerical method as the *generalised* false position method [281].

- *Resolution of special points.* The idea here is to first determine the special points of a curve or surface through standard symbolic processing techniques for resolution of equation systems. These special points then work as starting points for sampling the remaining singularity-free patches of a curve or surface through a Newton predictor-corrector.

In addition to special points, there is another problem underlying the predictor–corrector algorithms. They are not equipped with suited devices for sampling curves and surfaces with several components. To succeed on this we have first to find out a seeding point on each component, which may be quite difficult because there is no triangulation covering the domain to help us on this respect.

6.7 PC Algorithm for Nonmanifold Curves

This section describes a derivative-free continuation algorithm for nonmanifold implicit curves in $\mathbb{R}^2$, and is due to Morgado and Gomes [280]. It is also curvature-adaptive and suited for handling curves with singularities.

The basic idea behind this continuation algorithm is, given the previous and current points $\mathbf{x}_{i-1}$, $\mathbf{x}_i$ of a curve, to determine the next point $\mathbf{x}_{i+1}$ on the circle neighbourhood N_i centred at $\mathbf{x}_i$ (Figure 6.12). The algorithm uses numerical continuation to compute $\mathbf{x}_{i+1}$ in an arc of the frontier of N_i. This numerical method is inspired in the standard false position numerical method, and is called *angular false position method* (AFP) [280].

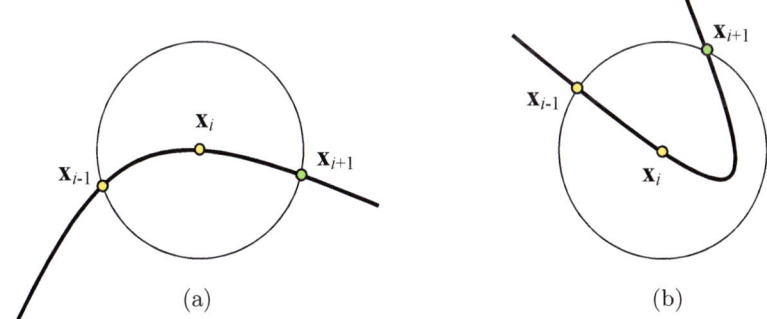

Fig. 6.12. Illustration of the basic idea behind Morgado-Gomes' algorithm for curves in $\mathbb{R}^2$.

6.7.1 Angular False Position Method

Traditionally, numerical methods operate on intervals in $\mathbb{R}$ or straight line segments in higher dimensions. Let us consider the initial interval $[\mathbf{p}_0, \mathbf{q}_0]$ bracketing the root $\mathbf{x}_{i+1}$, i.e. $[\mathbf{p}_0, \mathbf{q}_0]$ is transverse to the curve $\mathcal{C}$ at $\mathbf{x}_{i+1}$. For convenience, let us rewrite the standard interpolation formula of the false position method (see Section 5.6.2) as follows:

$$\mathbf{r}_k = \mathbf{q}_k - \frac{f(\mathbf{q}_k)}{f(\mathbf{q}_k) - f(\mathbf{p}_k)}(\mathbf{q}_k - \mathbf{p}_k) \qquad (6.28)$$

where $\mathbf{r}_k$ is the root of the secant line through $(\mathbf{p}_k, f(\mathbf{p}_k))$ and $(\mathbf{q}_k, f(\mathbf{q}_k))$. If $f(\mathbf{p}_k)$ and $f(\mathbf{r}_k)$ have identical signs, then we set $\mathbf{p}_{k+1} = \mathbf{r}_k$ and $\mathbf{q}_{k+1} = \mathbf{q}_k$; otherwise, we set $\mathbf{p}_{k+1} = \mathbf{p}_k$ and $\mathbf{q}_{k+1} = \mathbf{r}_k$. That is, the initial segment $[\mathbf{p}_0, \mathbf{q}_0]$ bracketing a root converges to a final smaller segment $[\mathbf{p}_n, \mathbf{q}_n]$ such that the next estimate $\mathbf{r}_n$ is a sufficiently good approximation of the next curve point $\mathbf{x}_{i+1}$, that is $\mathbf{r}_n \approx \mathbf{x}_{i+1}$.

Instead of using linear segments or intervals, the *angular* false position method uses arcs to find a bracketed root or curve point. The corresponding formula is then as follows:

$$\alpha(\mathbf{r}_k) = \alpha(\mathbf{q}_k) - \frac{f(\mathbf{q}_k)}{f(\mathbf{q}_k) - f(\mathbf{p}_k)}[\alpha(\mathbf{q}_k) - \alpha(\mathbf{p}_k)] \qquad (6.29)$$

where $\alpha(\mathbf{x})$ denotes the angle of the point $\mathbf{x}$ on the frontier of the circle neighbourhood N_i centred at a given origin $\mathbf{x}_i$. For example, in Figure 6.13, the angle of $\mathbf{p}_0$ is equal to $\frac{2\pi}{3}$.

6.7.2 Computing the Next Point

Morgado-Gomes' algorithm confines all computations to the neighbourhood N_i of the current curve point $\mathbf{x}_i$. The next curve point $\mathbf{x}_{i+1}$ results from the intersection of N_i and $\mathcal{C}$, and is numerically determined by the AFP method.

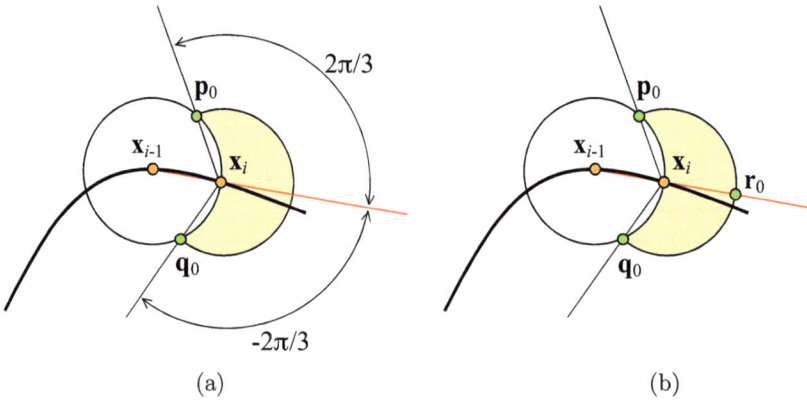

Fig. 6.13. Scanning neighbour circle arc in $[-\frac{2\pi}{3}, \frac{2\pi}{3}]$ for curve points in $\mathbb{R}^2$.

Recall that any point $\mathbf{p} = (x, y)$ on a circle with radius r and centred at $\mathbf{x} = (x_c, y_c)$ can be obtained from its angle α through the equations $(x, y) = (x_c + r\cos\theta, y_c + r\sin\theta)$, with $\alpha \in [-\pi, \pi[$, from which we can easily derive the 2×2 rotation matrix M_α, as usual in computer graphics.

To speed up our algorithm, and to prevent recomputing curve points on the previous circle neighbourhood, the circle points of the current circle are only computed for $\alpha \in [-\frac{2\pi}{3}, \frac{2\pi}{3}]$. These angles are those at which the current and previous circles intersect (Figure 6.13). So, the first step of the algorithm consists of determining these two intersection points, $\mathbf{p}_0$ and $\mathbf{q}_0$

$$\mathbf{p}_0 = \mathbf{x}_i + M_{2\pi/3} \cdot \overrightarrow{\mathbf{x}_{i-1}\mathbf{x}_i} \tag{6.30}$$

$$\mathbf{q}_0 = \mathbf{x}_i + M_{-2\pi/3} \cdot \overrightarrow{\mathbf{x}_{i-1}\mathbf{x}_i} \tag{6.31}$$

Then, we use (6.29) to determine the curve point on the arc $\widehat{\mathbf{p}_0\mathbf{q}_0}$. In practice, and to further speed up the root-finding process, we need to determine a preliminary estimate $\mathbf{r}_0 = \mathbf{x}_i + \overrightarrow{\mathbf{x}_{i-1}\mathbf{x}_i}$, and then we apply the AFP method to both arcs $\widehat{\mathbf{p}_0\mathbf{r}_0}$ and $\widehat{\mathbf{r}_0\mathbf{q}_0}$. Making an analogy to Newton's method, $\mathbf{r}_0$ works as a predicted point that is then corrected using the AFP method.

6.7.3 Computing Singularities

Morgado-Gomes' algorithm computes the singularities such as cusps and self-intersections using numerical approximation.

Cusps and Other High-curvature Points

As known, most curve continuation algorithms break down at singularities (see, e.g., [82]). But, Möller-Yagel's algorithm described in [277] copes with bifurcation points of curves by analysing sign changes of the partial derivatives in a rectangle neighbourhood. Nevertheless, it is a derivative-dependent algorithm, so it breaks down at other singularities (e.g. cusps and corners) which belong to the function domain, but not to domain of the partial derivatives. For example, it fails in rendering the diamond curve $|x| + |y| - 2 = 0$ which has four corners at $(0, 2)$, $(2, 0)$, $(0, -2)$ and $(-2, 0)$, where partial derivatives do not exist.

In contrast, Morgado-Gomes' algorithm does not compute derivatives at all. As a consequence, curves defined by differentiable and nondifferentiable functions can be, in principle, sampled and rendered in a straightforward manner. Another advantage of this derivative-free strategy is a shorter computation time for each sampled point. There is no need for computing derivatives, which in many cases are more time-consuming that functions themselves; for example, computing the partial derivatives of $F(x, y) = y(9 - xy)(x + 2y - y^2)((x - 10)^2 + (y - 4)^2 - 1)$ requires a bigger processing time overhead than the function itself.

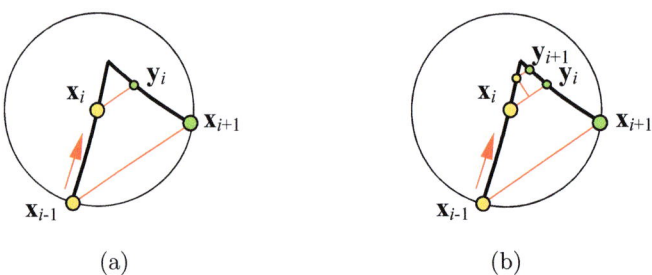

Fig. 6.14. Numerical approximation to a cusp point in $\mathbb{R}^2$.

Curvature flips at cusps and corners. Therefore, a cusp (or a quasi-cusp) is a point at or around which there is a high curvature variance within N_i (Figure 6.14). To make sure that there is a cusp (or a quasi-cusp) in N_i, we have to check whether the angle $\angle \mathbf{x}_{i-1}\mathbf{x}_i\mathbf{x}_{i+1}$ is small. If so, there is likely such a special point between $\mathbf{x}_i$ and $\mathbf{x}_{i+1}$. In addition, we have to check whether the mediatrix of $\overline{\mathbf{x}_{i-1}\mathbf{x}_{i+1}}$ inside N_i intersects the curve at exactly a single point. Under these two circumstances, we can say that a cusp exists within N_i. This means that $\mathbf{x}_{i+1}$ is not an appropriate point next to $\mathbf{x}_i$ (Figure 6.14(a)).

To compute a suited next point $\mathbf{x}_{i+1}$ we can use various strategies. One of them is to decrease the radius of N_i when the curvature increases. Another strategy is to fix the radius and sample the curve inside N_i somehow. A way to apply this second strategy is to assume that points after the cusp (or quasi-cusp) are image points of those before it in N_i (Figure 6.14(b)). For example, the former $\mathbf{x}_{i+1}$, now $\mathbf{y}_{i-1}$, is the image of $\mathbf{x}_{i-1}$. The image of $\mathbf{x}_i$ is $\mathbf{y}_i$ by tracing a line segment parallel to $\overline{\mathbf{x}_{i-1}\mathbf{y}_{i-1}}$. The next tentative point $\mathbf{x}_{i+1}$ is determined by intersecting the curve with the mediatrix of $\overline{\mathbf{x}_i\mathbf{y}_i}$ in N_i. This procedure is repeated for a few steps, stopping when the distance between the latest next point and its image is under a tolerance ϵ, meaning that our special point has been found, i.e. $d(\mathbf{x}_{i+n}, \mathbf{y}_{i+n}) \leq \epsilon$. At this point, set $\mathbf{x}_{i+1} = \mathbf{y}_{i+n}$ and label it as a cusp (or quasi-cusp).

Self-intersection Points

A self-intersection point can be viewed as a double cusp point, i.e. two cusps that come together (Figure 6.15). Therefore, no curve point on N_i can be the next point. In fact, with the exception of the point $\mathbf{x}_{i+2}$ in front of $\mathbf{x}_i$, the other two points form small angles with $\mathbf{x}_{i-1}$ and $\mathbf{x}_i$. But, $\mathbf{x}_{i+2}$ cannot be the next point either, because the segment from it to $\mathbf{x}_{i-1}$ intersects the curve at a nearer point. One solution is to use a procedure that converges to the self-intersection in a way similar to that one described above for cusps. This convergence process also stops when the distance between the latest next point and its image is under ϵ. This latest point will be the self-intersection point nearly, which will be set up as the next point $\mathbf{x}_{i+1}$.

6.7 PC Algorithm for Nonmanifold Curves   171

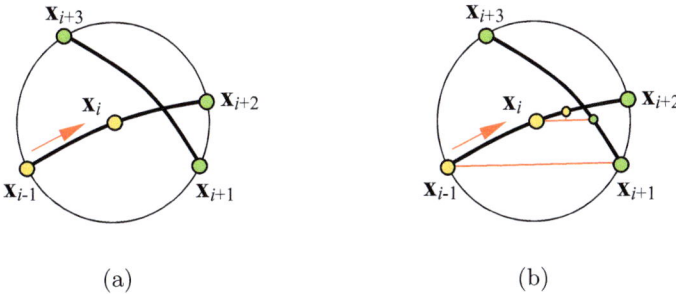

Fig. 6.15. Numerical approximation to a self-intersection point in $\mathbb{R}^2$.

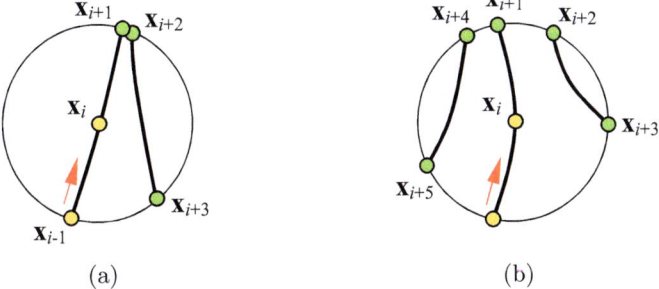

Fig. 6.16. Avoidance of the drifting phenomenon.

6.7.4 Avoiding the Drifting/Cycling Phenomenon

As seen above (cf. Figure 6.11), sometimes a curve comes close to itself so that algorithm goes cycling forever or just drifts away. In a way, this is similar to a self-intersection scenario because we have at least four curves points on N_i (cf. Figure 6.15). But, unlike the self-intersection neighbourhood, one of the curve points on N_i is the point next to $\mathbf{x}_i$ (Figure 6.16). Recall that the neighbourhood radius r is constant, even under ripples and undulations.

To determine the next point $\mathbf{x}_{i+1}$, we use two criteria: *angle criterion* (or curvature criterion) as above, and a new criterion, called *neighbour-branch criterion*. In Figure 6.16(a), $\mathbf{x}_{i+3}$ cannot be the next point because the angle $\angle \mathbf{x}_{i-1}\mathbf{x}_i\mathbf{x}_{i+3}$ is not approximately π within a given tolerance. But, both angles $\angle \mathbf{x}_{i-1}\mathbf{x}_i\mathbf{x}_{i+1}$ and $\angle \mathbf{x}_{i-1}\mathbf{x}_i\mathbf{x}_{i+2}$ are about π, and neither $\overline{\mathbf{x}_{i-1}\mathbf{x}_{i+1}}$ nor $\overline{\mathbf{x}_{i-1}\mathbf{x}_{i+2}}$ crosses the curve. Therefore, the next point will be either $\mathbf{x}_{i+1}$ or $\mathbf{x}_{i+2}$. To pick up the right next point, we use the neighbour-branch criterion. Basically, it is an elimination criterion, and can be described as follows:

1. Determine the midpoints of the segments $\overline{\mathbf{x}_{i-1}\mathbf{x}_{i+1}}$, $\overline{\mathbf{x}_{i+1}\mathbf{x}_{i+2}}$, and $\overline{\mathbf{x}_{i+2}\mathbf{x}_{i+3}}$ in Figure 6.16(a). The midpoint of $\overline{\mathbf{x}_{i-1}\mathbf{x}_{i+3}}$ is not calculated because $\mathbf{x}_{i+3}$ is not, by the angle criterion, a candidate to the next point.

2. Let us consider a segment with a midpoint M and P its projection on the frontier of N_i by prolonging the segment $\overline{\mathbf{x}_i M}$. If $\overline{\mathbf{x}_i P}$ intersects the curve inside N_i, then we discard the endpoints of the segment transverse to $\overline{\mathbf{x}_i P}$. This eliminates $\mathbf{x}_{i+2}$ as a candidate next point in Figure 6.16(a) because $\mathcal{C} \cap \overline{\mathbf{x}_i P} \neq \varnothing$, where P is the projection of the midpoint of $\overline{\mathbf{x}_{i+2} \mathbf{x}_{i+3}}$ on N_i. Therefore, the next point will be $\mathbf{x}_{i+1}$.

In Figure 6.16(b), the angle criterion eliminates $\mathbf{x}_{i+3}$ and $\mathbf{x}_{i+5}$ as candidates to the next point, while $\mathbf{x}_{i+2}$ and $\mathbf{x}_{i+4}$ are eliminated by the near-branch criterion. The remaining point $\mathbf{x}_{i+1}$ will be the next point. Note that the neighbourhood radius is constant independently of whether the curve oscillates or not.

The Algorithm

Morgado-Gomes' algorithm (Algorithm 16) essentially has the structure of a continuation algorithm (cf. Algorithm 15), but it does not use Newton's method to compute the next point of the curve. Instead, it uses a derivative-free method; hence its ability to cope with cusps (steps 7–8) and self-intersections (steps 12–13).

Besides, it avoids the drifting phenomenon nicely (step 11) using elimination criteria described above. However, this algorithm does not solve the global cycling problem. This global cycling phenomenon occurs when the current circle N_i overlaps a former or an intermediary circle neighbourhood previously calculated and processed. To prevent this global cycling phenomenon, we have

---

Algorithm 16 Derivative-free Predictor-Corrector Algorithm for Curves

1: **procedure** MORGADO-GOMES(f,Ω,r)
2:     Determine one point $\mathbf{x}_i$ on the curve.
3:     Set N_i as the circle neighbourhood with radius r centred at $\mathbf{x}_i$.
4:     Compute curve points $\mathcal{C} \cap N_i$ by means of the angular numerical method.
5:     **if** ($\#(\mathcal{C} \cap N_i) = 1$) **then**   ▷ a single candidate point
6:         $\mathbf{x}_{i+1} \leftarrow$ get such a single point from $\mathcal{C} \cap N_i$
7:         **if** ($\angle(\mathbf{x}_{i-1}\mathbf{x}_i\mathbf{x}_{i+1}) \not\approx \pi$) **then**
8:             $\mathbf{x}_{i+1} \leftarrow$ compute cusp
9:         **end if**
10:     **else**   ▷ two or more candidate points
11:         $\mathbf{x}_{i+1} \leftarrow$ get point from $\mathcal{C} \cap N_i$ by applying elimination criteria
12:         **if** ($\mathbf{x}_{i+1} = NULL$) **then**   ▷ there is a self-intersection point about $\mathbf{x}_i$
13:             $\mathbf{x}_{i+1} \leftarrow$ compute self-intersection point
14:         **end if**
15:     **end if**
16:     $\mathbf{x}_i \leftarrow \mathbf{x}_{i+1}$
17:     Go to step 3.
18: **end procedure**

to store circles in two vectors or lists. The first vector stores overlapping circles; typically, they are circles containing near-branch branches of the curve. The remaining circles are classified as nonoverlapping circles. So, if the next point $\mathbf{x}_{i+1}$ is in a nonoverlapping circle, the algorithm stops. But, if there is a not yet sampled branch coming out from a self-intersection, the algorithm restarts from this self-intersection point. Unfortunately, as usual for continuation algorithms, if the curve has various components inside the domain, some of them are likely missed out.

6.8 PC Algorithms for Manifold Surfaces

This section deals with predictor–corrector algorithms for implicit surfaces that are inspired in Newton's method. Therefore, these algorithms depend on the derivative or its higher-dimensional counterparts.

6.8.1 Rheinboldt's Algorithm

In computer graphics, while the *simplicial continuation* algorithms have been inspired by the Allgower–Schmidt algorithm [11], the class of *predictor-corrector* methods are rooted to Rheinboldt's work [340]. Rheinboldt proposes using a smoothly varying projection of the tangent plane onto the surface ("moving frame") to "wrap" a mesh onto the surface.

Therefore, Rheinboldt's algorithm for surfaces (Algorithm 17) in $\mathbb{R}^3$ ($n = 1$, $d = 2$) is very similar to that one for curves (Algorithm 15). However, the pseudocode of Algorithm 17 appears here simplified in order to highlight its similarities to Algorithm 15.

Note that, the predictor step (step 4) now outputs two points $\mathbf{x}_{i+1}$ and $\mathbf{x}_{i+2}$ close the surface $\mathcal{S}$ by computing two vectors with angle $\pi/3$ on the tangent plane T_i to $\mathcal{S}$ at $\mathbf{x}_i$. The angle $\pi/3$ aims at producing approximately equilateral triangles on the surface. The corrector step (step 5) places $\mathbf{x}_{i+1}$ and $\mathbf{x}_{i+2}$ on the surface. This allows us then to form a new triangle given by the vertices $\mathbf{x}_i$, $\mathbf{x}_{i+1}$ and $\mathbf{x}_{i+2}$.

Of course, the algorithm does not work in practice that way. Important issues like cycling, triangle overlapping, and triangle recomputation have to

---

Algorithm 17 Rheinboldt's Predictor-Corrector Algorithm for Surfaces

1: **procedure** RHEINBOLDT($\mathcal{S},\Omega,\delta$)
2:     Determine a point $\mathbf{x}_i$ on the surface $\mathcal{S}$ inside the bounding box Ω.
3:     Determine the tangent plane T_i to the surface $\mathcal{S}$ at $\mathbf{x}_i$.
4:     Step out a small amount δ along two vectors on T_i with angle $\pi/3$.
5:     Relocate the surface at both new points $\mathbf{x}_{i+1}$ and $\mathbf{x}_{i+2}$.
6:     Set $\mathbf{x}_i \leftarrow \mathbf{x}_{i+1}$ or $\mathbf{x}_i \leftarrow \mathbf{x}_{i+2}$.
7:     Go to step 3.
8: **end procedure**

be considered. We have also assumed that tessellation bricks are triangles. However, there are many ways of tessellating a surface in $\mathbb{R}^3$. For example, tessels can be triangles, squares or hexagons, though squares and hexagons are easily decomposed into two and six triangles, respectively.

6.8.2 Henderson's Algorithm

Henderson's algorithm is another predictor–corrector algorithm. Found a seeding point on the surface, the triangulation spirals away from it on the surface by attaching triangles beyond the current triangulation border. However, triangles are not determined directly. The Henderson triangulation results from covering the surface with an atlas of disks, being then their centres—which are points on the surface—used to triangulate it. This is illustrated in Figure 6.17. Therefore, we can say that Henderson's algorithm is inspired by devices commonly used in differential topology and geometry. The algorithm computes a set of points on the surface, and a set of mappings from the tangent space which cover the surface [183].

Henderson's algorithm (Algorithm 18) starts to differ from Rheinboldt's algorithm (Algorithm 17) at step 3. After determining the tangent plane T_i to the surface $\mathcal{S}$ at $\mathbf{x}_i$ (step 2), one determines a small disk D_i centred at $\mathbf{x}_i$ on T_i. Such a disk is here called *Henderson disk*.

At step 4, one determines a new point $\mathbf{y}_{i+1}$ on an non-overlapping arc of the Henderson disk D_i. For the initial disk, this boundary arc is the complete disk boundary. Non-overlapping arcs are those belonging to the border of the growing atlas.

Step 5 maps the new point $\mathbf{y}_{i+1}$ and its disk on the surface, merging it with the surface. The mapped point is now $\mathbf{x}_{i+1}$.

Before going to step 3, set $\mathbf{x}_i = \mathbf{x}_{i+1}$ or as any point on a nonintersecting disk arc. Note that when the covering of disks grows on the surface, its

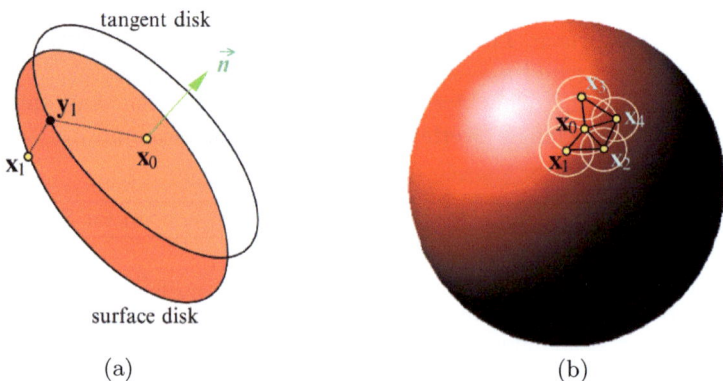

Fig. 6.17. (a) Mapping the Henderson disk on the surface; (b) covering the surface with disks followed by triangulation.

Algorithm 18 Henderson's Predictor–Corrector Algorithm for Surfaces

1: **procedure** HENDERSON($\mathcal{S},\Omega,\delta$)
2: Determine a point $\mathbf{x}_i$ on the surface $\mathcal{S}$ inside the bounding box Ω.
3: Determine the tangent plane T_i to the surface $\mathcal{S}$ at $\mathbf{x}_i$.
4: Determine the radius δ Henderson disk D_i at $\mathbf{x}_i$ on T_i.
5: Determine a point $\mathbf{y}_{i+1}$ on an non-overlapping arc of the boundary of D_i.
6: Map $\mathbf{y}_{i+1} \in \text{Bd}(D_i)$ onto a new point $\mathbf{x}_{i+1} \in \mathcal{S}$.
7: Set $\mathbf{x}_i \leftarrow \mathbf{x}_{i+1}$.
8: **if** $\text{Bd}(\text{Atlas}(\mathcal{S})) \neq \varnothing$ **then**
9:    Go to step 3.
10: **end if**
11: Triangulate $\mathcal{S}$.
12: **end procedure**

boundary consists of a set of connected nonintersecting arcs. The triangulation grows accordingly by connecting the centres of the disks mapped onto the surface $\mathcal{S}$.

Steps 4–6 are illustrated in Figure 6.17. The algorithm terminates when non-overlapping arcs run out (step 8) or, equivalently, when the boundary of the atlas of circles on the surface vanishes, i.e. $\text{Bd}(\text{Atlas}(\mathcal{S})) = \varnothing$. This guarantees that the whole surface is triangulated. The mechanism of non-overlapping arcs also ensures us that the algorithm does not loop locally and globally. Obviously, we are here assuming that the surface is manifold and thus closed, but the algorithm can be easily extended to manifold surfaces with boundary.

The triangulation produces a triangular mesh whose data structure consists of a vector of triangles and a vector of vertices. Each vertex is a surface point, so it stores its disk, which in turn must include data concerning overlapping and non-overlapping boundary arcs. The topological relations between disks are easily retrieved via their vertices and triangles.

6.8.3 Hartmann's Algorithm

Similar to Henderson's algorithm, Hartmann's algorithm only applies to surfaces in $\mathbb{R}^3$ without singularities [179]. Surfaces need not to be closed. But, instead of triangulating an atlas of disks on the surface, Hartmann directly triangulates each disk into an hexagon of six approximately equilateral triangles (Figure 6.18(a)).

The compatibility between overlapping hexagons is achieved by computing only the missing triangles of the new hexagon. This is illustrated in Figure 6.18(b), where we have three hexagons centred at $\mathbf{x}_0$, $\mathbf{x}_3$ and $\mathbf{x}_4$, respectively. To construct a new hexagon around $\mathbf{x}_2$, we have to take into account that three of its triangles are already attached to $\mathbf{x}_2$. Thus, we end up getting a triangulated surface consisting of a set of imaginary overlapping hexagons.

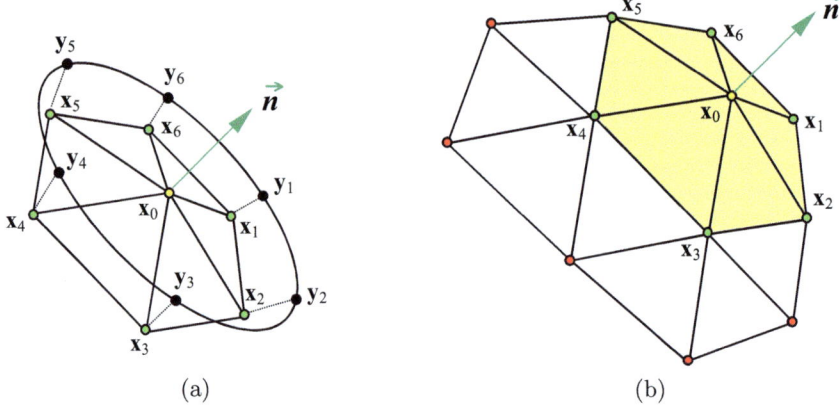

Fig. 6.18. (a) Hartmann's hexagon; (b) covering of three overlapping hexagons centred at $\mathbf{x}_0$, $\mathbf{x}_3$ and $\mathbf{x}_4$.

The idea is then to determine the tangent plane at each mesh border vertex $\mathbf{x}_0$, construct a small tangent disk—the Henderson disk—centred at $\mathbf{x}_0$, and inscribe an hexagon of triangles in such a disk. This works quite well for the first hexagon. Any subsequent hexagon already possesses at least one triangle in the growing mesh.

The construction of a new hexagon starts from a mesh border vertex $\mathbf{x}_0$ (Algorithm 19, step 2). This process consists of inserting the missing triangles around the hexagon centre $\mathbf{x}_0$ for completion; hence the need for compatibility between hexagons on the surface. This compatibility between hexagons is achieved through heuristics. Basically, we partition the external angle at $\mathbf{x}_0$ in order to obtain approximately regular triangles (i.e. approximately equilateral triangles). Hartmann [179], Karkanis and Stewart [208] and Raposo and Gomes [334] use the Henderson disk and similar heuristics, although Karkanis and Stewart also use the curvature criterion to produce good triangles in those surface regions where the curvature (and, consequently, the triangle size) changes quickly.

Hartmann's algorithm is described in Algorithm 19. Step 2 is the typical first step of any *continuation algorithm* which determines a seeding point $\mathbf{x}_0$ on the surface.

Step 3 leads to the construction of an orthonormal basis $(\mathbf{u}, \mathbf{v}, \mathbf{n})$ at $\mathbf{x}_0$.

The predictor step (steps 4–5) involves the definition of Henderson's disk centred at $\mathbf{x}_0$ and the computation of six vertices of its inscribed triangles as follows:

$$\mathbf{y}_j = \mathbf{x}_0 + \delta \cos(j\pi/3)\mathbf{u} + \delta \sin(j\pi/3)\mathbf{v} \qquad (6.32)$$

where $\mathbf{u}$ and $\mathbf{v}$ are unit base vectors in T_0.

To calculate the corresponding six sampling points on the surface we apply the Newton corrector to the points $\mathbf{y}_j$. The resulting points $[\mathbf{x}_{i+1}, \ldots, \mathbf{x}_{i+6}]$

Algorithm 19 Hartmann's Predictor–Corrector Algorithm for Surfaces

1: **procedure** HARTMANN($\mathcal{S},\Omega,\delta$)
2:   Determine a point $\mathbf{x}_0$ on the surface $\mathcal{S}$ inside the bounding box Ω.
3:   Determine a tangent plane T_0 to $\mathcal{S}$ at $\mathbf{x}_0$.
4:   Determine the radius δ Henderson disk D_i at $\mathbf{x}_0$ on T_0.
5:   Inscribe the Hartmann hexagon $[\mathbf{y}_1, \ldots, \mathbf{y}_6]$ in D_i, totally or partially.
6:   Map Hartmann's hexagon $[\mathbf{y}_1, \ldots, \mathbf{y}_6]$ onto $[\mathbf{x}_{i+1}, \ldots, \mathbf{x}_{i+6}]$ in $\mathcal{S}$.
7:   Triangulate $[\mathbf{x}_{i+1}, \ldots, \mathbf{x}_{i+6}]$ in $\mathcal{S}$, totally or partially.
8:   **if** $\{\mathbf{x}_k\} \neq \varnothing$ **then**
9:     Set $\mathbf{x}_0 \leftarrow \mathbf{x}_k$
10:    Go to step 3.
11: **end if**
12: **end procedure**

are the six vertices of the starting hexagon of the surface, that is, the first mesh boundary. This is valid for the first vertex, say hexagon, of the triangulation.

For other vertices, not all hexagon triangles need be determined, simply because some have been already determined. An easy way to determine the missing triangles around a border vertex $\mathbf{x}_0$ is to project its neighbour vertices back to tangent plane at $\mathbf{x}_0$ and fill in the pie slice of Henderson disk with the missing triangles. However, due to the varying curvatures of the surface, this strategy may originate thin triangles. To overcome this problem, we have sometimes to decompose the Henderson disk into 5-gons or 7-gons in order to keep the triangulation approximately regular.

The problem is then how many triangles are going to be inscribed in the missing Henderson slice defined by three consecutive border vertices $\mathbf{x}_{i-1}, \mathbf{x}_i, \mathbf{x}_{i+1}$, where $\mathbf{x}_i$ is the centre of the Henderson disk? We divide the external angle (i.e. outwards the triangulation border) $\theta = \angle \mathbf{x}_{i-1}\mathbf{x}_i\mathbf{x}_{i+1}$ into a number of angles with approximately $\frac{\pi}{3}$ radians. (See [334] for an elegant implementation of the external angle.)

To find the optimal number of triangles that fit θ, let us use the strategy described in [334]. First, we have to consider a range $[\theta_{\min}$ to $\theta_{max}]$ of acceptable angles around $\frac{\pi}{3}$, where $\theta_{\min} = \frac{\pi}{3} - \epsilon$ and $\theta_{\max} = \frac{\pi}{3} + \epsilon$, and where ϵ is a tolerance. It is necessary then to calculate the numbers of triangles $n_{\min}$ and $n_{\max}$ that result from dividing θ by $\theta_{\min}$ and $\theta_{\max}$, rounding them to the nearest integer. We select either $n_{\min}$ or $n_{\max}$ as the optimal number n_Δ of triangles depending on which one better approximates equilateral triangles, i.e whose angles are closest to $\frac{\pi}{3}$:

$$n_\Delta = \begin{cases} n_{\min} & \text{if } \left|\frac{\theta}{n_{\min}} - \frac{\pi}{3}\right| \leq \left|\frac{\theta}{n_{\max}} - \frac{\pi}{3}\right| \\ n_{\max} & \text{if } \left|\frac{\theta}{n_{\min}} - \frac{\pi}{3}\right| > \left|\frac{\theta}{n_{\max}} - \frac{\pi}{3}\right| \end{cases}$$

Once determined the number of triangles that fit θ around $\mathbf{x}_0$, the current mesh is ready to grow.

178   6 Continuation Methods

The external angle is a first device that guarantees that mesh grows beyond the border outwards. That is, it is the first condition to have a triangulation without overlapping triangles. However, this not enough to ensure re-triangulation of the surface because, when mesh grows on the surface, at some point its border or borders come too close to itself or to each other that they will overlap soon or later. To prevent this, Hartmann uses two proximity criteria as follows:

- *Two nonconsecutive vertices of the same border are near to each other.* If the distance d between two nonconsecutive border vertices is less that the Henderson disk radius δ, then they must be connected by a new edge to form a new triangle. This procedure splits the border into two. This *boundary splitting* operation is illustrated in Figure 6.19. Let Λ_m be a triangulation border, and let $\mathbf{x}_i$ and $\mathbf{x}_j$ (with $i < j$) be two of its vertices having at least two border vertices between them. If the Euclidean distance $d(\mathbf{x}_i, \mathbf{x}_j) < \delta$, one connects $\mathbf{x}_i$ to $\mathbf{x}_j$ by a new edge. As a consequence, Λ_m splits into two new borders, $\Lambda_m = \{\mathbf{x}_i, \mathbf{x}_{i+1}, \ldots, \mathbf{x}_j\}$ and $\Lambda_n = \{\mathbf{x}_1, \ldots, \mathbf{x}_i, \mathbf{x}_j, \ldots, \mathbf{x}_N\}$, where N is the number of vertices of the former Λ_m. In Figure 6.19(b), we can see these borders after attaching the triangle bounded by the border splitting edge $\overline{\mathbf{x}_i \mathbf{x}_j}$.
- *Two vertices of distinct borders are near to each other.* In this case, two vertices belonging to different borders are within the distance δ. Therefore, these borders are merged into a single one. This *merging* operation is illustrated in Figure 6.20. Let $\Lambda_m = \{\mathbf{x}_1, \ldots, \mathbf{x}_{m-1}, \mathbf{x}_i\}$ and $\Lambda_n = \{\mathbf{x}_j, \mathbf{x}_{m+2}, \ldots, \mathbf{x}_{m+n}\}$ two expansion borders. If the vertex $\mathbf{x}_i \in \Lambda_m$ and the vertex $\mathbf{x}_j \in \Lambda_n$ satisfy the condition $d(\mathbf{x}_i, \mathbf{x}_j) < \delta$, their host borders merge into a single one, say Λ_m, by attaching the new edge $\overline{\mathbf{x}_i \mathbf{x}_j}$ (Figure 6.20(b)). This implies deleting Λ_n after transferring its vertices into Λ_m, i.e.

$$\Lambda_m = \{\mathbf{x}_1, \ldots, \mathbf{x}_{m-1}, \mathbf{x}_i, \mathbf{x}_j, \mathbf{x}_{m+2}, \ldots, \mathbf{x}_{m+n}\}$$

where m and n denote the number of vertices of the former borders Λ_m and Λ_n, respectively.

This concludes the step 5 of the algorithm.

Step 6 is about mapping the vertices (and corresponding triangles) onto the surface. This is the corrector step. Every predicted vertex on the tangent plane is settled onto the implicit surface, its corrected position. As usual, the correction is usually done by a Newton-Raphson corrector (e.g. see [179] and [334]), but some researchers employ other numerical methods (e.g. Karkanis and Stewart use the bisection method in [208]).

Step 8 concerns the stopping criterion for the triangulation. The algorithm stops when the number of border vertices of the growing mesh goes to zero, i.e. $\{\mathbf{x}_k\} = \emptyset$. Recall that, we are here assuming that the surface is manifold.

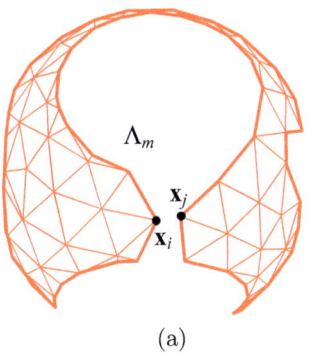 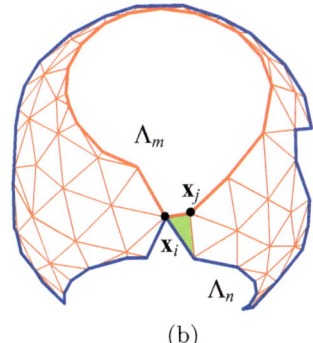

Fig. 6.19. Splitting a mesh border Λ_m (in thick red) into two smaller borders Λ_m (in thick red) and Λ_n (in thick blue).

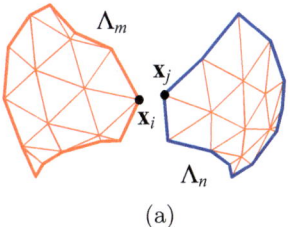 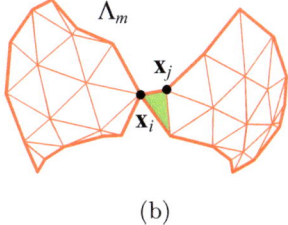

Fig. 6.20. Merging two mesh borders Λ_m (in thick red) and Λ_n (in thick blue) into a bigger border Λ_m (in thick red).

6.8.4 Adaptive Hartmann's Algorithm

In the context of implicit surfaces, an adaptive triangulation means a curvature-dependent triangulation. An adaptive hexagonal triangulation was proposed by Araújo and Jorge [20]. It is adaptive in the sense that the size of the triangles circumscribed by the Henderson disk depends on the local curvature of surface at each active or boundary vertex. That is, the radius of the Henderson disk is no longer constant; it depends on the local curvature.

By computing the local curvature of a surface at an active vertex, we are able to adapt the triangulation to shape variations of the surface, generating smaller triangles in regions of higher curvature and larger triangles where the curvature is smaller.

Thus, the adaptiveness of the Araújo-Jorge algorithm starts at the predictor step by changing the radius of Henderson's disk at an active vertex, depending on the curvature of the surface at such a vertex. The corrector step is identical to Hartmann's one, which is a Newton corrector for trivariate real functions. Mean curvature-based adaptive triangulations of Igea's implicit surface are shown in Figure 6.21.

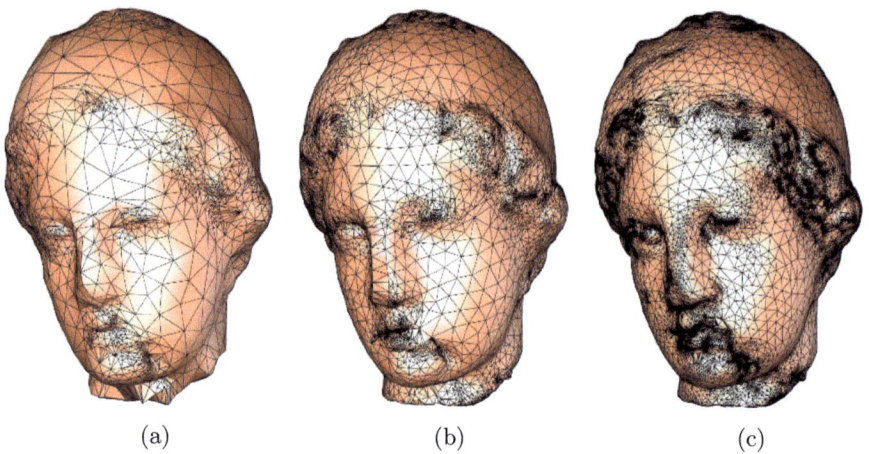

Fig. 6.21. Adaptive meshes of Igea's model using mean curvature heuristics: (a) 6908 triangles; (b) 25,596 triangles; (c) 82,511 triangles. Courtesy of B. Araújo [19].

As known, any PC algorithm for closed surfaces comprises two major stages: the *growing* stage and the *filling* stage. The growing stage consists of two steps, the predictor and corrector steps. The filling stage aims at filling the cracks with new triangles in order to close the surface. Note that both the original Hartmann algorithm and its adaptive counterpart due to Araújo and Jorge only apply to closed manifold surfaces. The cracks are just a result of avoiding that the growing surface mesh overlaps.

In the case of the Araújo-Jorge algorithm, the mesh overlapping is controlled by an octree data structure that stores all the points generated by the algorithm. Avoiding mesh overlapping is quickly done by first determining which octree cubes intersect or contain the sphere centred at an active boundary vertex with Henderson's radius (i.e. radius equal to the estimated edge length); then, one checks the distance between the current vertex and the boundary vertices associated to those intersecting cubes in order to prevent the mesh overlapping and fill eventual cracks. The efficiency of this overlapping avoidance procedure is reinforced by a cache mechanism that stores the last n visited octree cells to speed up the algorithm.

6.8.5 Marching Triangles Algorithm

In [187], Hilton et al. proposed a PC algorithm called *marching triangles* after the famous *marching cubes* algorithm (see the next chapter for further details). In a way, we can say that all the PC algorithms for implicit surfaces are based on the idea of marching triangles, i.e. triangles that are progressively attached to the mesh that approximates the surface. Algorithms differ in that they determine the triangles, i.e. how their triangulations are accomplished.

The marching triangles algorithm generates meshes with almost equilateral triangles. The approach proposed by Hilton et al. is also based on the prediction–correction step to compute new vertices of growing triangulation. A major problem is detecting the cycling phenomenon, say when the surface mesh starts to overlap. Even we succeed in controlling the cycling phenomenon, another problem comes up, which is the appearance of cracks in the mesh. Thus, unless we have a strategy to fill the cracks with new triangles, closed surface meshes cannot be generated at all.

The Delaunay Triangulation

The marching triangles algorithm relies on the *Delaunay triangulation* of a set $X = \{\mathbf{x}_0, \ldots, \mathbf{x}_n\}$ of points in $\mathbb{R}^3$. This 3-dimensional Delaunay triangulation is composed of tetrahedra such that each tetrahedron is inscribed in a sphere— i.e. the sphere passes through the vertices of the tetrahedron—which does not contain any other point of X.

In the case the points of X lie on a *manifold surface*, and according to Boissonnat [56], the surface triangulation in the Delaunay triangulation satisfies the condition that it is composed of triangles such that there exists a circumsphere that passes through the three vertices of each triangle, but it does not contain any other point of X. The result is the 2-dimensional analogue of the 3-dimensional Delaunay triangulation where the points of X lie on a manifold surface in $\mathbb{R}^3$ rather than $\mathbb{R}^2$.

The above definition of the manifold surface triangulation derived from the 3D Delaunay triangulation provides the incremental mechanism to construct the surface mesh by attaching triangle after triangle. This mechanism is based on the *3D Delaunay surface constraint*, which states that a triangle $\sigma = [\mathbf{x}_i, \mathbf{x}_{i+1}, \mathbf{x}_p]$ may only be attached to the mesh boundary at an edge $[\mathbf{x}_i \mathbf{x}_{i+1}]$ if no other triangle of the growing surface mesh intersects the sphere circumscribing $\sigma = [\mathbf{x}_i, \mathbf{x}_{i+1}, \mathbf{x}_p]$ with the same surface orientation. (Two triangles are said to have the same orientation if the dot product of their normals is positive.)

The 3D Delaunay surface constraint guarantees that each triangle uniquely defines the surface locally. In other words, the local surface does not over-fold or self-intersect [187]. Interestingly, this local Delaunay constraint also ensures that the triangulated surface is globally Delaunay [143].

The Algorithm

The marching triangles algorithm is described in Algorithm 20. Starting from a seeding triangle on the surface, the marching triangles algorithm spirals away on the surface by attaching new triangles to the border edges of the growing mesh. The new triangle edges are then appended at the rear of the list L of edges that form the border of the growing mesh.

Algorithm 20 Marching Triangles for Surfaces

1: **procedure** MARCHINGTRIANGLES($f,\mathcal{S},\Omega,\delta$)
2:     $\mathcal{T} \leftarrow \varnothing$ ▷ empty mesh of triangles
3:     $L \leftarrow \varnothing$ ▷ empty list of boundary edges
4:     Determine a seeding triangle σ_0 on $\mathcal{S}$ inside Ω.
5:     Add edges of σ_0 to L.
6:     $n \leftarrow \#L$ ▷ current size of L
7:     **for** $i = 0, n-1$ **do**
8:         Estimate a vertex position, $\mathbf{x}_p$, by stepping out a constant distance δ along a vector perpendicular to the boundary edge $e_i = [\mathbf{x}_k, \mathbf{x}_{k+1}]$ at the midpoint of e_i in the plane of the triangle $\sigma_i = [\mathbf{x}_i, \mathbf{x}_k, \mathbf{x}_{k+1}]$ that is bounded by e_i.
9:         Determine the nearest point $\mathbf{x}_P$ on $\mathcal{S}$ to $\mathbf{x}_p$, i.e. $f(\mathbf{x}_P) = 0$.
10:        **if** $\sigma_P = [\mathbf{x}_P, \mathbf{x}_k, \mathbf{x}_{k+1}]$ satisfies the 3D Delaunay constraint **then**
11:            Remove edge $[\mathbf{x}_k, \mathbf{x}_{k+1}]$ from L.
12:            $n \leftarrow n - 1$ ▷ delete one edge from L
13:            Add triangle $\sigma_P = [\mathbf{x}_P, \mathbf{x}_k, \mathbf{x}_{k+1}]$ to triangulation mesh $\mathcal{T}$.
14:            Add edges $[\mathbf{x}_P, \mathbf{x}_k]$ and $[\mathbf{x}_P, \mathbf{x}_{k+1}]$ to L.
15:            $n \leftarrow n + 2$ ▷ more two edges added to L
16:        **end if**
17:     **end for**
18: **end procedure**

The marching triangles algorithm iterates on the list L only once. When a new triangle at a boundary edge fails to satisfy the 3D Delaunay constraint, such an edge is left in L. At the end, L will accommodate an non-empty border of connected edges in the triangular mesh. This suggests that this algorithm is more adequate to polygonise open surfaces (also called surfaces with boundary) than closed surfaces (also called surfaces without boundary) simply because cracks will appear in the mesh.

Interestingly, the Delaunay constraint can also used to fix cracks in the polygonisation. In fact, as Akkouche and Galin noted in [2], the Delaunay constraint implies that the width of any crack in the surface does not exceed the length of the triangle edges. Therefore, to complete the triangulation we only need to connect the vertices of the boundary edges of L to create new triangles that fix the cracks. Similar solutions for filling cracks were proposed by Karkanis and Stewart [208] and Cermak and Skala [80].

In Algorithm 20, the *predictor* and *corrector* steps are steps 8 and 9, respectively. This original marching triangles algorithm has the disadvantage that the step length δ is constant. The triangulation tends to be regular, but it does not adapt to the curvature of the surface.

6.8.6 Adaptive Marching Triangles Algorithms

PC algorithms using curvature-dependent triangulations of implicit surfaces were proposed by Akkouche and Galin [2], Karkanis and Stewart [208], and

Cermak and Skala [79]. These *adaptive* marching triangles algorithms produce a mesh of approximately equilateral triangles with sizes dependent on the local surface curvature, although they do not use the Delaunay condition.

Karkanis-Stewart's algorithm estimates the curvature at a surface point x_p by computing the radius of curvature of several geodesics that pass through x_p, taking then the minimum. This algorithm is slower than the PL methods because it has the extra time overhead at computing local surface curvature for every new vertex x_p in order to generate triangles of the appropriate size. This is so because curvature computation of Karkanis-Stewart's algorithm assumes the second derivative is not directly available; consequently, the implicit function is invoked many times in the curvature calculation.

On the contrary, Akkouche-Galin's algorithm avoids the explicit computation of the local curvature of the surface; instead, Akkouche and Galin use a particular heuristics to speed up the algorithm.

6.9 Predictor–Corrector Algorithms for Nonmanifold Surfaces

As much as we know, there is no piecewise linear (PL) algorithm for nonmanifold implicit surfaces. However, there is a PC algorithm capable of polygonising non-manifold implicit surfaces under certain conditions. Such an algorithm extends Hartmann's algorithm in order to cope with self-intersections and multi-component implicit surfaces, and is due to Raposo and Gomes [334].

As known, continuation methods do not allow us to know *a priori* the number of topological components a surface possesses, i.e. its topological shape. Hence, the difficulties in finding a seeding point in each surface component to polygonise the whole surface. Raposo and Gomes proposed a function factorisation-based solution to overcome this problem. The idea is to factorise the implicit function f into function components $\{f_i\}$, also called *symbolic components* or *irreducible components*, before sampling the corresponding surface.

There are three sorts of symbolic components, namely:

- *One symbolic component matches a topological component.* In this case, a symbolic component corresponds to a topological component. For example, in Figure 6.22, the spherical surface $x^2 + y^2 + z^2 - 4 = 0$ is described by a single symbolic component $f(x,y,z) = x^2 + y^2 + z^2 - 4$, which embodies only one topological component. If we add another disconnected sphere $(x-9)^2 + (y-9)^2 + (z-9)^2 - 9 = 0$, we get a function $f(x,y,z) = (x^2 + y^2 + z^2 - 4).((x-9)^2 + (y-9)^2 + (z-9)^2 - 9)$ with two symbolic components, $f_1(x,y,z) = x^2 + y^2 + z^2 - 4$ and $f_2(x,y,z) = (x-9)^2 + (y-9)^2 + (z-9)^2 - 9$, each corresponding to a single topological component. Therefore, at least in this case, factorisation allows to know in advance the topological shape of the surface, i.e. the number of their topological components.

184   6 Continuation Methods

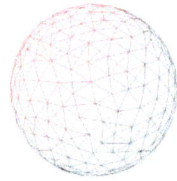 

Fig. 6.22. $f(x,y,z) = x^2 + y^2 + z^2 - 4 = 0$ has a single symbolic component, while its surface has a single topological component.

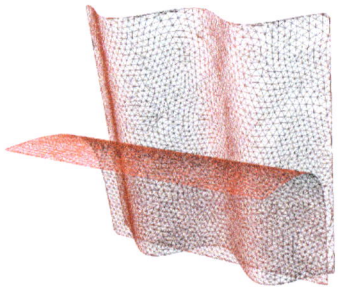 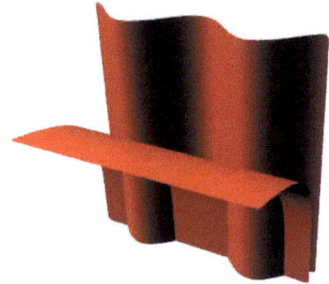

Fig. 6.23. $f(x,y,z) = x \ln x + \ln x \cos z - xy - y \cos z = 0$ has two symbolic components, but its surface has only one topological component.

- *Two or more symbolic components form a topological component.* This case is illustrated in Figure 6.23, where the surface $f(x,y,z) = x \ln x + \ln x \cos z - xy - y \cos z = 0$ has a single topological component with two intersecting symbolic components. These symbolic components are $f_1(x,y,z) = \ln x - y$ and $f_2(x,y,z) = \cos z + x$. Each symbolic component can be tessellated separately. Therefore, we do not need any particular procedure to treat self-intersections for this sort of surface during the triangulation stage, nor an exact symbolic algorithm to determine the intersection curve. A triangle-to-triangle algorithm suffices to polylinearise the intersection curve after triangulating all symbolic components; for example, the Möller algorithm is appropriate for this task [276].
- *One symbolic component possesses two or more topological components.* In this case, the algorithm only detects and tessellates one topological component of the symbolic component. For example, the paraboloid of two sheets $-x^2 - y^2 + z^2 = 1$ shown in Figure 6.24 consists of a single symbolic component with two topological components (say two sheets), but only one sheet is pictured. This is so because, in general, we do not know *a priori* the topological type of a symbolic component; consequently, finding a seeding point on the second topological component or sheet is only possible by chance. However, important results from the computation of the topological type of implicit curves and surfaces may be useful to solve this problem [163, 285, 355].

6.9 Predictor–Corrector Algorithms for Nonmanifold Surfaces

Fig. 6.24. $f(x, y, z) = -x^2 - y^2 + z^2 - 1 = 0$ has a single symbolic component, while its surface has two topological components.

Algorithm 21 Raposo-Gomes' Predictor–Corrector Algorithm for Surfaces
---
1: **procedure** RAPOSO-GOMES($f,\mathcal{S},\Omega,\delta$)
2:     Factorise f into irreducible function components f_i.
3:     **for** $i \leftarrow 0, n$ **do**   ▷ for each symbolic component
4:         HARTMANN($f_i,\mathcal{S},\Omega,\delta$)
5:     **end for**
6:     **for** $i \leftarrow 0, n-1$ **do**   ▷ for every two meshes M_i, M_j of f_i, f_j
7:         **for** $j \leftarrow 1, n$ **do**
8:             MÖLLER(M_i, M_j)   ▷ polyline of curve intersection
9:         **end for**
10:    **end for**
11: **end procedure**

Note that a symbolic component is not the same as a topological component. A function is said to be irreducible if it is non-constant and cannot be represented as the product of two or more nonconstant function components. Every function f can be factorised into irreducible function components $f_1, \ldots, f_n$, being factorisation unique up to permutation of the factors and the multiplication of constants. Thus, we use symbolic factorisation; that is, $f = f_1 \cdot \ldots \cdot f_n$, where each f_i represents a symbolic component of the surface.

Algorithm 21 describes Raposo-Gomes's algorithm. It is essentially a "divide-and-conquer" algorithm because the symbolic factorisation decomposes a function expression into subexpressions (or symbolic components). This way, one tessellates each symbolic component separately instead of the surface as a whole. After polygonising all irreducible components of the surface, one determines their intersection curve (step 3). This can be easily done by using Möller's algorithm [276] to find intersecting triangles of the irreducible components. Thus, neither analytic nor symbolic techniques are necessary to resolve self-intersections. Note that this resolution of singularities (i.e. self-intersections) is important for keeping a valid representation of the surface in the data structure, but it is not necessary for visualisation purposes.

6.10 Final Remarks

In this chapter we have dealt with the class of continuation algorithms, which includes both piecewise linear (PL) and predictor-corrector (PC) algorithms. PL algorithms were mainly developed in the field of numerical analysis using fixed triangulations of domain. In this context, the work of Allgower and colleagues possibly is the one with more impact in computer graphics and computational geometry. In turn, PC algorithms triangulate the surface directly. That is, no need exists for an intermediate triangulation of the domain. But, this intermediate triangulation of PL methods has the advantage that no concern is taken in relation to smoothness of functions. PL methods are thus inherently derivative-free methods.

7
Spatial Partitioning Methods

This chapter deals with spatial partitioning algorithms for rendering implicit surfaces. Typically, these algorithms start with a preliminary space decomposition of the domain (e.g. bounding box) into smaller subdomains or cells (e.g. cubic boxes), discarding those cells that do not intersect the surface. The surface is then polygonised or approximated by one or more polygons within each intersecting cell in order to render it on screen.

7.1 Introduction

Early spatial partitioning algorithms were developed by Wyvill et al. [421] for rendering soft and blobby objects, and Lorensen and Cline [247] who designed the marching cubes algorithm for generating human organ surfaces from medical image data sets. Depending on the input data, these algorithms can be classified as either continuous data-based algorithms or discrete data-based algorithms. Discrete data-based partitioning algorithms have been developed from Lorensen and Cline's algorithm. In this case, no function is known *a priori* so that the algorithm only operates on discrete data at the vertices of a grid, from which a surface is generated by interpolation. In contrast, continuous data-based partitioning algorithms can be viewed as a follow-up of the Wyvill et al. algorithm, as they operate on a given trivariate function such that continuous data can be evaluated at arbitrary points of the domain. In both cases, rendering an implicitly defined surface requires the computation of a polygonal mesh that approximates the surface.

Space partitioning algorithms subdivide (either *uniformly* or *adaptively*) the space into a lattice of cells to find those that intersect the implicit curve or surface. Usually, cells are either squares in $\mathbb{R}^2$ (respectively, cubes in $\mathbb{R}^3$ or n-cubes in $\mathbb{R}^n$) or triangles in $\mathbb{R}^2$ (respectively, tetrahedra in $\mathbb{R}^3$ or n-simplices in $\mathbb{R}^n$). The sign of the implicit function at the cell vertices determines a topological configuration (also known as topological type or pattern) that guides the polygonisation of the surface.

Cubes may lead to ambiguous configurations as more than one mesh can be created for the same configuration type; consequently, the mesh that approximates the surface may be generated with cracks. Some disambiguation strategies have been proposed in the literature, including simplicial decomposition, modified look-up table disambiguation, gradient consistency-based heuristics and quadratic fit, trilinear interpolation techniques, and recursive subdivision of space into smaller cells. Unlike cubes, tetrahedra tend to generate topologically consistent triangular meshes (i.e. without ambiguities), yet with distorted triangles. These distorted triangles require some kind of post-processing procedure to repair the resulting mesh.

7.2 Spatial Exhaustive Enumeration

This family of algorithms partition the space into axis-aligned n-cubes, sometimes called *voxels*. The well-known marching cubes (MC) algorithm belongs to this family. It was designed for the visualisation of the human anatomy. Medical 3D images are composed from uniform 2D slices taken from computerised tomography (CT) scanners—also called computerised axial tomography (CAT) scanners—magnetic resonance imagers (MRI), positron emission tomography (PET) scanners or even ultrasound scanners. These sliced 3D images contain detailed data about human organs that need be extracted and visualised for medical purposes. Many methods have been devised in last two decades and, in a way, explain the emergence of the research field of scientific visualisation.

We can ask ourselves, "Which is the relation between those sliced 3D images and spatial exhaustive enumeration algorithms in computer graphics and geometric modelling?" In fact, a sliced 3D image induces a space decomposition into voxels from which we can extract a cloud of points of a specific human organ. The points of different organs have distinct threshold values so that using a single threshold value over all slices we are able to extract a cluster of points for a particular organ. In other words, given a threshold value and a pack of 2D digital slices generated by some medical 3D scanner as input data source, the algorithm performs the exhaustive enumeration of a rectangular bounding box into voxels, from which points of a human organ surface can be extracted, interpolated and polygonised.

Thus, spatial exhaustive enumeration algorithms can be used to extract human organ surfaces from a pack of 2D pixel images. Note that these medical surfaces are not given *a priori* an algebraic or analytic expression. Such expression is given by the so-called *interpolants* as usual in scientific visualisation, which simply interpolate the medical data inside each voxel. It is convenient here to recall that spatial exhaustive enumeration algorithms apply not only to trilinear interpolants that approximate human organ surfaces, but also to general trilinear surfaces implicitly defined by level sets. The difference is that, instead of feeding the algorithm with a threshold value (also called isovalue) and a rectangular pack of 3D digital slices, the input consists of a real constant (or function value) and the trivariate expression of a real function.

7.2.1 Marching Squares Algorithm

Marching squares (MS) is the 2-dimensional version of marching cubes (MC) algorithm, i.e. the marching 2-cubes algorithm. It applies to only one digital 2D slice, while the MC algorithm applies to a pack of digital 2D slices. Therefore, given a threshold value and a single 2D slice, it generates a contour line by bilinear interpolation. That is, MS is a contour algorithm, and thus applies to many other scientific fields, namely: cartography, weather forecasting, fluid dynamics, etc. As known, the essential concept behind contouring is that of isolines, i.e. lines whose points are associated to equal values, the thresholds. For example, contours may represent lines with different temperatures (isotherms) or pressures (isobars) over the globe, as needed in weather forecasting. Obviously, there are several methods to generate contours, depending on the type of the grid, type of interpolation, and order of curve generation [66, 333, 344, 405]. In this section, the focus is on bilinear interpolation over a rectangular grid of squares.

Bilinear Interpolation

In contouring, we assume that data varies linearly between consecutive data points of a rectangular grid of squares. This assumption seems to be reasonable even when data does not vary linearly since we are able to guarantee a high data resolution in a preprocessing stage.

Bilinear interpolation extends linear interpolation to bivariate functions on a regular grid. The idea is to perform linear interpolation in two distinct directions, one after the other. So, let us determine the value of the unknown function f at the point $P = (x, y)$ of the square $[x_0, x_1] \times [y_0, y_1]$, assuming that the values of f at the corners $P_{00} = (x_0, y_0)$, $P_{01} = (x_0, y_1)$, $P_{10} = (x_1, y_0)$ and $P_{11} = (x_1, y_1)$ are known (Figure 7.1).

The linear interpolation in the x-direction on the horizontal square sides yields

$$f(x, y_0) \approx \frac{x_1 - x}{x_1 - x_0} f(P_{00}) + \frac{x - x_0}{x_1 - x_0} f(P_{10}) \qquad (7.1)$$

and

$$f(x, y_1) \approx \frac{x_1 - x}{x_1 - x_0} f(P_{01}) + \frac{x - x_0}{x_1 - x_0} f(P_{11}) \qquad (7.2)$$

where $x \in [x_0, x_1]$. Now, interpolating these two values in the y-direction we obtain

$$f(x, y) \approx \frac{y_1 - y}{y_1 - y_0} f(x, y_0) + \frac{y - y_0}{y_1 - y_0} f(x, y_1) \qquad (7.3)$$

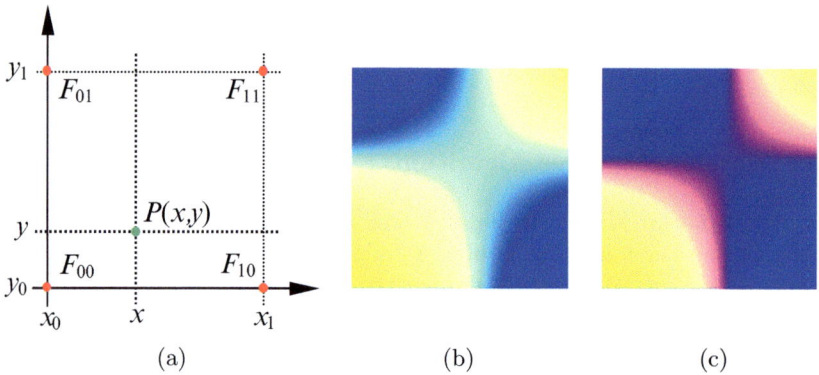

Fig. 7.1. Bilinear interpolation of a 2-cube $[x_1, x_2] \times [y_1, y_2]$ with $f_{00} = 0.0$, $f_{10} = 1.75$, $f_{01} = 1.75$ and $f_{11} = 0.5$: (a) interpolation scheme: first along $[x_1, x_2]$, then along $[y_1, y_2]$; (b) colour interpolation with $(R, G, B)_{\text{START}} = (0.5, 1.0, 0.75)$ and $(R, G, B)_{\text{END}} = (10.0, 5.0, 0.0)$; (c) colour interpolation with $(R, G, B)_{\text{START}} = (0.5, 0.1, 0.75)$ and $(R, G, B)_{\text{END}} = (10.0, 5.0, 0.0)$.

that is, the estimate of $f(x, y)$. Substituting Equation (7.1) and Equation (7.2) in Equation (7.3), we get the bilinear interpolant $F(x, y)$ that approximates $f(x, y)$, say $f(x, y) \approx F(x, y)$ with

$$F(x, y) = \frac{f(P_{00})}{(x_1 - x_0)(y_1 - y_0)}(x_1 - x)(y_1 - y)$$
$$+ \frac{f(P_{10})}{(x_1 - x_0)(y_1 - y_0)}(x - x_0)(y_1 - y)$$
$$+ \frac{f(P_{01})}{(x_1 - x_0)(y_1 - y_0)}(x_1 - x)(y - y_0)$$
$$+ \frac{f(P_{11})}{(x_1 - x_0)(y_1 - y_0)}(x - x_0)(y - y_0) \quad (7.4)$$

The isocontouring problem can be then rewritten as follows: given the values of a bilinear function $F(x, y)$ at the vertices of an axis-aligned square $D = [x_0, x_1] \times [y_0, y_1]$, determine and display isolines corresponding to threshold value c

$$C = \{(x, y) \in D \mid F(x, y) = c\} \quad (7.5)$$

Without loss of generality, we transform the square domain D into a unit square $I = [0, 1] \times [0, 1]$ for convenience, so that the bilinear interpolant is hereafter as follows:

$$F(x, y) = F_{00}(1 - x)(1 - y) + F_{10}x(1 - y) + F_{01}(1 - x)y + F_{11}xy \quad (7.6)$$

after labelling the function values as $F(x, y) = F_{xy}$ for simplicity. This bilinear function can be rewritten as

$$F(x,y) = Axy + Bx + Cy + D \tag{7.7}$$

where

$$\begin{aligned} A &= F_{00} - F_{10} - F_{01} + F_{11} \\ B &= F_{10} - F_{00} \\ C &= F_{01} - F_{00} \\ D &= F_{00} \end{aligned} \tag{7.8}$$

Note that the bilinear interpolant is *not* linear. On the contrary, it is the product $(ax+b)(cy+d) = Axy + Bx + Cy + D$ of two linear functions, where $A = ac$, $B = ad$, $C = bc$ and $D = bd$. The bilinear interpolant is quadratic along any straight line inside D, except along lines parallel to either in the x- or the y-direction where it is linear (simply because either y or x is constant, respectively).

Topological Configurations and Ambiguities

To correctly display the bilinear interpolant within a cell, we need to know its topological configuration inside such a cell. For that, we compute its partial derivatives as

$$\frac{\partial F}{\partial x} = Ay + B \quad \text{and} \quad \frac{\partial F}{\partial y} = Ax + C. \tag{7.9}$$

That is, these derivatives vanish at the stationary point $(-\frac{C}{A}, -\frac{B}{A})$. Besides, the eigenvalues of the Hessian matrix are of opposite signs, $\lambda = \pm A$; as a consequence the stationary point is a saddle point for $A \neq 0$, i.e. the intersection point of two hyperbola asymptotes. In this case, the contour curve is a hyperbola, which has to be approximated by straight line segments. But, if $A = 0$, the interpolant $F(x,y)$ is linear and the contours are just straight lines within the cell so that contouring is exact.

The possible topological configurations of the polylinearised contour curve are shown in Figure 7.2. Ambiguity appears when the contour curve is topologically equivalent to a hyperbola (configurations 5 and 10). We simply do not know how to connect pairs of hyperbola points on the boundary of the square. This happens when positive and negative vertices are diagonally opposed. A square point (x, y) is positive (respectively, negative) when its corresponding data is above (respectively, below) the threshold value c.

This hyperbola ambiguity can be solved by means of two methods. The first is known as the *four triangles method* and seems to be due to Dayhoff [96], Heap [180] and Wyvill et al. [422]. Basically, one computes the function value at the centre of the square. If this value is greater than the threshold value, the separation of the pairs of intersection points is done along the square diagonal that contains the positive vertices; otherwise, we use the diagonal defined by the negative vertices. These two diagonals divide the square into four triangles;

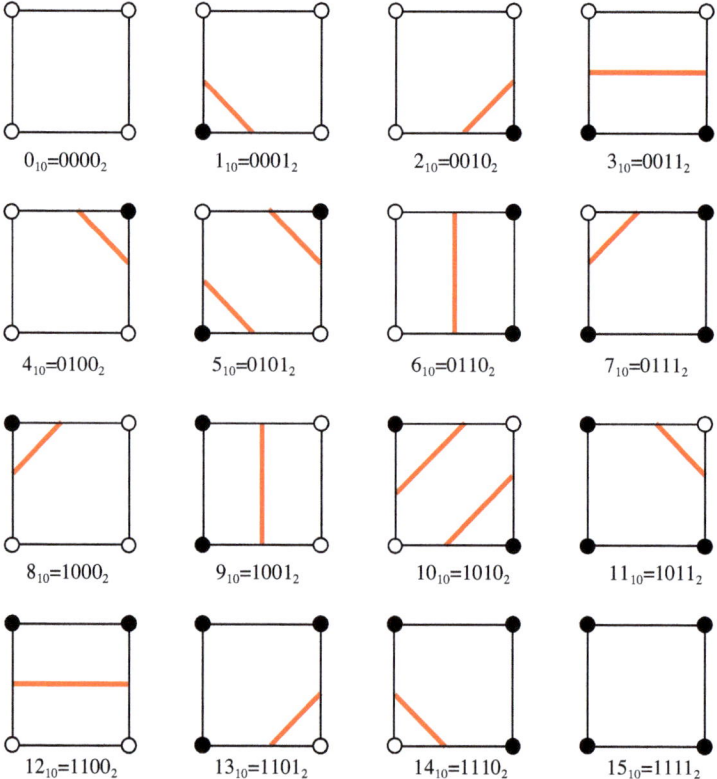

Fig. 7.2. Topological configurations for marching squares.

hence the name of four triangles. Unfortunately, this disambiguation method only works when the function values at the centre of the square and at the saddle point have identical signs.

The second disambiguation method is called *asymptotic decider* and was introduced by Nielson and Hamann [303] to solve ambiguities in the more general context of the MC algorithm. The pairwise connection is done after separating the two pairs of hyperbola points on the boundary of the cell. This is done by locating each of these points in relation to one of the asymptotes; for example, a point (x, y) in on the left of the asymptote $\frac{\partial F}{\partial y} = Ax + C$ if it satisfies $Ax + C < 0$; it is on the right if $Ax + C > 0$. This is an elementary space separation technique that works beautifully.

The Algorithm

Let us then describe the marching 2-cubes algorithm for implicitly defined curves. Recall that an implicit curve in $\mathbb{R}^2$ is defined by the zero set of a real bivariate function $f : \mathbb{R}^2 \rightarrow \mathbb{R}$, i.e. $f(x,y) = 0$. This is slightly different for isocontours because no input function $f(x,y)$ is given *a priori*; instead, one uses a bilinear interpolant $F(x,y) \approx f(x,y)$.

Marching squares algorithm essentially is a "divide and conquer" algorithm. It starts by splitting the axis-aligned rectangular domain or bounding box $\Omega = \Delta X \times \Delta Y$ into a grid of $n \times m$ squares of side length equal to δ. Then, each square is processed individually, that is, one evaluates the function $f(x,y)$ on its four vertices, stores these function values in the data structure, computes the intersection points between the curve and the square edges by using some root-finding method (see Part II), and then "marches" or moves onto the next square. The obvious data structure for this space decomposition is a 2-dimensional array $a[m,n]$ of $m \times n$ elements, in which each element stores the data corresponding to a square.

The crucial steps of the marching squares algorithm (Algorithm 22) are the steps 8 and 9 provided that they determine how the curve crosses each square, i.e. the accurate topological shape within each square. It is clear that this also depends on the square side length δ.

For computing the topological configuration within a square, we use a 4-bit code which encodes the state of each vertex with a single binary digit. If f evaluates negative at a vertex, its bit is set to 0; if f evaluates positive, the corresponding bit is set to 1. Therefore, each topological configuration corresponds to a specific 4-bit code. This code works as an index for a look-up table that stores all possible topological configurations. The data stored in this look-up table is used to correctly polylinearise the curve segment that crosses the square.

Taking into account that each square has four vertices and the function evaluates either positive or negative, we conclude that there are $4^2 = 16$ possible topological configurations within any square, i.e. a curve passes any square in up to different 16 ways. These 16 topological configurations form the lookup

---

Algorithm 22 The Marching Squares

1: **procedure** MARCHINGSQUARES($f,\Omega,\delta,a[m,n]$)
2:     $m \leftarrow \Delta X/\delta$
3:     $n \leftarrow \Delta Y/\delta$
4:     **for** $i \leftarrow 0, m-1$ **do**
5:        **for** $j \leftarrow 0, n-1$ **do**
6:           Create square $\square_{i,j}$.
7:           Evaluate f at each vertex of $\square_{i,j}$.
8:           Set up the topological configuration of the curve within $\square_{i,j}$.
9:           Find roots of f along edges of $\square_{i,j}$.
10:         Polylinearise the curve across $\square_{i,j}$.
11:         $a[i,j] \leftarrow \square_{i,j}$
12:        **end for**
13:     **end for**
14: **end procedure**

table used by the algorithm. These configurations are shown in Figure 7.2. The binary encoding of the vertices is counterclockwise, starting on the bottom-left vertex. For example, the pattern 1 (number base 10) in Figure 7.2 is encoded as 0001 (number base 2), while the pattern 9 (number base 10) is encoded as 1001 (number base 2). This encoding is used to index the pattern table. Once the correct pattern has been established, we can polylinearise the curve within the square with reasonable topological guarantees.

However, it is necessary to keep in mind that the topological configurations in Figure 7.2 are for bilinear interpolants. For more general polynomial functions of degree 3 or higher, the look-up table necessarily grows with new topological configurations and the assumption of linearity along the edges of a square is no longer valid. For example, it is possible to have more than one curve point on a single edge. In this case, a possible solution is to subdivide squares recursively until every square fits some of the those 16 configurations in Figure 7.2. Note that we have not considered the case that occurs when the curve crosses a vertex or the case of a curve self-intersection within a square.

7.2.2 Marching Cubes Algorithm

Marching cubes (MC) algorithm likely is the most used algorithm in scientific visualisation, including applications in medical imaging, bioinformatics, geographical information systems (GIS), weather forecasting, and many others. MC algorithm was introduced by Lorensen and Cline [247] in the context of medical imaging, though a similar algorithm due to Wyvill et al. [422] had been published before in the context of modelling soft objects. The main difference between these two algorithms lies in their spatial indexing data structures. The first uses a voxel-based data structure (i.e. a 3D array that mimics the partitioning of the bounding box into cubes), while the second uses a hash-table structure.

Trilinear Interpolation

Similar to the extraction of isocontours by using bilinear interpolation inside the unit 2-cube, we can extract isosurfaces by interpolating trilinearly values on eight vertices of the unit 3-cube. By generalisation of Equation (7.4), we obtain the trilinear interpolant

$$\begin{aligned}F(x,y,z) = &F_{000}(1-x)(1-y)(1-z) + F_{001}(1-x)(1-y)z \\ &+ F_{010}(1-x)y(1-z) + F_{011}(1-x)yz \\ &+ F_{100}x(1-y)(1-z) + F_{101}x(1-y)z \\ &+ F_{110}xy(1-z) + F_{111}xyz\end{aligned} \quad (7.10)$$

The trilinear interpolant is a cubic polynomial. As before, we are considering here the unit 3-cube $\mathbb{I}^3 = [0,1] \times [0,1] \times [0,1]$ because the extension to general case is done by using simple scaling factors. It is clear that inside each face of the unit 3-cube the values of F vary bilinearly because one of the coordinates remains constant. Along each edge of the 3-cube, F varies linearly as two out three coordinates do not vary.

Therefore, in addition to saddle points in faces (i.e. *face saddles*), there may be saddle points in the interior of each 3-cube, which are called *body saddles*. Recall that saddle points occur where the partial derivatives vanish simultaneously. For face saddles we use the two partial derivatives of the bilinear interpolant given by Equation (7.4) with the appropriate variables in place, while the three partial derivatives of the trilinear interpolant above are used to determine the body saddles of a 3-cube. As shown by Lopes and Brodlie [244] and Natarajan [295], extra topological configurations of face and body saddle points can be used to disambiguate the topological shape of a trilinear isosurface within each 3-cube correctly.

The algorithm

Similar to 2-dimensional contours and curves, a marching cubes algorithm involves a three-stages discretisation of the level set, namely:

- Partition of the bounding box.
- Sampling of the surface.
- Polygonisation of the surface.

The first stage partitions the bounding box into cubes (step 8 of Algorithm 23). In fact, this discretisation of the bounding box need not be done explicitly. There is no need to explicitly store edges and faces for each cube. It is enough to store the vertex data of each cube into a n-dimensional array, where n is the dimension of the cube (or of the space where the level set lies in).

MC algorithm creates and processes each cube at a time. After processing one cube, it moves (or *marches*) to the next one in an axis-aligned grid of equally sized cubes. The simplicity of this algorithm is a result of the one-to-one mapping between the cubes created inside the bounding box and the elements of the array data structure.

The second stage (steps 9–11 of Algorithm 23) concerns the sampling of the surface. Sampling consists in determining which cubes intersect the surface and where. This involves the following sequence of operations for each cube:

$$\text{evaluation} \rightarrow \text{classification} \rightarrow \text{interpolation}.$$

Sampling an implicit surface is the critical part of the algorithm because intersection points usually are determined by numerical interpolation (i.e. 2-point numerical methods), which may fail unless we use some of those certified techniques (e.g. interval arithmetic) described in Part II.

Algorithm 23 The Marching Cubes

1: **procedure** MARCHINGSQUARES($f,\Omega,\delta,a[m,n,o]$)
2:    $m \leftarrow \Delta X/\delta$
3:    $n \leftarrow \Delta Y/\delta$
4:    $o \leftarrow \Delta Z/\delta$
5:    **for** $i \leftarrow 0, m-1$ **do**
6:       **for** $j \leftarrow 0, n-1$ **do**
7:          **for** $k \leftarrow 0, o-1$ **do**
8:             Create cube $\square_{i,j,k}$.
9:             Evaluate f at each vertex of $\square_{i,j,k}$.
10:            Set up topological configuration of the surface within $\square_{i,j,k}$.
11:            Find roots of f along edges of $\square_{i,j,k}$.
12:            Polygonise the surface across $\square_{i,j,k}$.
13:            $a[i,j,k] \leftarrow \square_{i,j,k}$
14:          **end for**
15:       **end for**
16:    **end for**
17: **end procedure**

For implicit surfaces, one first proceeds to the *evaluation* (step 9 of Algorithm 23) of the function at each vertex of the current cube, whose values (either positive or negative) are stored in the corresponding data structures for vertices. Second, one encodes the topological configuration of the surface within the cube, a bit per vertex. Since there are two signs for function values and a 3-cube possesses 8 vertices, we readily come to the conclusion that there are $2^8 = 256$ possible shape configurations of the surface within a cube. In practice, we use a simplified lookup table with fourteen shape configurations (Figure 7.3). In fact, using cube symmetry operations (reflections and rotations), those 256 possible shape configurations are easily reduced to fourteen unique cases.

If a vertex has a function value equal or less than isovalue c of the surface, its bit is set to 0 (marked as ○ in Figure 7.3); otherwise, it is set as 1 (marked as ● in Figure 7.3). A cube edge crosses the surface if its vertices have distinct values (0 and 1). This bit encoding of vertices leads to the *classification* (step 10 of Algorithm 23) of the topological shape of the surface within each cube, as illustrated in Figure 7.3. This classification eases the *interpolation* (step 11 of Algorithm 23) of edges which do intersect the surface, reducing the processing workload to a minimum as non-transverse edges need not be processed.

Note that, for 3D medical images (e.g. MRI), no function is evaluated on the vertices because the data values are given by eight pixels, four each from two consecutive digital slices of a sliced volumetric data set. That is, sampling reduces to two stages: classification and interpolation. For that, we use a trilinear interpolant $F(x,y,z)$ to reconstruct the surface corresponding to a given threshold value. This interpolant is a cubic polynomial. Therefore,

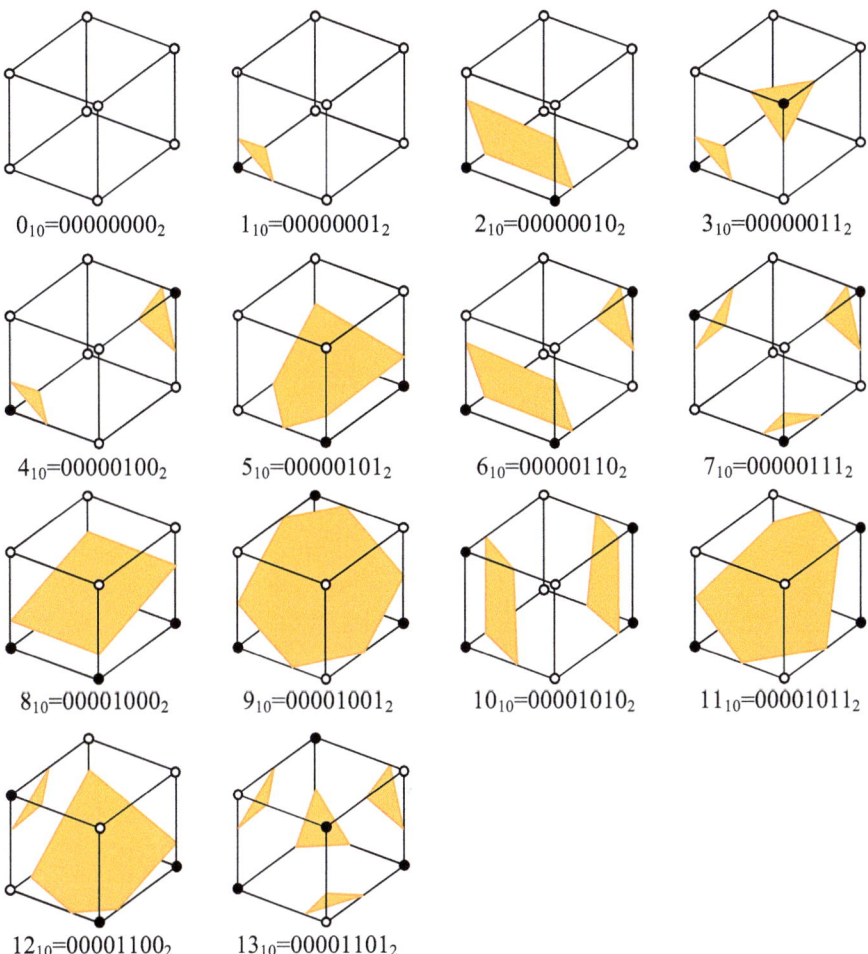

Fig. 7.3. Topological configurations for marching cubes.

intersection points between the surface (described by the interpolant) and voxel edges usually are found by linear interpolation because we only use the data values of the pixels of each slice to reconstruct the surface in a volumetric data set.

Found the interpolated points, only the polygonisation of the surface (third stage) remains to be done. The third stage reduces to triangulate the polygons within each cube before rendering the triangular mesh that approximates the surface (step 12 of Algorithm 23). For example, the quadrangles of the cases 2, 6, 8 and 10 can be easily decomposed into two triangles each. Thus, MC algorithm generates a mesh that approximates an implicit surface as described in Algorithm 23.

Ambiguities

MC algorithm does not offer topological guarantees. The polygonisation of the surface can be foiled up by eventual shape topological ambiguities that the algorithm cannot detect or solve. There are two types of ambiguities:

- Face ambiguities.
- Interior ambiguities.

Let us look at the 2D ambiguous configurations 5 and 10 in Figure 7.2. These configurations are ambiguous because they possess four intersection points on the boundary of a square. Disambiguation in 2D is then a matter of selecting the right pairs of intersection points to connect. In 3D, face ambiguities of a cube also occur when all its four edges intersect the surface. In this case, the triangulation procedure has to determine which pairs of intersection points to connect. If pairs are wrongly formed, "holes" or cracks may appear through the surface mesh when we try to merge the triangle edges of adjacent cubes. This first problem was pointed out by Dürst [122] and arises when the topological configurations of adjacent cubes do not match, as illustrated in Figure 7.4. In Figure 7.4(a), the adjacent cells possess matching topological configurations so that the surface will be polygonised without cracks in the shared face. Note that matching topological configurations mean matching polarity of vertices on the shared face. On the contrary, in Figure 7.4(b), there is not such a matching because the 3-type cell appears rotated 90 degrees in relation to its position in Figure 7.4(a); hence, the cracking phenomenon. That is, the lack of the shape continuity or matching between adjacent cells leads to the appearance of cracks in the final polygonised surface.

There are a couple of face disambiguation techniques. They are exactly those seen above for marching squares in 2D. The first, called the four triangles technique, was proposed by Wyvill et al. [422] in the context of isosurfacing in computer graphics. The second is due to Nielson and Hamann [303] and is called asymptotic decider. Recall that the asymptotic decider is based on the saddle point value of the bilinear interpolant to carry out the correct

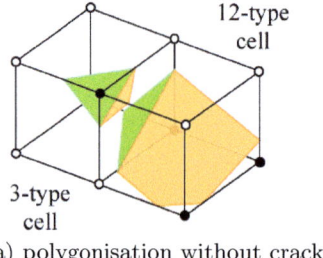
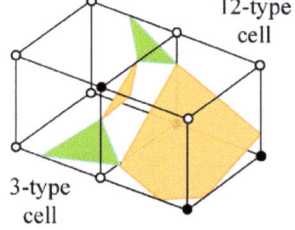

(a) polygonisation without cracks   (b) polygonisation with a crack

Fig. 7.4. Matching topological configurations of adjacent cubes.

7.2 Spatial Exhaustive Enumeration

connections between pairs of intersection points on an ambiguous face. Both techniques guarantee the continuity between cells. However, the four triangles technique not always guarantees the topological correctness of the surface on the domain boundary. Fortunately, the unlike the asymptotic decider does.

As van Gelder and Wilhelms [395] pointed out, there is continuity between cells if and only if each triangle edge is shared by exactly two triangles, except for those triangle edges lying on the bounding box boundary. Otherwise, cracks will appear through the surface mesh. The topological shape is certainly incorrect if not continuous. However, continuity is a necessary, but not sufficient, condition for getting a surface mesh with topological guarantees.

Recall that the asymptotic decider only aims to solve face ambiguities. But, other ambiguities may occur in the interior of a cube. In fact, Natarajan [295] and Chernyaev [84] independently noted that additional ambiguities may appear in the representation of the trilinear interpolant in the cube interior. This may even happen when the cube has no ambiguous faces.

For example, configuration 10 (Figure 7.3) has no ambiguous faces but it admits at least two different shapes, as shown in Figure 7.5 (recall that data varies trilinearly within the cell). The first shape is the usual one with two separate components, while the second is a simple component with a tunnel through it.

In [76], Natarajan proposes a similar method to the asymptotic decider to detect the existence of internal tunnels. For that, he uses the concept of body saddle point as an extension of 2D saddle point, i.e. a point at which all the three first derivatives of the trilinear interpolant vanish. So, for configuration 10, if the function evaluates negative at the body saddle point (i.e. it has opposite sign to the two marked positive vertices), the surface has two connected components inside the cube. If the body saddle point is positive, then there is a tunnel between those two marked positive vertices. Natarajan also indicates that internal tunnels may appear in configurations 4, 6, 7, 10, 12 and 13. Chernyaev [84] uses a different disambiguation strategy, but the results are essentially the same.

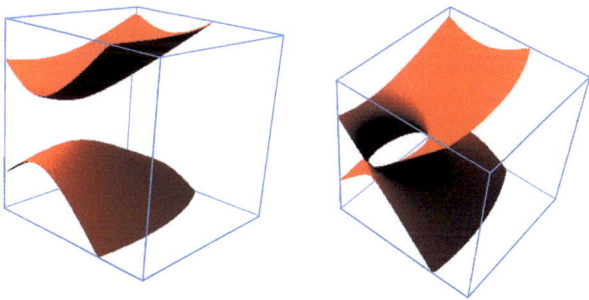

Fig. 7.5. Ambiguities in the cube interior.

Matveyev [261] also addresses the interior ambiguity problem, being the interior ambiguities resolved by inspecting the behaviour of the trilinear function along the cell diagonals. van Gelder and Wilhelms [395] propose a disambiguation technique in the interior of a cube, but this technique requires data beyond the extent of the cube itself, i.e. data from the surrounding cubes. It is a very time-consuming technique just to be used for disambiguation, with the further disadvantage that it must be applied to both nonambiguous and ambiguous cubes. This is troublesome because discontinuities may appear when one applies linear interpolation to an nonambiguous cube and cubic interpolation to an adjacent and ambiguous one. van Gelder and Wilhelms [395] propose other disambiguation techniques that use the gradient vector at the cube vertices to study how the function behaves across the domain. In fact, the gradient vector, which is normal to the surface, indicates the direction along which the function rises most rapidly, being its magnitude that determines how quickly the function rises in that direction.

Cignoni et al. [85] propose a disambiguation strategy based on an adaptive mesh refinement in order to get a very accurate representation for trilinear isosurfaces. For that, a new, exhaustive look-up table (ELUT) was designed to encode multi-entry patterns for each ambiguous configuration. Once again, in [301], Nielson extends his own work by presenting a more precise characterisation and classification of the isosurfaces of trilinear functions. Based on these results, he presents a new polygonisation algorithm that outputs a triangular mesh that approximates isosurfaces for data given on a 3D rectilinear grid. Lopes [243] and Lopes and Brodlie [244] also discusses and proposes accurate disambiguation techniques using additional points on the boundary and interior of the cube. Lewiner et al. [232] describes an efficient implementation of marching cubes with topological guarantees. Recently, Renbo et al. [337] have provided a robust and topologically correct MC algorithm without using the conventional look-up table.

In short, several techniques have been devised to overcome shape ambiguity problems on the boundary and interior of marching cubes. In addition to these local disambiguation techniques, various global solutions and algorithms have been proposed in the literature to solve these ambiguity problems. Two of these algorithms are the *dividing-cubes* and the *marching-tetrahedra*, which can be viewed as variants of the marching cubes.

7.2.3 Dividing Cubes

Dividing cubes algorithm was proposed by Cline et al.[86] and is a variant of marching cubes. It was introduced in the context of the production of 3D medical images, i.e. surface reconstruction and rendering, in order to bypass the scan conversion step of polygonal rendering algorithm.

Dividing cubes differs from marching cubes in that each cube is divided into pixel-sized cubes, also called pixel-sized voxels. This division depends on both image and data resolution. Each voxel is classified as being inside, outside or intersecting the surface with reference to the threshold value of the isosurface. As usual, it is this threshold value that determines which human organ will be visualised. But, unlike marching cubes, the sampling stage reduces to check whether each of those pixel-sized cubes belong to the isosurface or not. There is no need for setting topological configurations for cubes, neither applying numerical interpolation to find surface points between vertices. This explains why there is no concern about the shape ambiguities over cells.

After extracting the surface through this pixel-sized sampling, the visualisation of the surface is straightforward. The algorithm generates a single surface point for each pixel-sized voxel that intersects the surface. Such a point is the centre of this pixel-sized voxel. Then, one computes the gradient vector at the voxel centre point by interpolating the gradients on its eight vertices in order to display it according to the Phong shading model.

Thus, the idea is to approximate the surface by a cloud of points instead of a mesh of triangles. Displaying point primitives is more efficient than triangles in terms of memory and time because point primitives can be displayed on the raster directly. This means that the polygonisation stage of the marching cubes is no longer necessary. This is particularly adequate in high-resolution medical imaging, as the density of triangles increases in such a way that the size of each triangle decreases and tends to the pixel size. Consequently, rendering points instead of triangles pays off in terms of computational cost.

7.2.4 Marching Tetrahedra

Marching tetrahedra (MT) is an algorithm for computing a triangular mesh that approximates an isosurface in a 3D volume. It is another attempt to solve the ambiguities of the marching cubes. Where the marching cubes algorithm decomposes the 3D volume into cubic cells, the marching tetrahedra algorithm performs such a decomposition into tetrahedral cells or tetrahedra.

Tetrahedral Decompositions of a Cube

There are exactly 74 triangulations of the 3-cube, which fall into six classes of combinatorially different types [44, 107, 225]. Representatives of three of these classes are shown in Figures 7.6–7.8. All these triangulations of the 3-cube are regular. But, for higher-dimensional cubes, d-cubes ($d \geq 4$), not all triangulations are regular [107].

Interestingly, the smallest size of a triangulation of the 3-cube that slices off its vertices is 5 [198]. Such a minimal triangulation is shown in Figure 7.6. Recall that the size of a triangulation is the number of its higher-dimensional simplices. It is also known that the maximum size of a triangulation of the d-cube is $d!$ [225]. Therefore, any maximal triangulation of the 3-cube has size 6, as those depicted in Figure 7.7 and Figure 7.8.

202   7 Spatial Partitioning Methods

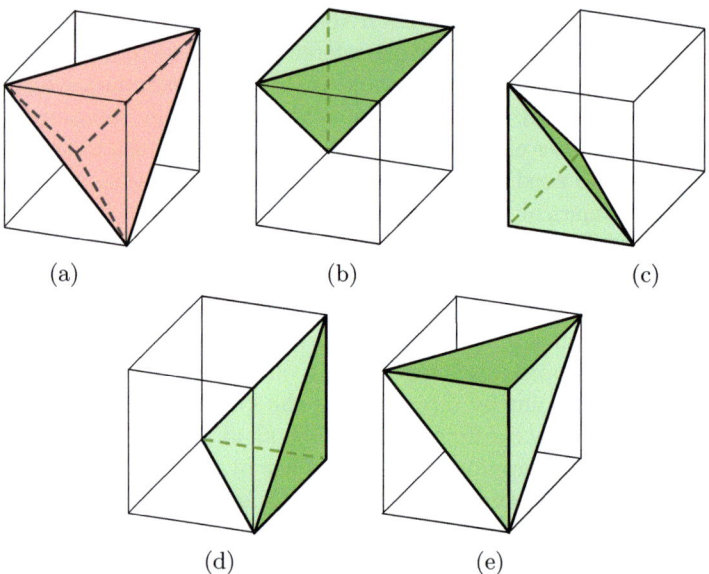

Fig. 7.6. 5-decomposition of a cube into tetrahedra: (a) one equilateral tetrahedron and (b)-(e) four cubic tetrahedra.

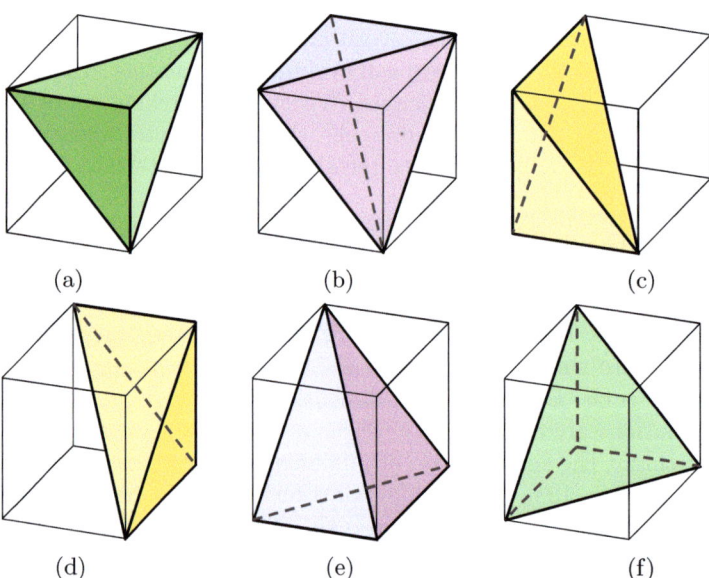

Fig. 7.7. 6-decomposition of a cube into tetrahedra.

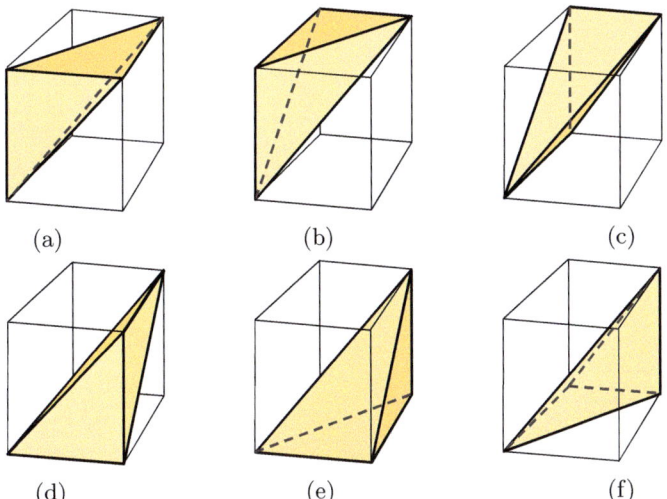

Fig. 7.8. Kuhn decomposition of a cube into tetrahedra.

Piecewise Linear Interpolation

Using piecewise linear interpolation aims at solving ambiguities within n-dimensional hypercubes. Each hypercube is divided into $d!$ smaller pieces, called simplexes. Recall that a simplex is a convex region bounded by hyperplanes of lower dimension.

The interpolation varies linearly over each simplex. Again, for simplicity, we only consider here the unit hypercube $\mathbf{I}^d = [0, 1] \times [0, 1] \times \ldots \times [0, 1]$; the general case simply requires scaling factors. Let us first consider the 2-dimensional case, i.e. the unit square in Figure 7.9(a). Let us also label the values of the linear interpolant at each of the vertices as $F(0,0) = F_{00}$, $F(1,0) = F_{10}$, $F(1,1) = F_{11}$, and $F(0,1) = F_{01}$.

The diagonal line in Figure 7.9(a) slices the square into $2! = 2$ triangles: the lower triangle is the region $0 \leq y \leq x \leq 1$, and the upper triangle is the region $0 \leq x \leq y \leq 1$. The linear interpolant over the lower triangle is given by the following expression:

$$F(x,y) = F_{00} + (F_{10} - F_{00})x + (F_{11} - F_{10})y. \tag{7.11}$$

Analogously, in the upper triangle, the linear interpolant is

$$F(x,y) = F_{00} + (F_{01} - F_{00})y + (F_{11} - F_{01})x. \tag{7.12}$$

Let us consider now the linear interpolation over the unit cube $\mathbf{I}^3 = [0, 1]^3$, as shown in Figure 7.9(b). In this case, the cube is sliced into $3! = 6$ tetrahedra, as for example the Kuhn triangulation shown in Figure 7.8. Let us consider,

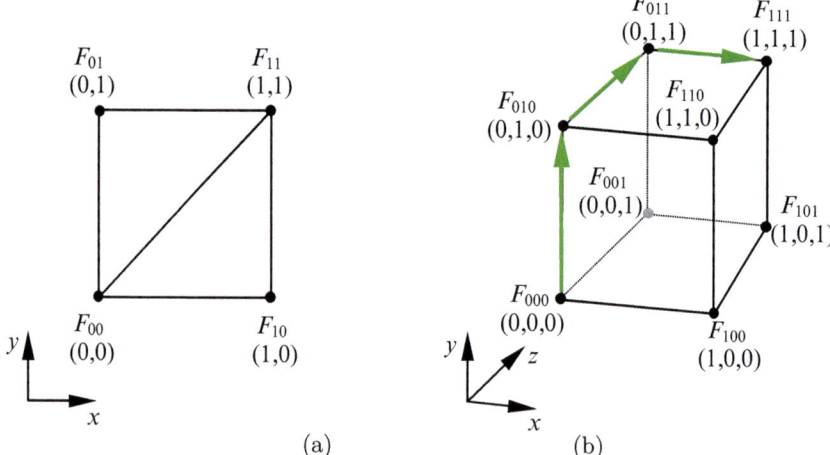

Fig. 7.9. Piecewise interpolation of the unit 2-cube and the 3-cube.

for example, the tetrahedron $0 \leq x \leq z \leq y \leq 1$ (Figure 7.8(b)). As illustrated by the arrows on the edges of the cube depicted in Figure 7.9(b), we follow the path by sorting the variable values in ascending order, say $y \geq z \geq x$, so that the interpolation formula for every point within this tetrahedron is:

$$F(x,y,z) = F_{000} + (F_{010} - F_{000})y + (F_{011} - F_{010})z + (F_{111} - F_{011})x. \quad (7.13)$$

Note that the function values F_{000}, F_{010}, F_{011}, and F_{111} are picked up by following the ascending order: first y, second z, and then x. Note that every point within a tetrahedron has the same ascending order. Therefore, it is straightforward to obtain interpolation formulas for the remaining five tetrahedra inside the cube. And, more importantly, this easily generalises to higher dimensions.

Recall that the main problem with cubic grids—whose scalar data is stored at the vertices—is that the linear interpolation over a cube may produce ambiguous surface configurations inside such a cube. The idea of splitting cubes into tetrahedra aims at solving such ambiguities because the linear interpolation over a tetrahedron becomes unique and the isosurfaces between any two neighbouring tetrahedra are conformal [409]. Note that we are here assuming that the interpolation scheme acts on scalar values at the vertices so that the isovalues inside a tetrahedron correspond to isolevel planes, as shown in Figure 7.10. This means that the transition between two neighbouring tetrahedra is not differentiable, but it is scalar-value conformal, as typical for piecewise linear interpolation.

As noted above, up to symmetry, there are six possible ways of decomposing a regular cube into tetrahedra. These six combinatorial classes fall into two

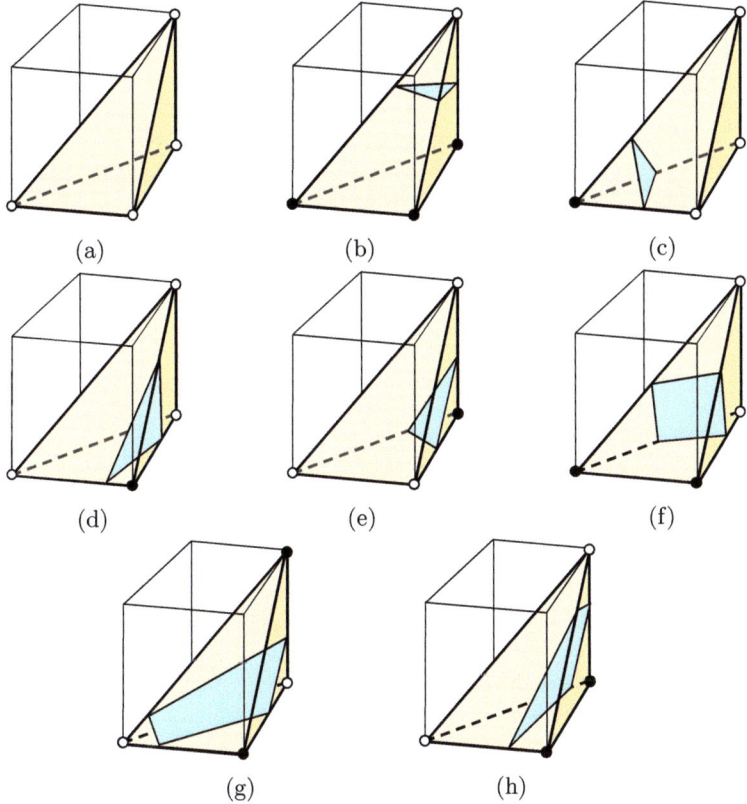

Fig. 7.10. Shape configurations inside a tetrahedron.

families with reference to their size: the 5-tetrahedral decomposition and the 6-tetrahedral decomposition. The 5-tetrahedral decomposition corresponds to the minimal triangulation of a cube into five tetrahedra (Figure 7.6), which yields an orientation switch of two opposite diagonal face edges of the cube. Consequently, the tessellation of the isosurface inside a given cube forces a particular tessellation of all neighbouring cubes in order to guarantee a conformal mesh. To be more specific, if such 5-tetrahedral cubes are stacked together to a chain, the mesh in each cube must be rotated by an angle of $\pi/2$.

In the case of a 6-tetrahedral decomposition (i.e. a maximal tetrahedralisation) of a cube, opposite face edges have identical orientations. This means that we can use the same procedure and direction of cutting a cube into tetrahedra so that non-conformal mesh tessellations vanish completely. Looking at Figure 7.8, we see that the Kuhn decomposition of a cube into six tetrahedra is obtained by splitting diagonally through the three pairs of opposing faces. In addition to the twelve edges of the cube, we now have more six face diagonals, and the main diagonal. Similar to marching cubes, the intersections of these 19 edges with the isosurface are approximated by linear interpolation of the values at the grid corners. Note that all faces of the original cube are now divided into two triangles, so that adjacent cubes share all edges in the common face. This nice property prevents the appearance of cracks in the rendered surface. This is very important for maintaining topological consistency because interpolation of the two distinct diagonals of a face usually produces different intersection points. Another advantage is that up to five of computed intersection points (including their surface normals and other graphics attributes) can be reused when it comes the turn of processing the neighbour cube.

The Marching Tetrahedra Algorithm

Using tetrahedra has the following advantages:

- *Generality*. It works on both unstructured and structured meshes. This makes the marching tetrahedra a generic solution for isosurface extraction on all grid types. Recall that a structured mesh admits the standard decomposition of a cube into five tetrahedra [3, 49, 115, 169, 172, 262, 302, 304, 305, 320, 366] or six tetrahedra [3, 10, 302, 305] without adding supplementary points. On the contrary, an unstructured mesh results from decomposing a cube into tetrahedra with reference to some supplementary point. By adding the cube centroid as a supplementary point, we can produce a 12-tetrahedral subdivision after splitting each cube face by a single diagonal [3, 49, 75, 81]. We can then progressively add centroids to faces, splitting each face into four triangles in order to produce 14, 16, 18, 20, 24 or 48 tetrahedra [3, 34, 427]. In particular, the 24-tetrahedral subdivision of a tetrahedron appears in a number of works [3, 10, 156, 407], and is known as barycentric subdivision (BCS). In general, the BCS of an n-dimensional simplex consists of $(n+1)!$ simplices; hence a tetrahedron or 3-simplex is BCS-decomposed into 24 tetrahedra.
- *Disambiguation*. The second advantage is that tetrahedra are less prone to shape ambiguities. The main reason behind this is that the number of surface configurations in a tetrahedron is far less than the number of configurations inside a cube. In fact, taking into account that each tetrahedron has only four vertices, we can say that there are only 16 topological configurations, which can be reduced to eight by symmetry, as illustrated in Figure 7.10. These eight cases can be even reduced down to three cases by using rotations. The first case is the topological pattern 0 (Figure 7.10(a)), where no surface intersects the tetrahedron. Note that the filled and hollow circles at the vertices indicate that the vertices are on different sides of the surface. The cases 1, 2, 3 and 4 shown in Figure 7.10(b)-(e), respectively, represent the same topological pattern of a surface triangle defined on three faces of the tetrahedron. Finally, the cases 5–7 shown in Figure 7.10(d)-(g) represent the topological pattern of a convex quadrangle

defined on the four faces of the tetrahedron; each quadrangle is usually divided into two triangles for polygonisation purposes.

The tetrahedral decomposition of the cube ends up with a set of tetrahedra within which the isosurface is correctly drawn as a plane. Note that we are here assuming that a linear model is being used. However, if the data vary trilinearly within the cubic cell, as is the case in the MC, then such a tetrahedral decomposition may be not free of ambiguities. Therefore, it is not correct to assume linear variation of data along the edges of a tetrahedron [243]. That is, no claim can be made about the automatic removal of ambiguities of MC by simply decomposing cubes into tetrahedra.

The marching tetrahedra algorithm was first suggested by Shirley and Tuchman [366]. See also Bloomenthal [50] for an elegant implementation of a tetrahedral polygoniser. This algorithm is essentially the marching cubes algorithm (Algorithm 23) with the 5-tetrahedral decomposition step for each cube. Obviously, the look-up table has now three unique entries for topological configurations, and the surface within each tetrahedron is approximated by two triangles at maximum.

7.3 Spatial Continuation

Spatial continuation is a hybrid scheme that combines exhaustive enumeration and continuation. The spatial partitioning is driven by a continuation scheme, as that one presented by Wyvill et al. [422]. Continuation consists of producing new transverse cubes (i.e. cubes intersected by the implicit surface) incrementally from a seeding cube which straddles the surface (Figure 7.11);

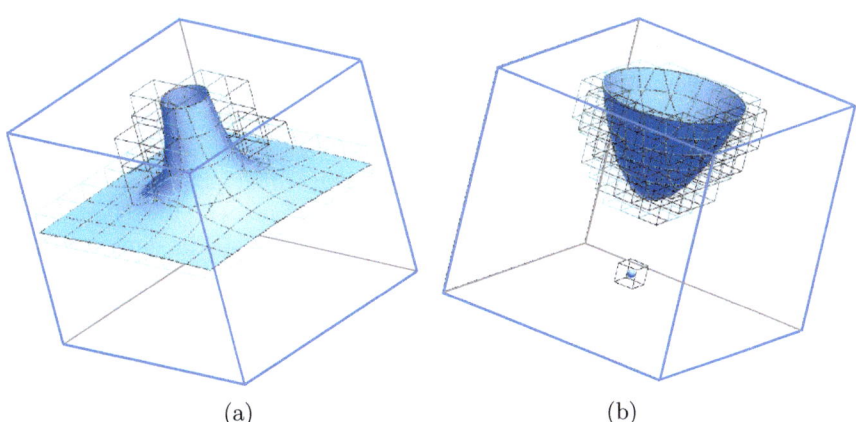

(a)   (b)

Fig. 7.11. Spatial continuation on two implicit surfaces: (a) $z - \frac{1}{x^2+y^2} = 0$; (b) $(z - (x^2 + y^2))(x^2 + y^2 + (z+2)^2) = 0$.

this process continues until the entire surface is enclosed by the collection of cubes [50, 52]. The surface within each cube is then polygonised, i.e. the surface patch inside each cube is approximated by one or more polygons, as described [49]. As expected, this "marching cubes" method also produces cracks in the surface because of the ambiguities described above. An implementation in C of this method is presented by Bloomenthal [50].

These algorithms combine the principles of both spatial exhaustive enumeration and continuation. As a consequence, the main problem of continuation algorithms—i.e. the computation of at least one seeding point on each component of the surface—may be then solved by applying interval arithmetic to axially aligned rectangular boxes belonging to the complement of the union of surface-straddling cubes inside the bounding box.

7.4 Spatial Subdivision

Subdivision is an *adaptive* space partitioning technique. It is another attempt to solve the ambiguity problems resulting from the use of regular space grids. Before proceeding, let us recall that an *implicit object* is defined as the zero set of a real function $f : \Omega \subseteq \mathbb{R}^n \to \mathbb{R}$, i.e. it is the solution set of an equation $f(\mathbf{p}) = 0$. For well-behaved functions, this zero set is a $(n-1)$-dimensional variety in $\mathbb{R}^n$; in particular, such a zero set is an implicit curve in $\mathbb{R}^2$ or an implicit surface in $\mathbb{R}^3$.

7.4.1 Quadtree Subdivision

This section shows how to achieve an *adaptive* polygonal approximation to a curve implicitly defined in $\mathbb{R}^2$ as follows:

$$\mathcal{C} = \{(x,y) \in \Omega \subseteq \mathbb{R}^2 : f(x,y) = 0\} \qquad (7.14)$$

where Ω is the domain given by an axis-aligned bounding box. Following Lopes et al. [245], what we mean by *adaptive* is twofold: first, the subdivision of the bounding box Ω into smaller boxes is more intensive or finer near the curve $\mathcal{C}$; second, the polygonal approximation is curvature-adaptive, i.e. the higher the curvature of $\mathcal{C}$, the finer is the quadtree subdivision (Figure 7.12).

The advantage of the quadtree subdivision over the spatial enumeration is that the size of the boxes is shape-adaptive so that eventual shape ambiguities are resolved by further subdivision. For example, to make sure that the curve depicted in Figure 7.12(b) does not self-intersect on the positive x-axis and near to the origin, the quadtree has been subdivided down to a finer resolution around there.

However, even so, if the resolution of the subdivision—i.e. the minimum size of boxes—is not enough, some small components and isolated points of the curve may remain undetected and are missed. In other words, the topological shape of the curve may be not preserved. The obvious solution for

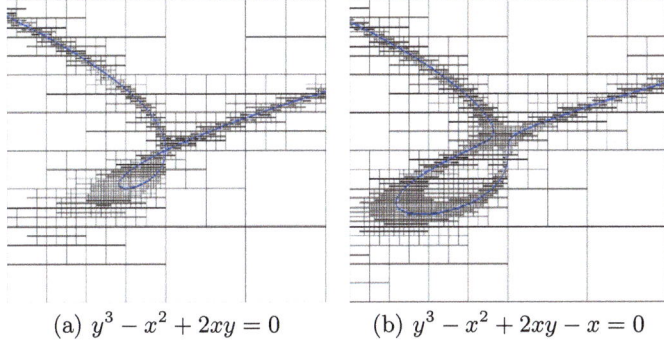

Fig. 7.12. Quadtree subdivision for two implicit curves.

this problem is to use interval arithmetic or affine arithmetic (see Chapter 4 for more details). Doing so, we add robustness to the curve polylineariser, in which the interval arithmetic plays the role of *curve locator* within each box. The curve exists inside a box if f takes on the value 0 over two perpendicular intervals or sides of a square. These boxes are called zero boxes. However, as seen in Chapter 4, there may be false zero boxes, and the results are even worse if floating-point computations are involved.

Note that a curve locator (e.g. interval arithmetic, affine arithmetic, or any of their variants) is not used to sample the curve because that would require to recursively subdivide a box down to a nearly infinitesimal resolution. Instead, we use a *root finder* to compute the curve points that result from the intersection between the curve and the edges of each zero box. Usually, such a root finder builds on some classical numerical method (e.g. bisection method, false position method or Newton's method), but there is no impediment to the usage of a symbolic root finder (e.g. Bézier root finder) based on the Descartes rule. However, most symbolic root finders only apply to polynomials, not to generic real functions.

The quadtree subdivision algorithm for implicit curves is described in Algorithm 24. This algorithm has three subdivision stopping conditions:

- *Inexistence of curve components.* Testing the existence of any curve segment inside a box $\Box_i$ is done through interval arithmetic. This criterion appears at the step 4 and discards the boxes of the quadtree that do not contain any curve component or segment. In fact, the box exclusion test $0 \notin \text{Image}(\Box_i)$ guarantees that only empty boxes (i.e. boxes without any segment of the curve $f^{-1}(0)$) are immediately discarded. However, it may happen that—as explained in Chapter 4—not only true zero boxes, but also some false zero boxes will be considered for subdivision, i.e. there may be redundant and unnecessary subdivision of some boxes.
- *Maximum resolution.* The maximum resolution is the admissible minimum size of the boxes. When the size of a box falls below a given threshold Δ,

210   7 Spatial Partitioning Methods

Algorithm 24 Quadtree Subdivision Algorithm for Implicit Curves
---
1: **procedure** QUADTREE-BASEDIMPLICITCURVE($f,\mathcal{C},\Omega,\Delta,\tau$)
2:    Subdivide Ω into 4 equally sized boxes $\square_i$
3:    **for** $i \leftarrow 0, 3$ **do**
4:       **if** $0 \notin \text{Image}_f(\square_i)$ **then**   ▷ box exclusion test
5:          Discard $\square_i$
6:       **else**
7:          **if** $(\text{size}(\square_i) < \Delta) \vee (\text{curvature}(\mathcal{C}) < \tau)$ **then**
8:             Find roots of f along edges of $\square_i$ ▷ curve points $\mathcal{C} \cap \text{Fr}(\square_i)$
9:             Polylinearise the curve across $\square_i$
10:          **else**
11:             QUADTREE-BASEDIMPLICITCURVE($f,\mathcal{C},\square_i,\Delta,\tau$)
12:          **end if**
13:       **end if**
14:    **end for**
15: **end procedure**

---

the recursive subdivision stops and the box becomes a leaf box of the quadtree.

- *Minimum curvature.* The minimum curvature τ of the curve inside a given box works as a threshold below which the subdivision also stops. The idea is to stop subdividing a box when a curve segment inside it is approximately flat.

Note that Algorithm 24 also has the classical structure of a space partitioning algorithm, namely: partitioning (step 2), sampling (step 8), and polylinearisation (step 9). It is a robust algorithm because it uses interval arithmetic as a fast and robust discarder of empty boxes (i.e. boxes that do not contain any curve segment). However, it is not strictly necessary to use the interval arithmetic as a discarder of empty boxes. By evaluating f at the vertices of a given box $\square_i$, we are able, in principle, to check whether a box is empty or not. In fact, if f does not change sign at the vertices of $\square_i$, we conclude that $\square_i$ is an empty box. But, this alternative technique for checking the transversality of the curve within a box fails if a small component of the curve lies entirely in a box; hence the use of interval arithmetic.

It seems that Suffern [377] was who first tried to use adaptive enumeration, instead of full enumeration, to approximate implicit curves. Shortly afterwards, Suffern and Fackerell [379] introduced interval arithmetic as a robust support for the enumeration of implicit curves, whose algorithm is essentially the Algorithm 24. Nevertheless, the credit of the first application of interval arithmetic in computer graphics is due to Mudur and Koparkar [287]. In [370, 371], Snyder describes a geometric modelling system based on interval arithmetic, which includes an approximation algorithm for implicit curves, but the corresponding quadtree decomposition is not adapted to the curvature. These pioneering works on interval methods in computer graphics have given rise to interesting research results, in particular to the development of several variants of interval arithmetic (e.g. affine arithmetic [17, 90, 102]).

7.4.2 Octree Subdivision

Similar to 2D implicit curves, a 3D implicit surface $\mathcal{S}$ is defined as a zero set of some real function f, whose domain in now in $\mathbb{R}^3$:

$$\mathcal{S} = \{(x, y, z) \in \Omega \subseteq \mathbb{R}^3 : f(x, y, z) = 0\} \tag{7.15}$$

where Ω is the domain given by an axis-aligned bounding box.

As for quadtrees, the idea behind the octree subdivision is to provide an *adaptive* approximation to implicit surfaces. That is, a cubic box through which the surface passes is subdivided into eight smaller boxes. These smaller cubes are stored into an octree data structure (see Chapter 2 for further details). Therefore, the first stopping criterion for an octree approximation to an implicit surface is the emptiness of a given octree box (i.e. the box exclusion criterion). Also, the curvature of the surface within a box and box resolution (i.e. a box has reached its minimum size) can work as stopping criteria for the octree subdivision (step 7 of Algorithm 25). Other criteria appear listed in [49]. Their importance come from the fact that they reinforce the *adaptivity* of the approximation to the implicit surface.

However, these adaptive criteria are not sufficient to resolve all the topological ambiguities. Note that a leaf box is topologically unambiguous if the surface can be approximated by a single polygon within such a box. Topological unambiguity may work as the fourth stopping criterion of the octree subdivision. But, resolving all ambiguities may not be an easy task. For example, it is not easy to distinguish a surface consisting of two touching spheres

---

Algorithm 25 Octree Subdivision Algorithm for Implicit Surfaces

1: **procedure** BLOOMENTHAL(f,$\mathcal{S}$,Ω,Δ,τ)
2:     Subdivide Ω into eight equally sized boxes $\square_i$
3:     **for** $i \leftarrow 0, 7$ **do**
4:         **if** $0 \notin \text{Image}_f(\square_i)$ **then**   ▷ box exclusion test
5:             Discard $\square_i$
6:         **else**
7:             **if** (size($\square_i$) < Δ) $\vee$ (curvature($\mathcal{S}$) < τ) **then**
8:                 Find roots of f along edges of $\square_i$ ▷ surface points $\mathcal{S} \cap \text{Fr}(\square_i)$
9:                 Polygonise the surface $\mathcal{S}$ across $\square_i$
10:            **else**
11:                BLOOMENTHAL(f,$\mathcal{S}$,$\square_i$,Δ,τ)
12:            **end if**
13:         **end if**
14:     **end for**
15: **end procedure**

from another having two *almost* touching spheres inside a given box; the first has only one component, while the second possesses two components. The problem here is that likely the polygonisation of the surface will be messed-up inside the box where the surface-touching point lies in.

Satisfied the subdivision stopping conditions (step 7 of Algorithm 25), one starts the polygonisation (or triangulation) of the surface within each leaf cube (step 9), after which the surface mesh is ready to be rendered. Algorithm 25 is essentially Bloomenthal's algorithm [49]. Possibly, this is the earliest work on adaptive approximation of implicit surfaces through cubic boxes. Bloomenthal's algorithm does not use interval methods for robustness; instead, Bloomenthal's algorithm simply inspects the function at the vertices of a box to check whether the box intersects the surface. This function inspection technique is not obviously robust because small components entirely inside a box are certainly missed. The box emptiness test (steps 4 of Algorithm 25), as opposed to the transversality test (i.e. the intersection between a box and the surface), is generically written in order to comprise both these robust and non-robust solutions. A robust and accurate computation of transversality can be carried out using not only interval arithmetic, but also Lipschitz constants [185, 207], or even derivative bounds [177].

In the line of the Bloomenthal algorithm, other octree-based recursive space subdivision have been developed and proposed in the literature [24, 316, 325, 352, 376, 378, 379]. In [379], Suffern and Fackerell proposed the first robust implementation of Bloomenthal's algorithm by using interval arithmetic. In [24], Balsys and Suffern improved the crack removal algorithm proposed by Bloomenthal [49]. With adaptive subdivisions, cracks may appear in the polygonal mesh that approximates the surface. In fact, adjacent boxes with different sizes (or different subdivision depth levels) mean that the surface is approximated with different resolutions; consequently, the surface is approximated by nonmatching polygons on overlapping back-to-back faces of adjacent boxes with different sizes. Putting it differently, the reason behind the cracking problem is that the polygonisation of each cell is carried out independently of its adjacent cells. Suffern and Balsys [378] also proposed an algorithm to compute the intersections of implicit surfaces, having them argued that this algorithm could be extended to polygonise self-intersecting surfaces.

More recently, and following the principle of polygonising with topological guarantees, Paiva et al. [316] introduced another adaptive algorithm for implicit surfaces; its robustness stems from the fact that all topological components of the surface are located using interval arithmetic; hence, the topological guarantees. In addition to the box emptiness test, which is a topological criterion to locate surface components, Paiva et al. also used a second topological criterion for locating tunnels and enabling the corresponding box subdivision. Interestingly, they also use a third subdivision geometric criterion that has to do with the curvature-based adaptivity; the curvature is estimated from the variation of the gradient. The polygonal mesh that approximates the surface is generated from the dual grid of the octree

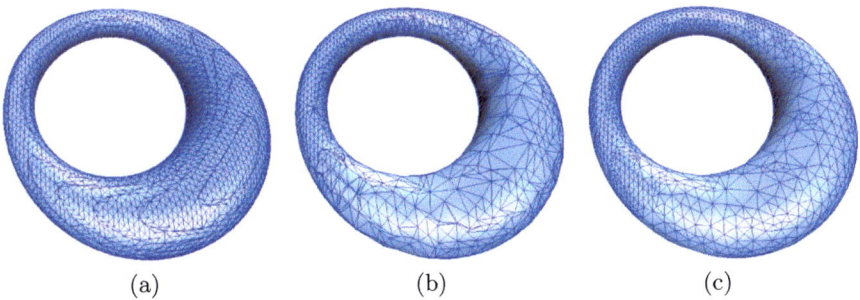

Fig. 7.13. (a) The marching cubes generates 11,664 triangles; (b) the dual marching cubes generates 5396 triangles; (c) the quality–improved dual marching cubes also generates 5396 triangles after using a simple mesh processing. (Figure kindly provided by Dr. Afonso Paiva and his colleagues.)

using the an enhanced Schaefer-Warren method [351]. Thus, unlike the uniform tessellation generated by the marching cubes, the algorithm of Paiva et al. tessellates an implicit surface adaptively, i.e. according to the value of the surface curvature. This is illustrated in Figure 7.13, where the cyclide $(x^2+y^2+z^2)^2-2(x^2+r^2)(a^2+b^2)-2(y^2-z^2)(a^2-b^2)+(a^2-b^2)^2+6abrx = 0$ with $a = 10$, $b = 2$ and $r = 2$ appears tessellated using the marching cubes (Figure 7.13(a)) and the dual octree technique of Paiva et al. (Figures 7.13(b) and (c)). Paiva et al. call their algorithm *dual marching cubes*.

7.4.3 Tetrahedral Subdivision

As argued by Hall and Warren [172], one major drawback of applying the methods of Wyvill et al.[422] and Lorensen and Cline [247] to contour a trivariate function is that these methods must sample the function uniformly at a cubic grid of points. As a consequence, to accurately approximate the contour, the function must be sampled closely, and thus heavily, even in regions where the function is nearly linear. One solution to this problem is to use an adaptive subdivision scheme, sampling more closely near high-curvature regions of the surface. There are two adaptive subdivision schemes: (a) the adaptive octree subdivision scheme or, alternatively, (b) the adaptive tetrahedral subdivision scheme. In the previous section, an adaptive octree subdivision scheme has been described. This section deals with adaptive tetrahedral subdivision schemes.

The first adaptive tetrahedral subdivision scheme to polygonise implicit surfaces was proposed by Hall and Warren [172]. Hall-Warren's algorithm performs an adaptive partition of space into tetrahedra. Interestingly, and regardless of the subdivision level, this tetrahedral subdivision enjoys the honeycomb property, i.e. the collection of all tetrahedra forms a *honeycomb*, the 3D analogue of a tessellation [93]. A honeycomb is a polyhedral partition of

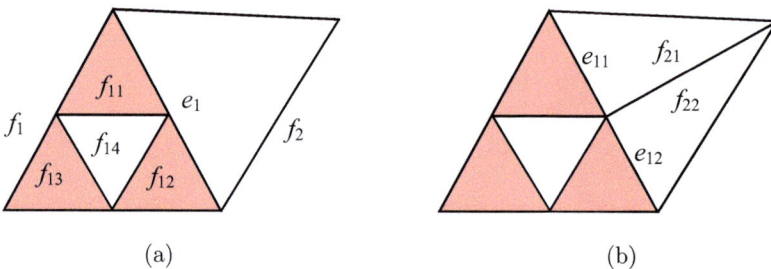

Fig. 7.14. Keeping the honeycomb in 2D.

space in which the i-dimensional face of each polyhedron meets only one other i-dimensional face on a $(i-1)$-dimensional face. Figure 7.14 illustrates this in 2D for triangles. The equilateral triangles f_1 and f_2 share a common edge e_1, but after splitting f_1 into f_{11}, f_{12}, f_{13} and f_{14} the honeycomb rule is violated (Figure 7.14(a)). In order to maintain the honeycomb rule, e_1 must be split into e_{11} and e_{12}, and f_2 into f_{21} and f_{22}. Note that the partition of f_2 is partial. The reader is referred to Hall and Warren [172] to observe the subdivision patterns of equilateral and isosceles triangles, as needed to decompose equilateral and cubic tetrahedra, respectively, of the 3D honeycomb.

By maintaining a honeycomb, the algorithm guarantees that the surface will be approximated by a polygonal mesh without cracks. Possibly, this is a major advantage of adaptive tetrahedral subdivisions over the octree subdivision of space, yet not all types of tetrahedral subdivisions maintain a honeycomb. In the rest of this section, we only deal with honeycombs.

Tetrahedral Honeycombs

Hall-Warren's algorithm uses an adaptive subdivision of a tetrahedron into twelve smaller tetrahedra. The result is an unstructured tetrahedral subdivision, as opposed to a Kuhn subdivision. The subdivision of a regular tetrahedron is performed in two steps (Figure 7.15).

First, one subdivides such a regular tetrahedron into four regular tetrahedra (Figure 7.15(a)) and one regular octahedron (Figure 7.15(b)) by cutting off each corner of the original tetrahedron. This cut is done in a way that each face of the tetrahedron, an equilateral triangle, is subdivided into four smaller equilateral triangles (Figure 7.15(a)). Second, the remaining regular octahedron (Figure 7.15(b)) left in the middle of the original tetrahedron is then split into eight similar tetrahedra by creating a vertex at the centre of the octahedron and projecting edges to each of its corners (Figure 7.15(c)). These eight tetrahedra are called cubic tetrahedra, since each of these tetrahedra may be formed by cutting a corner off a cube.

7.4 Spatial Subdivision   215

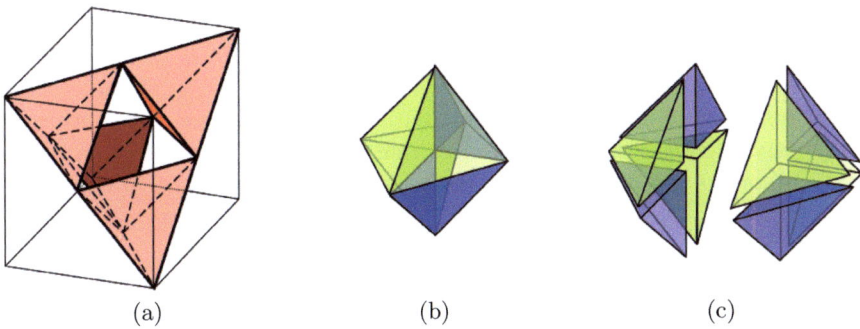

Fig. 7.15. Subdivision of an equilateral tetrahedron into four equilateral tetrahedra (in red) and eight cubic tetrahedra (in green and blue).

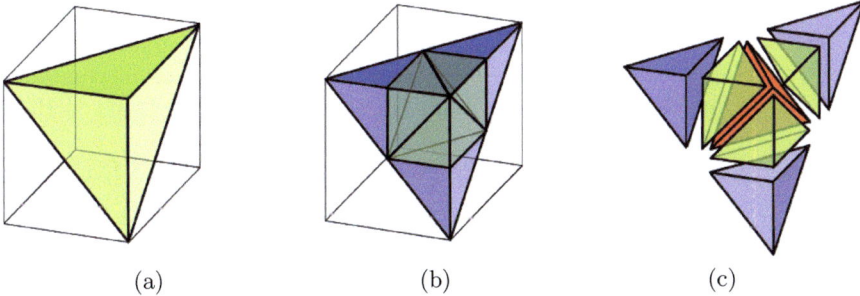

Fig. 7.16. Subdivision of a cubic tetrahedron into one regular tetrahedron (in red) and six cubic tetrahedra (in green and blue).

The 12-tetrahedra subdivision described above involves two types of tetrahedra: regular tetrahedra and cubic tetrahedra. To keep the honeycomb we need a matching subdivision for cubic tetrahedra. This subdivision is illustrated in Figure 7.16.

First, one subdivides *the* equilateral face of a cubic tetrahedron (the posterior tetrahedron face in Figure 7.16(a)) into four equilateral triangles (Figure 7.16(b)). Then, one subdivides the remaining three isosceles triangles as shown in Figure 7.16(b). The resulting tetrahedron decomposition consists of one regular tetrahedron and three pairs of cubic tetrahedra as shown in Figure 7.16(c). The small regular tetrahedron (in red) in Figure 7.16(c) can be alternatively obtained by projecting from the vertex opposite the equilateral face of the original cubic tetrahedron.

These two mutually recursive subdivisions involving only two types of tetrahedra (regular tetrahedra and cubic tetrahedra) allow us to construct arbitrarily fine honeycombs. Similar to the 2D case, the recursive subdivision of a single tetrahedron may cause the honeycomb property to be lost, unless the neighbours of that tetrahedron are partially subdivided to maintain the property. This way, it is possible to construct a *continuous* (i.e. without discontinuities) piecewise linear function that interpolates the values at the vertices of the tessellation, as usual in discrete data-based algorithms. For continuous data-based algorithms, this guarantees that the mesh that approximates the implicit surface is formed without cracks.

The Algorithm

The algorithm may start with either a regular tetrahedron or with a tetrahedral mesh on the domain, though a nonregular tetrahedron could be also used by treating it as if it were regular. The nonregularity of a tetrahedron does not break the honeycomb property of the subdivision scheme because the subdivision of an irregular original tetrahedron is just the image under a linear transformation of the subdivision of a regular tetrahedron [172].

Velho [397, 398, 399] and Hall and Warren [172] should be given the credit of introducing the first adaptive tetrahedralisation algorithms in computer graphics. In general terms, they are similar, so we are going to focus on Hall-Warren's algorithm.

The Hall-Warren algorithm (see Algorithm 26) consists of five major stages:

- Decomposition of the bounding box into cubes.
- 5-tetrahedral decomposition of cubes.
- Uniform subdivision of tetrahedra.
- Adaptive subdivision of tetrahedra.
- Polygonisation of transverse tetrahedra

The first stage decomposes an axially aligned bounding box into a grid of cubes (step 2 of Algorithm 26). At the second stage (steps 6–9 of Algorithm 26), these cubes are partitioned by using, for example, a 5–*tetrahedral decomposition*. Recall that the 5-tetrahedron decomposition consists of one regular tetrahedron and four cubic tetrahedra (Figure 7.6). The third stage (step 10 of Algorithm 26) performs a *uniform subdivision* of the tetrahedra down to a given minimum level l_{MIN}, regardless of whether the surface crosses a tetrahedron or not. Therefore, l_{MIN} works as a stopping condition for the first stage of the algorithm. The fourth stage (step 12 of Algorithm 26) concerns the *adaptive subdivision* of tetrahedra. Only transverse tetrahedra are subdivided, but the subdivision of each tetrahedron depends on the curvature of the surface. Finally, the algorithm performs the polygonisation (steps 14–15) of the surface inside each transverse tetrahedron.

It is worthy noting that Hall-Warren's algorithm comprises three tetrahedral decompositions. After applying a 5-tetrahedral decomposition to each cube of the original cube grid (steps 6 and 7 in Algorithm 26), more two tetrahedral decompositions take place: a uniform decomposition (step 10) and an adaptive decomposition (step 12).

7.4 Spatial Subdivision

Algorithm 26 Tetrahedral Subdivision Algorithm for Implicit Surfaces
1: **procedure** HALLWARREN($f,\mathcal{S},\Omega,l_{MIN},l_{MAX}$)
2:     Subdivide Ω into a grid of equally sized boxes $\{\square_i\}$
3:     $\{\Delta_j^{ON}\} \leftarrow \varnothing$ ▷ list of ACTIVE tetrahedra
4:     $\{\Delta_j^{newON}\} \leftarrow \varnothing$ ▷ list of new ACTIVE tetrahedra
5:     $n \leftarrow \#\{\square_i\}$
6:     **for** $i \leftarrow 0, n-1$ **do**
7:         Subdivide $\square_i$ into five tetrahedra $\{\Delta_k\}$ ▷ 5-tetrahedron decomposition
8:         $\{\Delta_j^{newON}\} \leftarrow \{\Delta_j^{newON}\} \cup \{\Delta_k\}$
9:     **end for**
10:    UNIFORMSUBDIVISION($\{\Delta_j^{ON}\}, \{\Delta_j^{newON}\},l_{MIN}$)   ▷ uniform subdivision
11:    $\{\Delta_k^{OFF}\} \leftarrow \varnothing$ ▷ list of PASSIVE tetrahedra
12:    ADAPTIVESUBDIVISION($\{\Delta_k^{ON}\}, \{\Delta_k^{OFF}\},l_{MAX}$)   ▷ adaptive subdivision
13:    $N \leftarrow \#\{\Delta_k^{OFF}\}$
14:    **for** $i \leftarrow 0, N-1$ **do**   ▷ polygonisation
15:        Polygonise Δ_k^{OFF}
16:    **end for**
17: **end procedure**

Algorithm 27 Uniform Tetrahedral Subdivision
1: **procedure** UNIFORMSUBDIVISION($\{\Delta_j^{ON}\}, \{\Delta_j^{newON}\},l_{MIN}$)
2:     **if** $l_{MIN}=0$ **then**
3:         $\{\Delta_j^{ON}\} \leftarrow \{\Delta_j^{ON}\} \cup \{\Delta_j^{newON}\}$
4:         **return**
5:     **end if**
6:     $n \leftarrow \#\{\Delta_j^{newON}\}$
7:     **for** $i \leftarrow 0, n-1$ **do**
8:         Subdivide Δ_j^{newON} into $\{\Delta_k\}$
9:         UNIFORMSUBDIVISION($\{\Delta_j^{ON}\}, \{\Delta_k\},l_{MIN}-1$)
10:    **end for**
11: **end procedure**

The uniform decomposition subdivides regular and cubic tetrahedra down for a number l_{MIN} of subdivisions, as described in Algorithm 27. This uniform subdivision of each tetrahedron terminates after completing a number l_{MIN} of recursion cycles (step 2 of Algorithm 27), being then the resulting terminal tetrahedra inserted into the list $\{\Delta_j^{ON}\}$ of active tetrahedra (step 3 of Algorithm 27). These tetrahedra are called "active" in the sense that they still need be processed to check whether they contain the surface or not.

At the beginning of the adaptive subdivision stage, the list $\{\Delta_j^{ON}\}$ then contains all terminal tetrahedra generated at the uniform subdivision stage. Algorithm 28 takes as input this list of active tetrahedra, and outputs the list $\{\Delta_i^{OFF}\}$ of passive tetrahedra. These passive tetrahedra are transverse to the surface and are ready to polygonisation stage. The active tetrahedra Δ_j^{ON} are subject to two stopping subdivision criteria (see Algorithm 28):

Algorithm 28 Adaptive Tetrahedral Subdivision

1: **procedure** ADAPTIVESUBDIVISION(f,$\mathcal{S}$,$\{\Delta_j^{ON}\}$, $\{\Delta_i^{OFF}\}$,l_{MAX})
2:     $n \leftarrow \#\{\Delta_j^{ON}\}$
3:     **if** ($n = 0$) **or** ($l_{MAX}=0$) **then**
4:         $\{\Delta_j^{OFF}\} \leftarrow \{\Delta_i^{OFF}\} \cup \{\Delta_i^{newOFF}\}$
5:         **return**
6:     **end if**
7:     **for** $i \leftarrow 0, n-1$ **do**
8:         **if** $0 \notin \text{Im}_f(\Delta_j^{ON})$ **then**   ▷ exclusion box test
9:             Discard Δ_j^{ON}
10:         **else**
11:             **if** $\mathcal{S}$ is approx. flat in Δ_j^{ON} **then**   ▷ curvature test
12:                 $\{\Delta_i^{newOFF}\} \leftarrow \text{insert}(\Delta_j^{ON})$
13:             **end if**
14:         **end if**
15:     **end for**
16:     $m \leftarrow \#\{\Delta_j^{newOFF}\}$
17:     **for** $i \leftarrow 0, m-1$ **do**
18:         **if** Δ_i^{newOFF} is adjacent to any Δ_j^{ON} **then**
19:             Subdivide Δ_i^{newOFF} into $\{\Delta_k\}$ to maintain the honeycomb
20:             $\{\Delta_i^{OFF}\} \leftarrow \{\Delta_i^{OFF}\} \cup \{\Delta_k\}$
21:         **else**
22:             $\{\Delta_i^{OFF}\} \leftarrow \text{insert}(\Delta_i^{newOFF})$
23:         **end if**
24:     **end for**
25:     $N \leftarrow \#\{\Delta_j^{ON}\}$
26:     **for** $j \leftarrow 0, N-1$ **do**
27:         Subdivide Δ_j^{ON} into $\{\Delta_k\}$
28:         $\{\Delta_j^{newON}\} \leftarrow \{\Delta_j^{newON}\} \cup \{\Delta_k\}$
29:     **end for**
30:     $\{\Delta_j^{ON}\} \leftarrow \emptyset$
31:     $\{\Delta_j^{ON}\} \leftarrow \{\Delta_j^{ON}\} \cup \{\Delta_j^{newON}\}$
32:     ADAPTIVESUBDIVISION(f,$\mathcal{S}$,$\{\Delta_j^{ON}\}$, $\{\Delta_i^{OFF}\}$,$l_{MAX} - 1$)
33: **end procedure**

- *Exclusion test.* Interestingly, Hall and Warren do not use the intermediate value theorem to exclude empty (i.e. nontransverse) tetrahedra. Instead, they use Descarte's rule applied to Bézier formulation of the polynomial that defines the surface. This test changes the state of a tetrahedron from "active" to "discarded" (steps 8–9) so that nontransverse tetrahedra are deleted.
- *Curvature test.* This test evaluates the curvature of the surface within each active tetrahedron. If the curvature falls below a given threshold, i.e. the surface is approximately flat therein, the active tetrahedron is re-labelled as a new passive tetrahedron (steps 11–12). Therefore, such a tetrahedron is inserted into the collection Δ_i^{newOFF} of new passive tetrahedra.

In order to maintain the honeycomb (steps 17–24 of Algorithm 28), one proceeds as follows. If a new passive tetrahedron is adjacent to any active tetrahedron, it is subdivided into smaller tetrahedra, which will be then labelled as passive tetrahedra and inserted into $\{\Delta_i^{\text{OFF}}\}$, ready for polygonisation; otherwise, the new passive tetrahedron is simply relabelled as a passive tetrahedron and inserted into $\{\Delta_i^{\text{OFF}}\}$ also for polygonisation.

The remaining active tetrahedra—those containing surfaces patches with significant curvature—are then subdivided into smaller active tetrahedra, as illustrated in steps 25–28 of Algorithm 28. Then, of course, the algorithm recurses on this set of active tetrahedra (step 32 of Algorithm 28). The adaptive subdivision terminates when at least one of the following conditions is satisfied (step 3 of Algorithm 28):

- A predefined level l_{MAX} of adaptive subdivision is reached;
- The number n of active tetrahedra is zero.

Finally, Hall–Warren's algorithm uses the honeycomb consisting of passive tetrahedra, as well as any remaining active ones, to create a piecewise planar approximation (steps 14–15 of Algorithm 26).

Various adaptive tetrahedralisation algorithms are based on Hall-Warren's approach, as those described by Hui and Jiang [199] and Müller and Wehle [289]. Hui-Jiang's algorithm extends Hall-Warren's algorithm in that it uses a heuristics based on Schmidt's work [352] to avoid intersection in highly curves surfaces, surfaces with self-intersections or multiple components; a process called compensate-subdivision is also used to eliminate cracks in the tessellated surface.

The reader is still referred to Ning and Blomenthal [305] and Zhou et al. [428] for more details on tetrahedralisations, as well as resolution of ambiguities.

7.5 Nonmanifold Curves and Surfaces

So far, in this chapter, we have studied implicit curves and surfaces—in general, varieties—that are manifolds. By definition, an n-dimensional manifold ($n \in \mathbb{N}$) is, everywhere, locally homeomorphic to $\mathbb{R}^n$. This means that, an infinitesimal neighbourhood of any point on an n-manifold is topologically equivalent to an n-disk. Putting this differently, exactly n independent directions can be defined as the axes of a local coordinate system at each point of an n-manifold. On the contrary, a variety V is said to be *not* an n-manifold if and only if the number of independent directions we can follow from a point is different from n.

For example, in Figure 7.17, the curve (a) is not manifold because it has an isolated point at the origin so that no 1-dimensional local coordinate system can be defined; the curve (b) is not manifold either because at the self-intersection we can define two up to four 1-dimensional local systems. In

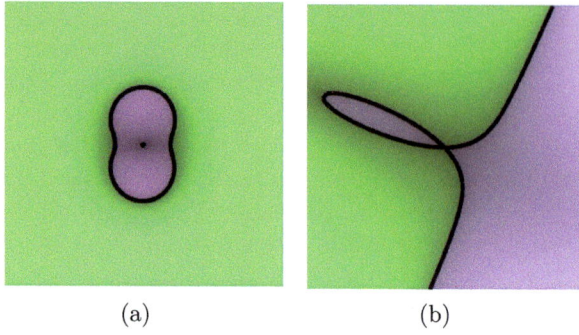

Fig. 7.17. Nonmanifold algebraic curves: (a) $x^4 + 2x^2y^2 - x^2 + y^4 - 4y^2 = 0$ and (b) $2y^3 - (3x - 3)y^2 - (3x^2 - 3x)y - x^3 = 0$.

respect to nonmanifold surfaces, the Steiner surface, the Whitney umbrella surface with handle (Figure 1.12(a)), and the Kummer surfaces (Figure 7.18(b)) are some examples we can find in the literature. In other words, a manifold is homogeneous in dimension and does not possess self-intersections.

7.5.1 Ambiguities and Singularities

Nonmanifold features of implicit curves and surfaces arise a series of problems to polygonisers. The main problem comes from the fact that the dimension may not be homogeneous. Dimension is a topological invariant, so if the dimension of a surface is not uniform, the polygoniser will face serious difficulties in keeping topological guarantees. As far as the authors know, there is no algorithm, at least in the computer graphics literature, to resolve isolated singularities of an n-variety, i.e. k-directional singularities ($0 \leq k < n$); these singularities include isolated points of curves, and isolated points and lines for surfaces.

As shown in Chapter 6, many m-directional singularities ($m > n$) or self-intersections can be resolved through symbolic factorisation, in particular when a topological component has two or more symbolic components. But, when a topological component has a single symbolic component (e.g. Whitney umbrella surface) that self-intersects, such a procedure is no longer possible. It is true that the mathematics behind the resolution of singularities (see, e.g., Lu [249]) is wellknown and there are some symbolic techniques (see, e.g., Bodnár and Schicho [55]) to compute them, but they are computationally expensive and only apply to polynomial functions.

Thus, the integration of a singularity solver into a polygoniser remains an open issue in computer graphics, regardless of whether the nature of the polygoniser, either continuation-based polygoniser or space partitioning-based polygoniser. In fact, no many articles have been published on the polygonisation of non-manifold implicit surfaces.

7.5.2 Space Continuation

Similar to conventional polygonisers, Bloomenthal and Ferguson [53] use a uniform partitioning of the bounding box into smaller cubic boxes, but these subsidiary boxes are obtained by spatial continuation. That is, with the exception of the seeding box, every zero box (i.e. surface-intersecting box) is obtained from a previously formed, adjacent zero box. To prevent cyclic propagation—inherent to continuation algorithms—the location of each visited box is stored in a hash table, as described by Wyvill et al. in [422]. Each zero box is then decomposed into six tetrahedra. This tetrahedralisation aims to resolve eventual ambiguities. But, as seen above, tetrahedralisation-based disambiguation is not sufficient to ensure that the surface polygonisation is performed with topological guarantees, even for manifold surfaces.

Unlike manifold implicit surfaces, the implicit scalar fields underlying nonmanifold surfaces are no longer defined by real functions that bisect space into interior and exterior regions. To solve this problem, Bloomenthal and Ferguson [53] introduced a multiple space classification as a generalisation of the binary space classification into positive and negative regions. But, the use of multiple regions rather complicates the polygoniser, in particular the polygonisation of surface borders and surface intersections. Recall that Bloomenthal and Ferguson's algorithm only applies to nonmanifold surfaces with homogeneous dimension, i.e. surfaces with boundaries and surfaces with intersections.

7.5.3 Octree Subdivision

In [352], Schmidt proposes an octree subdivision-based polygoniser for self-intersecting surfaces. Using an octree data structure means that the polygonisation is in principle adaptive. Tetrahedralisations of the leaf zero boxes (i.e. boxes that intersect the surface at the maximum subdivision depth) also take place in hope of resolving topological ambiguities caused by eventual surface self-intersections. Unfortunately, as shown before, the resolution of ambiguities through the tetrahedralisation of zero boxes may fail.

Balsys and Suffern also proposed an adaptive polygonisation for self-intersecting surfaces [25, 378], but their polygoniser does not use tetrahedralisations of the zero leaf boxes. Two examples of implicit surfaces rendered by the Balsys-Suffern polygoniser are shown in Figure 7.18. Following their previous own work on manifold surfaces [378], Balsys-Suffern's method uses an octree spatial data structure and a box exclusion test for adaptivity, interval arithmetic for robustness, and uses a numerical root-finder for point sampling over box edges. Balsys and Suffern's polygoniser is capable of rendering a number of important non-manifold implicit surfaces, but even so it cannot be considered a general polygoniser because the surface is limited to intersect any box edge twice at most. It is not able to handle isolated points and dangling lines properly either. Despite its limitations, Balsys-Suffern's polygoniser is the most general polygoniser for nonmanifold implicit surfaces we can found in the literature.

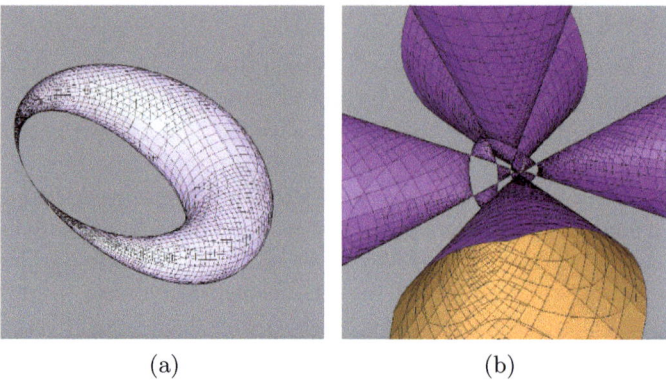

(a)   (b)

Fig. 7.18. Two surfaces generated by Balsys-Suffern's polygoniser: (a) the Dupin cyclide $(x^2+y^2+z^2-r^2)^2-2(x^2+r^2)(a^2+b^2)-2(y^2-z^2)(a^2-b^2)+8abrx+(a^2-b^2)^2 = 0$, with $a = 10, b = 2$, and $r = 2$; (b) the Kummer surface $(x^2+y^2+z^2-u^2)^2-\lambda pqrs = 0$, with $u \in \mathbb{R}^+$ controlling the number of double points of the surface, $\lambda = \frac{3u^2-1}{3-u^2}$ is a scaling factor taken here as 1, and $p = 1-z-\sqrt{2}x$, $q = 1-z+\sqrt{2}x$, $r = 1+z+\sqrt{2}y$, $s = 1+z-\sqrt{2}y$ are the tetrahedral coordinates. A Kummer surface has sixteen double points, i.e. the maximum number of double points for a surface of degree 4 in 3D space. By using the default value $u = 1.3$, all these double points are real and displayed as the vertices of five tetrahedra. (Figure kindly provided by Dr. Ron Balsys and Dr. Kevin Suffern.)

Balsys-Suffern's Algorithm

Pseudo-code describing Balsys-Suffern's algorithm appears in Algorithm 29. This adaptive algorithm recursively partitions the given cubic bounding box Ω into eight equally sized boxes (step 2), which are stored into the nodes of an octree data structure. An interval arithmetic-based exclusion test is used to discard cubic boxes that do not intersect the surface (steps 4 and 5). Each zero box (i.e. nonexcluded box) is then subdivided recursively down (step 32) until the minimum subdivision depth Δ_{MIN} is reached (step 7). The minimum depth works as landmark that indicates the beginning of the polygonisation. However, the polygonisation of a box only takes place if the topological type of the surface inside such a box is valid (step 11) and the polygons pass the flatness test (step 12), after which eventual mesh cracks are repaired. Note that the algorithm forces the polygonisation of the surface in a zero box when the maximum depth Δ_{MAX} is reached (step 19), regardless of its local flatness.

The admissible topological configurations of the surface inside a zero box are depicted in Figure 7.19, namely: (a) a single surface patch, (b) two non-intersecting patches, and (c) two intersecting patches. Similar to conventional polygonisers, the topological pattern of the surface inside a box is determined by first computing the intersection points between the surface and the edges of such a box. For this task, Balsys and Suffern use a numerical root finder

7.5 Nonmanifold Curves and Surfaces

Algorithm 29 Balsys-Suffern Algorithm for Nonmanifold Implicit Surfaces

```
 1: procedure BALSYSSUFFERN(f,S,Ω,Δ,τ)
 2:     Subdivide Ω into eight equally sized boxes □_i
 3:     for i ← 0, 7 do
 4:         if 0 ∉ Im_f(□_i) then                  ▷ box exclusion test
 5:             Discard □_i
 6:         else
 7:             if depth(□_i) ≥ Δ_MIN then         ▷ minimum depth test
 8:                 if depth(□_i) ≤ Δ_MAX then     ▷ maximum depth test
 9:                     Find roots of f on edges of □_i  ▷ sampling surface points
10:                     Determine topological type τ of S inside □_i
11:                     if τ is valid then         ▷ topological type test
12:                         if τ is flat then      ▷ surface flatness test
13:                             Polygonise S across □_i
14:                             Fix cracks
15:                         else                   ▷ not flat enough yet
16:                             if depth(□_i) < Δ_MAX then
17:                                 BALSYSSUFFERN(f,S,□_i,Δ,τ)
18:                             else
19:                                 Polygonise S across □_i
20:                                 Fix cracks
21:                             end if
22:                         end if
23:                     else                       ▷ type τ not valid
24:                         if depth(□_i) < Δ_MAX then
25:                             BALSYSSUFFERN(f,S,□_i,Δ,τ)
26:                         else
27:                             ;                  ▷ polygonisation fails
28:                         end if
29:                     end if
30:                 end if
31:             else
32:                 BALSYSSUFFERN(f,S,□_i,Δ,τ)
33:             end if
34:         end if
35:     end for
36: end procedure
```

based on the false position method, combined with binary interval subdivision. But, these sampled surface points on the box edges do not allow us to distinguish the topological pattern in Figure 7.19(b) from the topological pattern in Figure 7.19(c). A gradient-based criterion is used to disambiguate these two cases. If the angle between the gradient vectors at the sampled surface points is within a small range, then the topological pattern is that one shown in Figure 7.19(b); otherwise, we have the pattern in Figure 7.19(c). Unfortunately, the polygonisation fails for other topological configurations (steps 11, 23 and

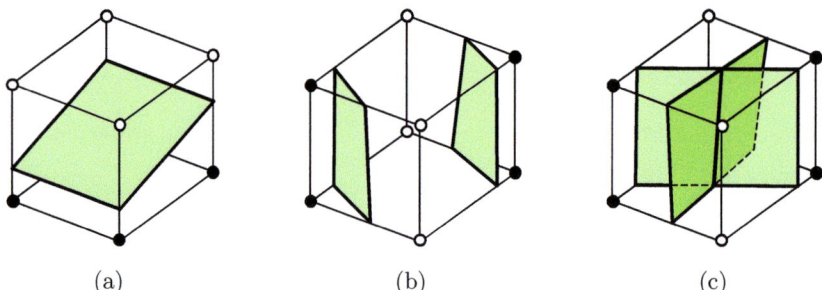

Fig. 7.19. Admissible surface's topological configurations in a zero box according to Balsys-Suffern's polygoniser.

27); in particular, if the surface crosses an edge of a zero box more than twice, there are not any guarantees that the surface will be polygonised correctly.

Balsys-Suffern's algorithm is a curvature-driven adaptive subdivision algorithm. In the literature, we find several criteria to estimate the curvature of a surface, namely:

- the planarity of the surface in the box;
- the divergence of surface normals;
- the chord distance of the surface.

Balsys and Suffern use the first two in the flatness test (step 12), whereby, if the surface patch is not flat enough inside the box, or the surface normals is beyond a certain threshold, the box is further subdivided. The maximum depth Δ_{MAX} works as the principal stopping criterion (step 12); in particular, it is used to stop subdivision in boxes where the surface has extreme curvature (step 18).

7.6 Final Remarks

A number of space subdivision-based algorithms have been devised for rendering implicit curves, surfaces, and even high-dimensional varieties. They all are based on locating a series of boxes in space that intersect the surface. For rendering purposes, the surface is usually polygonised, i.e. the surface is approximated by one or more polygons in each cube. Early algorithms were developed by Wyvill et al. [421] for rendering soft or blobby objects and Lorensen and Cline [247] who designed the marching cubes algorithm for generating human organ surfaces from medical image data sets. These algorithms have given rise to two major families for defining implicit surfaces through space subdivisions: continuous data-based algorithms and discrete data-based algorithms. Discrete data-based subdivision algorithms have been developed from Lorensen and Cline's algorithm. No function is known *a priori* so that

7.6 Final Remarks

the algorithm only operates on discrete data at the vertices of a grid, from which a surface is generated by interpolation. In contrast, continuous data-based subdivision algorithms operate on a given function such that continuous data can be evaluated at arbitrary points of the domain.

Conventional manifold polygonisers are based on the principle that the implicit function is continuous. They also assume that the function evaluates positively on one side and negatively on the other side of the surface. Thus, they perform a binary partitioning of space.

Implicit surfaces in 3D geometric modelling are limited to two manifolds because the corresponding implicit fields are usually defined by real-valued functions that bisect space into interior and exterior. We present a novel method of modelling nonmanifold surfaces by implicit representation. Our method allows discontinuity of the field function and assesses the special meaning of the locus where the function is not differentiable. The enhancement can yield a nonmanifold surface with such features as holes and boundaries. The discontinuous field function also enables multiple classification of the field, which makes it possible to represent branches and intersections of the implicit surfaces. The implicit field is polygonised by the algorithm based on the marching cubes algorithm, which is extended to treat discontinuous fields correctly. We also describe an efficient implementation of converting a surface model into a set of discrete samples of field function, and present the result of the non manifold surfaces reproduced by our method. The implicit surfaces are directly visualised at interactive frame rates independent of surface complexity by the hardware-accelerated volume rendering method. We also developed a system for visualising the implicit surfaces and have confirmed that it can render surfaces at sufficient quality and speed.

Implicitly defined surfaces $f(\mathbf{x}) = 0$ are usually displayed after computing a polygonal mesh which approximates it. Space partitioning algorithms just partition the bounding box surrounding the surface into a 3D polyhedron mesh in order to sample the surface. The algorithms differ from each other in how spatial partitioning is done.

8
Implicit Surface Fitting

Surface reconstruction has become an important research topic in part due to the appearance of 3D range scanners on the market. These scanners are capable of acquiring unstructured 3D point datasets from the surface of a given physical object. Digital scans allow for high-quality surface reconstructions, but this requires a particular care in recovering sharp features such as ridges, corners, spikes, etc. Surface reconstruction has many applications in science and engineering, in particular, geometric modelling, computer graphics, virtual reality, computer animation, computer vision, computer-assisted surgery, and reverse engineering.

8.1 Introduction

This chapter deals with surface-fitting algorithms that reconstruct surfaces from clouds of points. There are several surface reconstruction techniques depending on the representation in hand: *simplicial, parametric, implicit* surfaces. Even though they are different representations, they share various issues and problems. This chapter starts with a brief review on these surface reconstruction techniques, after which the focus will be on the implicit ones.

8.1.1 Simplicial Surfaces

Sometimes, simplicial surfaces are also called triangulated surfaces. There are two main classes of techniques to reconstruct simplicial surfaces from a scattered set of points: *Delaunay-based* and *region-growing techniques*.

With Delaunay-based approach, we end up having a *space partitioning* into tetrahedra. To be more specific, the typical Delaunay-based surface reconstruction algorithm consists of two steps:

- *Delaunay triangulation.* First, one constructs the Delaunay triangulation (or Voronoi diagram) from such a cloud of points, which consists of a partition of the convex hull of these sample points into a finite set of tetrahedra. The main advantage of the Delaunay triangulation comes from its uniqueness; it is unique if no five sample points can be found on a common sphere. Several algorithms for constructing a Delaunay triangulation can be found in the literature, in particular those given by Bowyer[60], Watson [404], Avis and Bhattacharya [23], Preparata and Shamos [328], Edelsbrunner [126], and, more recently, Hjelle and Dæhlen [191], just to mention a few.

- *Triangulated surface extraction.* Terminated the triangulation of the cloud of points, it only remains to identify which simplices belong to the surface. Thus, the reconstruction of simplicial surfaces consists in finding the subgraph of Delaunay triangulation of the initial set of points [127]. The identification of surface simplices varies from an algorithm to another.

The first Delaunay-based surface reconstruction method seems to be due to Boissonnat [56]. In the class of Delaunay-based surface reconstruction methods, we also find the α-shapes of Edelsbrunner and Mücke [127], the crust and the power crust algorithms of Amenta et al. [12, 13, 14], the co-cone algorithm of Dey et al. [110, 111], and more recently the reconstruction algorithms due to Yau et al. [218, 424]. Amongst these algorithms, the algorithm of Yau et al. [218] preserves sharp features, but only in convex regions, while the algorithm of Amenta et al. [14] is capable of reconstructing sharp edges and corners by steering poles, a subset of circumcentres of tetrahedra. In both cases, multiple Delaunay computations are required, so that the corresponding reconstruction algorithms are rather time-consuming to use.

The second class of algorithms to reconstruct simplicial surfaces is based on the concept of *continuation* (see Chapter 6 for more details about continuation). In the context of surface reconstruction, continuation algorithms are known as *region-growing algorithms*. Starting from a seed triangle, say initial region, the algorithm iterates by attaching new triangles to the region's boundaries. The early surface-based algorithm due to Boissonnat [56], the graph-based algorithm of Mencl and Müller [267], the ball-pivoting algorithm of Bernardini et al. [43], the projection-based triangulating algorithm of Gopi and Krishnan [165], the interpolant reconstruction algorithm of Petitjean and Boyer [323], the advanced-front algorithm of Hung and Menq [196], and the greedy algorithm of Cohen-Steiner and Da [88], all fall into the class of region-growing algorithms.

In the literature, we also find hybrid algorithms that combine Delaunay-based and region-growing approaches; for example, the algorithm of Kuo and Yau [218, 219] is a representative of these hybrid algorithms. These algorithms were later improved in order to reconstruct surfaces with sharp features [220].

8.1.2 Parametric Surfaces

Parametric surface fitting algorithms, also called spline-based surface reconstruction algorithms, are quite common in numerical analysis and computer graphics. In the context of parametric representations, the problem of surface reconstruction involves the computation of a surface S that *approximates* as much as possible each point of a given cloud of points in $\mathbb{R}^3$. The goal is then to find a parametric surface S, defined by a function $F(u,v)$, that closely approximates a given cloud of points, where F belongs to a specific linear space of functions. Examples of such parametric surfaces are Bézier and B-spline surfaces [132].

Traditionally, the parametric surface reconstruction algorithm consists of four main steps, namely:

- *Mesh generation* from the unorganised point cloud. This can be done by, for example, using the marching cubes [247], Delaunay triangulations [26] (see the previous section for further references), and α-shapes [127].
- *Mesh partitioning* into patches homeomorphic to disks. These patches are also known as charts. The surface mesh partitioning becomes mandatory when the surface is closed or has genus greater than zero. Roughly speaking, there are two ways of cutting surfaces into charts: segmentation techniques and seam generation techniques. For parametrisation purposes, segmentation techniques divide the surface into several charts in order to keep as short as possible the parametric distortion resulting from the cuts. Unlike segmentation, seam cutting techniques are capable of reducing the parametric distortion without cutting the surface into separate patches. For that purpose, they use seams (or partial cuts) to reduce a surface of genus greater than zero to a surface of genus zero. For more details about surface mesh partitioning, the reader is referred to Sheffer et al. [361] and the references therein.
- *Parametrisation.* For each mesh patch, one constructs a local parametrisation. These local parametrisations are made to fit together continuously such that they collectively form a globally continuous parametrisation of the mesh. In computer graphics, this method was introduced by Eck et al. [125], who used harmonic maps to construct a (local) parametrisation of a disk over a convex polygonal region. Nevertheless, before that, Pinkall and Polthier had already used a similar method for computing piecewise linear minimal surfaces [324]. For more details on this topic, the reader is referred to Floater and Hormann [141] and the references therein.
- *Surface fitting.* Terminated the parametrisation step, which outputs a collection of pairs of parameters (u_i, v_i) associated to the points (x_i, y_i, z_i) of the cloud, it remains the problem of surface fitting. Surface fitting consists in minimising the distance between each point (x_i, y_i, z_i) and its corresponding point of the surface $F(u_i, v_i)$.

The standard approach of surface fitting reduces to the following minimisation problem:

$$\min \sum_i ||\mathbf{x}_i - F(u_i, v_i)||^2 \qquad (8.1)$$

where $\mathbf{x}_i$ is the ith input cloud point (x_i, y_i, z_i) and $||\cdot||$ is the Euclidean distance between $\mathbf{x}_i$ and the corresponding point on the surface $F(u_i, v_i)$ in the above mentioned linear space of functions. The objective function of this minimisation problem is then the squared Euclidean norm. Its computation can be done easily by the least squares method; hence the least-squares (LS) fitting for parametric surface reconstruction [87, 132]. As argued in [326], this is the main approach to approximating an unstructured cloud of points by a B-spline surface.

Alternatives to LS fitting using parametric surfaces are:

- *Active contours.* This approximation approach is borrowed from computer vision and image processing, and is due to Kass et al. [209] who introduced a variational formulation of parametric curves, called snakes, for detecting and approximating contours in images. Since then various variants of snakes or active contours have appeared in the literature [45]. In the context of parametric surface reconstruction, the active contour technique was introduced by Pottmann and Leopoldseder [326], which uses local quadratic approximants of the squared distance function of the surface or point cloud to which we intend to fit a B-spline surface. Interestingly, this approach avoids the parametrisation problem, i.e. the third step of the standard procedure described above.
- L_p *fitting.* The use of L_p norms in fitting curves and surfaces to data aims at finding a member of the family of surfaces in $\mathbb{R}^n$ which gives a best fit to N given data points. The least squares or L_2 norm is just an example of a fitting technique that minimises the orthogonal distances from the data points to the surface. Note that the least squares norm is not always adequate, in particular when there are wild points in the data set. This leads us to look at other L_p norms for surface fitting [22]. For example, Marzais and Malgouyres [260] uses a linear programming fitting which is based on the L_∞ norm, also called the uniform or Chebyshev norm. The L_∞ fitting outputs a grid of control points of a parametric surface (e.g. Bézier or B-spline surface).

8.1.3 Implicit Surfaces

Most implicit surface reconstruction algorithms from clouds of points are based on Blinn's idea of blending local implicit primitives [47], called *blobs*. This blending effect over blobs fits the requirements of modelling a molecule from an union of balls that represent atoms. Muraki [293] combines Gaussian blobs to fit an implicit surface to a point set. Lim et al. [239] use the blended union of spheres in order to reconstruct implicit solids from scattered data; the spheres are obtained from a previous configuration of spheres given by the Delaunay tetrahedralisation of the sample points.

In computer graphics literature, in 1987, Pratt [327] was who first called attention to fitting implicit curves and surfaces to data, since parametric curves and surface had received the most attention in the fitting literature, creating the misleading idea that implicit curves and surfaces are less suitable for fitting purposes. Also, Pratt affirms that none treatment of least squares fitting of implicit surfaces to data was found in the literature. In 1991, Taubin [383] noted that there was no previous work on fitting implicit curves in 3D, having found only a few references on fitting quadric surfaces to data in the literature of pattern recognition and computer vision.

Since then, two major classes of implicit fitting methods have been introduced in the literature:

- *Global methods*. These methods aim to construct a single function such that its zero set interpolates or approximates the cloud of points globally.
- *Local methods*. In this case, the global function results from blending local shape functions, each one of which interpolates or approximates a subcloud of points.

Now, there is an extensive literature on global implicit surface fitting that uses a single polynomial to fit a point cloud. Taubin [383] introduced algorithms to reconstruct algebraic curves and surfaces based on minimising the approximate mean square distance from the cloud points to the curve or surface, which is a nonlinear least squares problem. In certain cases, this problem of implicit polynomial fitting leads to the generalised eigenvector fit, i.e. the minimisation of the sum of squares of the function values that define the curve or surface. Also, Hoppe et al. [192] proposed an algorithm based on the idea of determining the zero set of a locally estimated signed distance function, say the distance to the tangent plane of the closest point; such a zero set is then used to construct a simplicial surface that approximates the actual surface. Similarly, Curless and Levoy [94] use a volumetric approach to reconstruct shapes from range scans that is based on estimating the distance function from a reconstructed model. As Curless and Levoy noted, the isosurface of this distance function can be obtained in an equivalent manner by means of least squares (LS) minimisation of squared distances between range surface points and points on the desired reconstruction. Other surface reconstruction algorithms based on signed distance are due to Bernardini et al. [42] and Boissonnat and Cazals [57].

An important representation of implicit surfaces is the moving least squares (MLS) surfaces [229, 263]. Roughly speaking, a MLS surface is a LS surface with local shape control. The main shortcoming of MLS (and also LS) is that this approach transforms sharp creases and corners into rounded shapes. To solve the problem of reconstructing sharp features, Kobbelt et al. [213] proposed an extended marching cubes algorithm, Fleishman et al. [139] designed a robust algorithm based on the moving least squares (MLS) fitting, and Kuo and Yau [220] proposed a combinatorial approach based on the Delaunay to produce a simplicial surface with sharp features from a point cloud.

Another family of implicit surface reconstruction algorithms use radial basis functions (RBFs). Some algorithms employ *globally* supported radial basis functions, namely those due to Savchenko et al. [349], Turk and O'Brien et al. [392, 393], and Carr et al. [76]. Unfortunately, because of their global support, RBFs fail to reconstruct surfaces from large datasets, i.e. point sets having more than a few thousands points. This fact led to the development of reconstruction algorithms that use Wendland's *compactly* supported RBFs [408]; for example, the algorithms proposed by Floater and Iske [142], Morse et al. [282], Kojekine et al. [214], and Ohtake et al. [311, 312] fall into this category. These algorithms are particularly suited to reconstruct *smooth* implicit surfaces from large and incomplete datasets.

Another yet family of implicit surface reconstruction algorithms is the partition of unity (PoU). This approach uses the divide-and-conquer paradigm. The idea is to adaptively subdivide the box domain into eight subsidiary boxes recursively. A necessary but not sufficient condition to subdivide a box is the existence of data points in such a box. Then, one uses locally supported functions that are blended together by means of the partition of unit. This partition of unit is simply a set of smooth, local weights (or weight functions) that sum up to one everywhere on the domain. Ohtake et al. [308] use the multilevel partition of unity (MPU) together with three types of local approximation quadratic functions (i.e. local shape functions) to reconstruct implicit surfaces from very large sets of points, including surfaces with sharp features (e.g. sharp creases and corners). Interestingly, Ohtake et al. [312] and Tobor et al. [389] combine RBFs and PoU as a way of getting a more robust method against large, non-uniform data sets, i.e. sets with variable density of points, but the algorithm due to Tobor et al. [389] has the advantage that it also works in the presence of noisy data.

In the remainder of the present chapter, we will focus on the most used or significant methods in implicit surface reconstruction, namely: blob functions, moving least squares, radial basis functions, and partition of unity implicits.

8.2 Blob Surfaces

In order to break away from the conventional ball-and-stick and space-filling models, Blinn [47] introduced the *blobby model* in computer graphics for visualising molecules. This model represents a surface of an object as an isosurface of a global scalar field built from local scalar fields associated to subsidiary or constituent primitives (Figure 8.1).

Instead of using the traditional implicit quadrics, we use electron density functions to model atoms and molecules. Recall that a molecule is an aggregate of atoms. For example, Figure 8.2 shows the trypsin molecule—with the identifier 4PTI in the protein data bank (PDB)—as (a) an aggregate of atoms and as (b) an isosurface of a molecular scalar field.

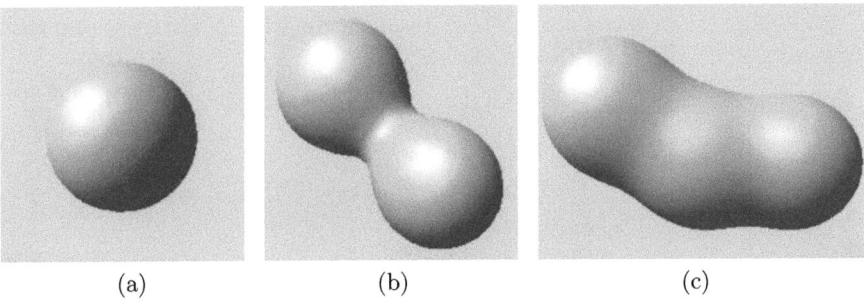

Fig. 8.1. Blobby models with (a) one primitive, (b) two primitives, and (c) three primitives. (Courtesy of Paul's Projects.)

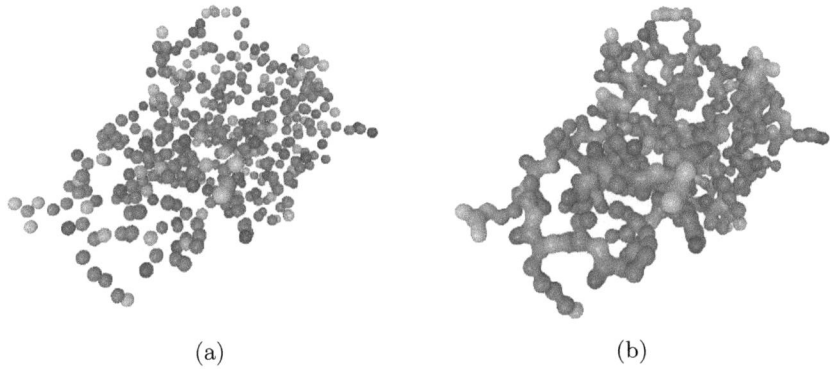

Fig. 8.2. The trypsin molecule as (a) an aggregate of atoms and as (b) an isosurface of a molecular scalar field.

As said above, the blobby model represents a 3-dimensional object in $\mathbb{R}^3$ as an isosurface of a scalar field generated by composition of local scalar fields, each generated by a geometric primitive (e.g. point or sphere). This means that a field value at a point $\mathbf{x} = (x, y, z)$ generated by a primitive or atom A_i centred at a point $\mathbf{x}_i$ is given by

$$f_i(\mathbf{x}) = b_i\, e^{-a_i\, d_i(\mathbf{x})} \qquad (8.2)$$

where the $d_i(\mathbf{x})$ dictates the shape of the scalar field. Equation (8.2) is known as Blinn's Gaussian function. In fact, the exponential term is nothing more than a Gaussian bump centred at $\mathbf{x}_i$ which has height b_i and standard deviation a_i. If $d_i(\mathbf{x})$ is the square of the Euclidean distance between $\mathbf{x}$ and $\mathbf{x}_i$, that is

$$d_i(\mathbf{x}) = (x - x_i)^2 + (y - y_i)^2 + (z - z_i)^2 \qquad (8.3)$$

then the field is spherically symmetric.

The global density function of a given molecule with N atoms is obtained by summing up the contribution of each atom

$$f(\mathbf{x}) = \sum_{i=1}^{N} f_i(\mathbf{x}) \qquad (8.4)$$

or, equivalently,

$$f(\mathbf{x}) = \sum_{i=1}^{N} b_i\, e^{-a_i\, d_i(\mathbf{x})} \qquad (8.5)$$

Now, we can define an implicit surface as the zero set of points where f equals a given threshold T

$$F(\mathbf{x}) = f(\mathbf{x}) - T = 0. \qquad (8.6)$$

Although Blinn's implicit model has been primarily designed to represent molecules, many other applications have been described and discussed in the literature. This is particularly true since the appearance of alternative implicit blob models as generic shape representations, namely: the metaballs [306], the soft objects [422], and the blobby model [293]. All these models rely on the same global implicit function, but the subsidiary local functions differ slightly. As explained in Chapter 9, these local functions are similar to Blinn's exponential density function.

8.3 LS Implicit Surfaces

The first comprehensive treatment of the least squares (LS) method was published in 1805 and is due to Legendre [226]. In 1809, Gauss [157] published a book in which he also describes the LS method. Gauss mentioned that he had been using the LS method since 1795, thus starting an anteriority dispute about the discovery of the method with Legendre, in a way similar to the Leibniz-Newton controversy about the invention of Calculus.

8.3.1 LS Approximation

The LS method is an approximation method, and thus it results in smoothing rather than interpolating the scattered data. The LS approximation starts from the formulation of the following *problem*. Given a set of N observations or scalar values $\{f_i\}_{i=1}^{N}$ on a set of points located at positions $\{\mathbf{x}_i\}_{i=1}^{N}$ in $\mathbb{R}^n$, the problem is to find an unknown, *globally* defined function $f(\mathbf{x}) : \mathbb{R}^n \to \mathbb{R}$ that fits the given observed values f_i as near as possible with respect to some metric. If such a metric is the sum of the squares of the errors, also called *residuals*, at the data points, we come up with the well-known least squares solution. That is, we minimise the LS error

8.3 LS Implicit Surfaces

$$E_{LS} = \sum_{i=1}^{N} r_i^2 \qquad (8.7)$$

where the residuals r_i are given by

$$r_i = f(\mathbf{x}_i) - f_i \qquad (8.8)$$

The residuals are thus differences between the function values (or theoretical values) at the data points and the observed values, so that the LS best fit is obtained when the LS error of Equation (8.7) is reduced to a minimum.

As explained in the remainder of this section, the minimisation of the LS error is crucial to achieve the unknown function $f(\mathbf{x})$. This will allow us to define the implicit surface given by $f^{-1}(0)$ that fits the given observed data points $\mathbf{x}_i$. Such an unknown function $f(\mathbf{x})$ can be designed as a linear combination of K basis functions $p_j(\mathbf{x})$

$$f(\mathbf{x}) = \sum_{j=1}^{K} c_j p_j(\mathbf{x}) \qquad (8.9)$$

so that the "best" fit—which depends on the criterion that one uses in a specific context—to data set is obtained by adjusting and determining the real parameters c_j.

The functions $p_j(\mathbf{x})$ usually are *polynomial* basis functions, i.e. they form a polynomial basis. This means that $f(\mathbf{x})$ belongs to the space $\prod_d^n$ of n-variate polynomials of total degree less or equal to d. Examples of polynomial bases are:

(i) $\mathbf{p}(\mathbf{x}) = [1]$ for a constant fit in arbitrary dimensions;
(i) $\mathbf{p}(\mathbf{x}) = [1, x, y, z]^T$ for a linear fit in $\mathbb{R}^3$ ($d = 1$ and $n = 3$);
(i) $\mathbf{p}(\mathbf{x}) = [1, x, y, x^2, xy, y^2]^T$ for a quadratic fit in $\mathbb{R}^2$ ($d = 2$ and $n = 2$).

Thus, in vector-vector notation, Equation (8.9) can be written as follows

$$f(\mathbf{x}) = \mathbf{p}^T(\mathbf{x})\,\mathbf{c} \qquad (8.10)$$

where $\mathbf{c}$ stands for the vector of the real coefficients c_j associated to the basis functions p_j.

Thus, to achieve the fit function $f(\mathbf{x})$ given by Equation (8.9), it only remains to determine the parameter values c_j that minimise the least squares error E_{LS}. Such a minimum is found by setting the corresponding gradient to zero. This means that we have K gradient equations for K parameters

$$\frac{\partial E_{LS}}{\partial c_k} = 2 \sum_{i=1}^{N} r_i \frac{\partial r_i}{\partial c_k} = 0, \quad k = 1, \ldots, K \qquad (8.11)$$

or, equivalently,

8 Implicit Surface Fitting

$$\sum_{i=1}^{N}(f(\mathbf{x}_i)-f_i)\frac{\partial f(\mathbf{x}_i)}{\partial c_k}=0,\quad k=1,\ldots,K \tag{8.12}$$

Since $\frac{\partial f(\mathbf{x}_i)}{\partial c_k}=p_k(\mathbf{x}_i)$, it follows that

$$\sum_{i=1}^{N}\left(\sum_{j=1}^{K}c_j p_j(\mathbf{x}_i)-f_i\right)p_k(\mathbf{x}_i)=0,\quad k=1,\ldots,K \tag{8.13}$$

or, equivalently,

$$\sum_{i=1}^{N}\sum_{j=1}^{K}c_j p_{ij} p_{ik}=\sum_{i=1}^{N}f_i p_{ik},\quad k=1,\ldots,K \tag{8.14}$$

where $p_{ij}=p_j(\mathbf{x}_i)$ and $p_{ik}=p_k(\mathbf{x}_i)$. These K simultaneous linear equations are called the *normal equations*, which are written in matrix notation as follows:

$$\sum_{i=1}^{N}\sum_{j=1}^{K}c_j p_{ij} p_{ik}=\sum_{i=1}^{N}f_i p_{ik},\quad k=1,\ldots,K \tag{8.15}$$

From Equations (8.7) and (8.8), we obtain the system of equations in matrix-vector notation

$$(\mathbf{P}^T\mathbf{P})\mathbf{c}=\mathbf{P}^T\mathbf{f} \tag{8.16}$$

hence

$$\mathbf{c}=(\mathbf{P}^T\mathbf{P})^{-1}\mathbf{P}^T\mathbf{f} \tag{8.17}$$

where $\mathbf{c}=[c_1,\ldots,c_K]^T$ is a vector of real coefficients, $\mathbf{f}=[f_1,\ldots,f_N]^T$, and

$$\mathbf{P}=\begin{pmatrix} p_{11} & p_{12} & \cdots & p_{1K} \\ p_{21} & p_{22} & \cdots & p_{2K} \\ \vdots & \vdots & & \vdots \\ p_{N1} & p_{N2} & \cdots & p_{NK} \end{pmatrix} \tag{8.18}$$

is the $N\times K$ matrix of basis functions $p_{ij}=p_j(\mathbf{x}_i),\, i=1,\ldots,N,\, j=1,\ldots,K$.

Using matrix calculus, it can be proved that $\mathbf{c}$ is a unique solution of the system of equations (8.17), i.e. the unique solution of the least squares problem. So, if the square matrix $M_{LS}=\mathbf{P}^T\mathbf{P}$ is nonsingular or, equivalently, $\det(M_{LS})\neq 0$, we can substitute Equation (8.17) into Equation (8.10) to obtain the fit function $f(\mathbf{x})$.

Example 8.1. Let us consider the following set of sixteen data points in the Euclidean space $\mathbb{R}^2$

$$\{\mathbf{x}_i\}=\left\{\begin{array}{l}(-1,-1),\,(-1,0),\,(-1,1),\,(-1,2),\\ (0,-1),\,(0,0),\,(0,1),\,(0,2),\\ (1,-1),\,(1,0),\,(1,1),\,(1,2),\\ (2,-1),\,(2,0),\,(2,1),\,(2,2)\end{array}\right\} \tag{8.19}$$

8.3 LS Implicit Surfaces

as well as the corresponding set of associated observed function values in $\mathbb{R}$

$$\{f_i\} = \begin{Bmatrix} 1.0,\ 0.0,\ 1.0,\ 1.0, \\ 0.0,\ 2.0,\ 0.0,\ 0.0, \\ 1.0,\ 0.0,\ \text{-}0.5,\ \text{-}0.5, \\ 0.5,\ 0.5,\ 0.5,\ 0.5 \end{Bmatrix}. \tag{8.20}$$

These sample data points $\{\mathbf{x}_i, f_i\}$ in the product space $\mathbb{R}^2 \times \mathbb{R}$ are pictured in Figure 8.3 as marked as • (say, bullets). The four graphs (in green) in Figure 8.3 were produced using this sample data set, but their LS approximating functions are obviously different.

Figure 8.3(a) shows the graph of the fit function

$$f(\mathbf{x}) = 1 + A\,x + B\,y$$

with the linear basis $\mathbf{p}(\mathbf{x}) = [1, x, y]^T$ and the coefficient vector $\mathbf{c} = [A, B]^T = [-0.133, -0.190]^T$.

Figure 8.3(b) shows the graph of the fit function

$$f(\mathbf{x}) = 1 + A\,x + B\,y + C\,x^2$$

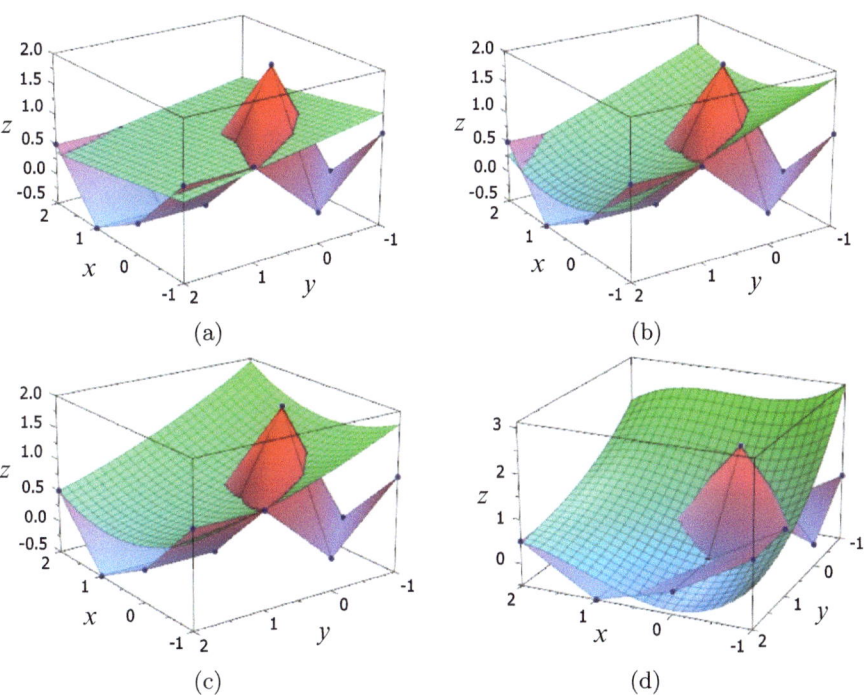

Fig. 8.3. Four LS surfaces (in green) approximating the same input data set.

with the linear basis $\mathbf{p}(\mathbf{x}) = [1, x, y, x^2]^T$ and the coefficient vector $\mathbf{c} = [A, B, C]^T = [-0.286, -0.239, 0.144]^T$.

Figure 8.3(c) shows the graph of the fit function

$$f(\mathbf{x}) = 1 + A\,x + B\,y + C\,x^2 + D\,xy + E\,y^2$$

with the linear basis $\mathbf{p}(\mathbf{x}) = [1, x, y, x^2, xy, y^2]^T$ and the coefficient vector $\mathbf{c} = [A, B, C, D, E]^T = [-0.179, -0.440, 0.176, -0.051, 0.058]^T$.

Figure 8.3(d) shows the graph of the fit function

$$f(\mathbf{x}) = 1 + A\,x + B\,y + C\,x^2 + D\,xy + E\,y^2 + F\,x^3 + G\,x^2y + H\,xy^2 + I\,y^3$$

with the linear basis $\mathbf{p}(\mathbf{x}) = [1, x, y, x^2, xy, y^2, x^3, x^2y, xy^2, y^3]^T$ and the coefficient vector

$$\mathbf{c} = [A, B, C, D, E, F, G, H, I]^T$$
$$= [0.035, 0.287, -0.656, -0.229, -0.306, 0.215, 0.207, -0.007, 0.044]^T.$$

Recall that the coefficient vectors of the example above were all determined using Equation (8.17). Besides, as shown in Chapter 1, a graph of each function $f(\mathbf{x}) : \mathbb{R}^2 \to \mathbb{R}$ is the zero set of another function $F(\mathbf{x}, z) : \mathbb{R}^2 \times \mathbb{R} \to \mathbb{R}$ given by

$$F(\mathbf{x}, z) = f(\mathbf{x}) - z = 0 \qquad (8.21)$$

since $f(\mathbf{x})$ is C^1-differentiable, as it is the case. This means that the four graphs in green shown in Figure 8.3 are also zero sets of $F(x, y, z)$ as given by Equation (8.21), i.e. they are implicit LS surfaces.

8.3.2 WLS Approximation

The method of weighted least squares (WLS) is a generalisation of the method of least squares. Similar to ordinary least squares, the unknown values of the coefficients (or parameters) $c_1, \ldots, c_K$ are estimated by finding their corresponding numerical values that minimise the sum of the squared deviations between the observed values f_i and the true values $f(\mathbf{x}_i)$. Unlike ordinary least squares, however, each term of the approximant includes a weight, w_i, that determines how much each observed value influences the final coefficients c_i. The WLS approximation that is minimised to obtain the unknown coefficients c_i is given by the minimisation of the error

$$E_{WLS} = \sum_{i=1}^{N} w_i\, r_i^2 \qquad (8.22)$$

where each *residual* r_i is weighted by the corresponding weight value w_i.

The ordinary LS is just a particular case of the WLS with all weights equal to 1. The WLS method is thus adequate when it may seem reasonable to assume that not all data points should be treated equally. This allows us, in principle, to get a better approximation to the point dataset than that one produced by the ordinary LS approximation.

To determine the parameter values c_j that minimise E_{WLS}, we follow the same procedure we used to minimise E_{LS}, i.e. using K null derivatives. Now, the gradient equations for the sum of squares in Equation (8.22) are

$$\frac{\partial E_{WLS}}{\partial c_k} = 2 \sum_{i=1}^{N} w_i \, r_i \frac{\partial r_i}{\partial c_k} = 0, \quad k = 1, \ldots, K. \tag{8.23}$$

from which, and using the same method employed in the ordinary least squares, we can easily derive the coefficients c_k. The coefficient vector is now given by

$$\mathbf{c} = \mathbf{A}^{-1}(\mathbf{x}) \, \mathbf{B}(\mathbf{x}) \, \mathbf{f} \tag{8.24}$$

with

$$\mathbf{A}(\mathbf{x}) = \mathbf{P}^T \mathbf{W}(\mathbf{x}) \, \mathbf{P} \quad \text{and} \quad \mathbf{B}(\mathbf{x}) = \mathbf{P}^T \mathbf{W}(\mathbf{x}),$$

and where the *weight matrix* $\mathbf{W}(\mathbf{x})$ is an $N \times N$ diagonal matrix given by

$$\mathbf{W}(\mathbf{x}) = \begin{pmatrix} w_1(\mathbf{x}) & 0 & 0 & \cdots & 0 \\ 0 & w_2(\mathbf{x}) & 0 & \cdots & 0 \\ 0 & 0 & w_3(\mathbf{x}) & \cdots & 0 \\ \vdots & \vdots & \vdots & \ddots & \vdots \\ 0 & 0 & 0 & \cdots & w_N(\mathbf{x}) \end{pmatrix} \tag{8.25}$$

Note that, the matrix notation has been reformulated in order to include weighting of the scattered data points. More specifically, there is a weight $w_i(\mathbf{x}) = w(||\mathbf{x} - \mathbf{x}_i||)$ for each data point $\mathbf{x}_i$, where $||\mathbf{x} - \mathbf{x}_i||$ stands for the Euclidean distance d_i between $\mathbf{x}$ and $\mathbf{x}_i$; hence the diagonal matrix $W(\mathbf{x})$, with $w_i = W_{ii}$, $i = 1, \ldots, N$.

8.3.3 MLS Approximation and Interpolation

Moving least squares (MLS) is a mesh-free approximation method. Therefore, it is often understood as an alternative to the traditional finite element and finite difference methods to scattered node configurations with no predefined connectivity [135, 233, 234]. In approximation theory, the MLS method seems to be due to McLain [263], later developed by Lancaster and Salkauskas [221] for approximating (or smoothing) and interpolating scattered data, though a particular case goes back to Shepard [363]; see also Fasshauer [136] and references therein. In statistics, the MLS method is known as mesh-free *local regression* and has been used by statisticians for the last 100 years approximately [105, 167, 415] (also, see Loader [240] and references therein).

The MLS Surfaces

The MLS method is essentially a WLS method. Accordingly, the MLS method uses the fit function given by Equation (8.9), here rewritten for our convenience

$$f(\mathbf{x}) = \sum_{j=1}^{K} c_j p_j(\mathbf{x}) \qquad (8.26)$$

or, in matrix notation, as

$$f(\mathbf{x}) = \mathbf{p}^T(\mathbf{x})\,\mathbf{c} \qquad (8.27)$$

where $\mathbf{p}(\mathbf{x})$ is the vector of basis functions, and $\mathbf{c}$ is the vector of real coefficients given by (8.24).

The MLS method differs from the WLS method in that the fit is allowed to change depending on where we evaluate the function. The WLS approximation, as well as the LS approximation, is global because the coefficients c_j, and the corresponding fit function $f(\mathbf{x})$ given by Equation (8.26), are evaluated only once. On the contrary, the MLS approximation is local because:

- First, the fit function (8.26) is evaluated for each fixed point $\mathbf{x}_i$. Therefore, the coefficient vector $\mathbf{c}$ is computed for each $\mathbf{x}_i$. Usually, but not necessarily, the set of fixed points is a proper subset of the input data set. Such fixed points are also called *nodes*, *evaluation points*, or *centres*.
- Second, and more importantly, the approximation is local because each centre $\mathbf{x}_i$ is associated with a compactly supported weight function $w_i(\mathbf{x}) = w(||\mathbf{x} - \mathbf{x}_i||)$, that is $w_i(\mathbf{x})$ rapidly decays to zero with the distance $d_i = ||\mathbf{x} - \mathbf{x}_i||$.

The weight function is a common feature to all mesh-free methods: MLS, kernels, and partitions of unity (PoU) [39]. Each centre $\mathbf{x}_i$ is associated with a *domain of influence*, called the support of the weight function $w_i(\mathbf{x})$. The support of $\mathbf{x}_i$ may be compact or not. Using a compact support (Figure 8.4(f)), we have $w_i(\mathbf{x}) > 0$ inside a subdomain that is small in relation to the domain, and $w_i(\mathbf{x}) = 0$ outside it.

There are infinitely many possibilities for the weight function $w(d)$, namely:

- *Thin-plate weight functions*. Thin-plate functions are radial basis functions (RBFs) of the form:

$$w(d) = \begin{cases} d^k & \text{if } k = 1, 3, 5, \ldots \\ d^k \ln(d) & \text{if } k = 2, 4, 6, \ldots \end{cases} \qquad (8.28)$$

These functions have not compact support. Two examples of thin-plate weight functions are pictured in Figure 8.4(a)–(b). To avoid problems with thin-plate functions of even dimension k at $d = 0$ (where $\ln(0) = -\infty$), we set up $w(0) = 0$. Alternatively, for even k, the weight function may be redefined by implementing the natural logarithm as follows:

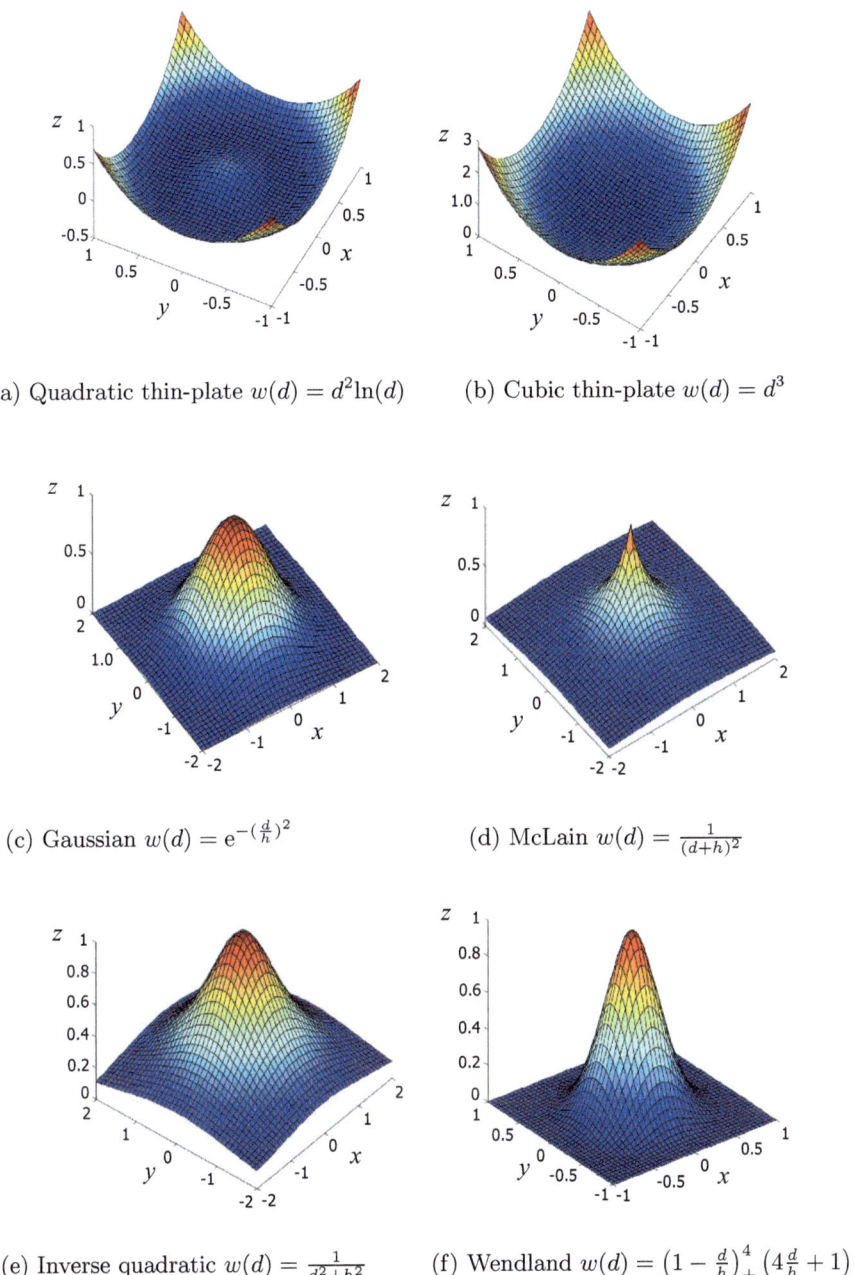

Fig. 8.4. Weight functions.

$$w(d) = \begin{cases} d^{k-1}\ln(d^d) & \text{if } d < 1 \\ d^k \ln(d) & \text{if } d \geq 1. \end{cases} \qquad (8.29)$$

- *Gaussian weight function.* This weight function has not compact support either (Figure 8.4(c)). It is given by

$$w(d) = e^{-(\frac{d}{h})^2} \qquad (8.30)$$

where h is a parameter that can be used to smooth out small features in the data [230, 5]. This parameter h is related to the full width at half maximum of the Gaussian peak, and corresponds to the position of the circle of inflection points on the Gaussian.

- *McLain weight function.* This function appears in Figure 8.4(d) and is given by

$$w(d) = \frac{1}{(d+h)^2}. \qquad (8.31)$$

- *Inverse quadratic weight function.* This is a variant of McLain's function (Figure 8.4(e)). It is as follows

$$w(d) = \frac{1}{d^2 + h^2}. \qquad (8.32)$$

In this case, there is a singularity at $d = 0$ if the parameter h vanishes.

- *Wendland weight function.* This function is shown is Figure 8.4(f). Unlike the previous weight functions, the Wendland function [408] does have compact support

$$w(d) = \left(1 - \frac{d}{h}\right)_+^4 \left(4\frac{d}{h} + 1\right), \qquad (8.33)$$

where h stands for the radius of support, and $d \in [0, h]$. Note that $w(0) = 1$, $w(h) = 0$, $w'(h) = 0$ and $w''(h) = 0$, i.e. this function is C^2 continuous. The first factor of the Wendland weight function (8.51) is a truncated quartic power function.

By choosing an adequate weight function, the fit function behaves as either an approximant or interpolant, even with lower-degree basis functions. For example, Shen et al. [362] use the inverse quadratic weight function (8.32). In this case, and assuming that $d_i = ||\mathbf{x} - \mathbf{x}_i||$ is the Euclidean distance between the centre $\mathbf{x}_i$ and some data point $\mathbf{x}$, we have:

- The weight function $w(d_i)$ quickly, monotonically decreases to zero as $d_i \to \infty$.
- The weight function $w(d_i)$ tends to infinity near the corresponding input data point $\mathbf{x}_i$ (where $d_i \to 0$) when h is very small, which forces the fit function to interpolate the corresponding function values. Thus, the parameter h determines how the MLS fit function behaves, either as an interpolant or as an approximant.

The inverse quadratic weight function rapidly decays to zero, but it has not compact support. This ensures *a priori* that we end up having a smooth assembly of the local fits into the global fit. On the contrary, using compactly supported weight functions (e.g. the Wendland function) does not guarantee such a smooth assembly, unless we use some extra algorithmic machinery to ensure that every point of the domain Ω is covered by at least one fitting polynomial. In other words, we have to make sure that the compact supports of weight functions $w(d_i)$ centred at $\mathbf{x}_i$ cover Ω or, equivalently,

$$\Omega = \bigcup_{i=1}^{N} \mathrm{supp}(w(d_i)). \tag{8.34}$$

Thus, since the coefficient vector $\mathbf{c}$ varies with $\mathbf{x}$, the polynomial coefficients have to be recomputed for every centre. We do so by evaluating a new weight matrix $\mathbf{W}_i$ for each centre $\mathbf{x}_i$, which is seen as a new location of the *moving* $\mathbf{x}$; hence the *moving* least squares (MLS). That is, one computes a local MLS fit for each translate of $\mathbf{x}$ individually. This leads us to slightly change Equation (8.26) for each local MLS function as

$$f_i(\mathbf{x}) = \sum_{j=1}^{K} c_{ij}(\mathbf{x}) p_j(\mathbf{x}) \tag{8.35}$$

or, using the matrix notation,

$$f_i(\mathbf{x}) = \mathbf{p}^T(\mathbf{x}) \mathbf{c}_i(\mathbf{x}) \tag{8.36}$$

which give us the approximation to the surface in the neighbourhood of the ith fixed point $\mathbf{x} = \mathbf{x}_i$, and where $\mathbf{c}_i$ is the corresponding coefficient vector obtained by minimising the WLS error (8.22). Thus,

$$\mathbf{c}_i = \mathbf{A}_i^{-1}(\mathbf{x}) \mathbf{B}_i(\mathbf{x}) \mathbf{f} \tag{8.37}$$

with

$$\mathbf{A}_i(\mathbf{x}) = \mathbf{P}^T \mathbf{W}_i(\mathbf{x}) \mathbf{P} \quad \text{and} \quad \mathbf{B}_i(\mathbf{x}) = \mathbf{P}^T \mathbf{W}_i(\mathbf{x}),$$

We can then say that there is local degree-d polynomial reproduction in the neighbourhood of each data point $(\mathbf{x}_i, f_i)$. Obviously, computing each local polynomial $f_i(\mathbf{x})$ that fits the surface in the neighbourhood $N(\mathbf{x}_i)$ of the centre $\mathbf{x}_i$ requires that we determine which data points are inside $N(\mathbf{x}_i)$ by means of, for example, k-d trees [362].

Finally, by summing up the local MLS fit functions given by Equation (8.35), we obtain a *global* MLS approximation written as follows:

$$F(\mathbf{x}) = \sum_{i=1}^{N} f_i(\mathbf{x}) \tag{8.38}$$

By substituting Equation (8.35) into Equation (8.38), we get the global MLS approximation written in the form of Equation (8.26), i.e.

$$F(\mathbf{x}) = \sum_{j=1}^{K} c_j(\mathbf{x}) p_j(\mathbf{x}) \quad \text{with} \quad c_j(\mathbf{x}) = \sum_{i=1}^{N} c_{ij}(\mathbf{x}). \tag{8.39}$$

Also, substituting Equation (8.37) into Equation (8.36), and Equation (8.36) into Equation (8.38), and rearranging the equation in relation to the observed values f_i, Equation (8.38) can be written in the following form:

$$F(\mathbf{x}) = \sum_{i=1}^{N} \mathbf{p}^T(\mathbf{x}) \mathbf{A}_i^{-1}(\mathbf{x}) \mathbf{B}_i(\mathbf{x}) \mathbf{f}$$

$$= \sum_{i=1}^{N} \left(\sum_{j=1}^{K} p_j(\mathbf{x}) [\mathbf{A}_i^{-1}(\mathbf{x}) \mathbf{B}_i(\mathbf{x})]_j \right) f_i$$

that is

$$F(\mathbf{x}) = \sum_{i=1}^{N} \phi_i(\mathbf{x}) f_i \tag{8.40}$$

where

$$\phi_i(\mathbf{x}) = \sum_{j=1}^{K} p_j(\mathbf{x}) [\mathbf{A}_i^{-1}(\mathbf{x}) \mathbf{B}_i(\mathbf{x})]_j \tag{8.41}$$

is known as the MLS shape function associated to jth node; $[\mathbf{A}_i^{-1}(\mathbf{x}) \mathbf{B}_i(\mathbf{x})]_j$ is the jth column of the $K \times N$ matrix $\mathbf{A})_i^{-1}(\mathbf{x}) \mathbf{B}_i(\mathbf{x})$. Remarkably, the MLS shape functions (8.41) form a partition of unity since there exists a p_j such that $p_j(\mathbf{x}) = 1$, a result due to Duarte and Oden [118].

The MLS Algorithm

A naïve computation of the global MLS function would involve the computation of a local MLS function in the neighbourhood of each data point $\mathbf{x}_i$ ($i = 1, \ldots, N$). As argued by Shen et al. [362], this naïve computation is impracticable for large data sets.

A solution for this problem is to use a rather smaller number of centres, say $n < N$, in the neighbourhood of each one estimates the corresponding local MLS fit. This centre reduction can be achieved by using an adaptive octree subdivision of domain Ω into leaf subdomains Ω_k ($k = 1, \ldots, n$), and then defining a centre for each leaf subdomain. Of course, the recursive subdivision of the domain stops when the number of data points inside each subdomain falls below a given threshold.

Stopped the subdivision of the domain, one determines the local MLS fit $f_i(\mathbf{x})$ given by (8.35) within each leaf subdomain. To be more precise, and

8.3 LS Implicit Surfaces

Algorithm 30 The MLS Implicit Surface Reconstruction

1: **procedure** MLS($\{\mathbf{x}_i\},\{\mathbf{f}_i\},P(\mathbf{x})$)   ▷ $i \leftarrow 1, N$
2:     $F \leftarrow 0$ ▷ initialise the global MLS function
3:     Partition Ω into n leaf domains Ω_k ▷ $n < N$
4:     **for** $k \leftarrow 1, n$ **do**
5:       Evaluate the local MLS function f_k ▷ local MLS function (8.35)
6:       $F \leftarrow F + f_k$ ▷ global MLS function (8.38)
7:     **end for**
8:     Ray-trace or polygonise the implicit surface F
9: **end procedure**

in order to guarantee the continuity of the global reconstruction surface, the local MLS approximations (8.35) are determined within overlapping spherical neighbourhoods enclosing leaf subdomains, instead of the leaf subdomains themselves.

Several surface reconstruction algorithms use this adaptive space subdivision scheme, namely those due to Ohtake et al. [308], Tobor et al. [389], and Yang et al. [423]. Algorithm 30 concerns the MLS algorithm with centre reduction to reconstruct implicit surfaces from scattered datasets of points.

The partition (step 3) of the domain Ω into subdomains Ω_k involves the computation of the K nearest neighbours of $\mathbf{c}_k$, the centre of the box subdomain Ω_k. These K nearest neighbour points are easily retrieved from a K-d tree and constitute a spherical neighbourhood $N(\mathbf{c}_k)$ that encloses Ω_k. Assuming that the local reconstruction functions f_k contribute equally to the global reconstruction function F, we end up having a partition of unity for MLS surfaces (step 8).

It is worthy noting that point sets obtained from range scanners are usually noisy and do not include normals, which poses some difficulties in rendering MLS surfaces. Ohtake et al. [308] implemented a technique to generate pseudo-normals from surface point datasets, which was originally suggested by Turk and O'Brien [392]. This technique takes advantage of the definition of a closed implicit surface as a zero set such that one attaches a zero constraint to a point on the surface, a positive offset constraint to a point just outside the surface, and a negative one just inside. Shen et al. [362] improved on this technique not only for rendering purposes, but also to reduce undesirable oscillations in the process of fitting a surface to points of the dataset. Interestingly, Sun-Jeong Kim and Chang-Geun Song [211] use the MLS approximation itself not only to approximate the surface in the neighbourhood of each point, but also to obtain normals and relevant differential data.

Levin's MLS Surfaces

A different MLS approach to reconstruct an implicit surface $\mathcal{S}$ in $\mathbb{R}^3$ from a unstructured point dataset $\{\mathbf{x}_i\}_{i=1}^N$ on $\mathcal{S}$ or nearly on $\mathcal{S}$ was originally proposed by Levin [230]. This method, also known as PMLS (projection moving

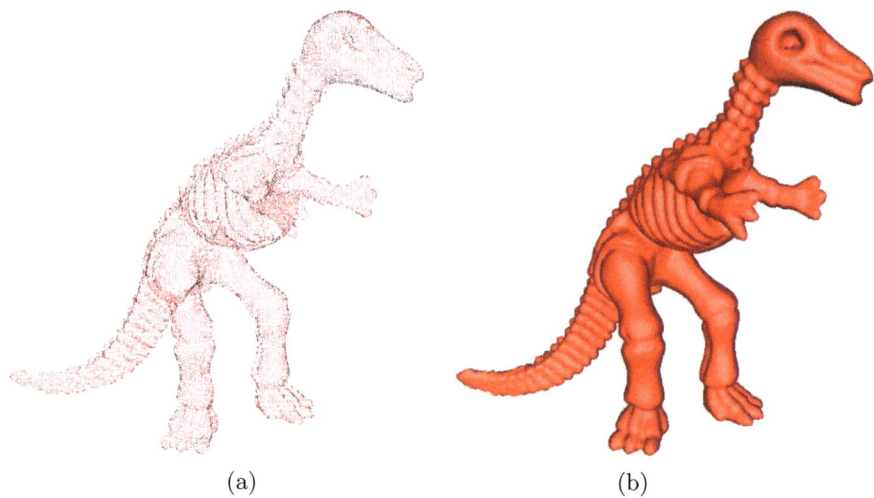

Fig. 8.5. A MLS surface reconstructed from a cloud of points using Pointshop 3D.

least squares), is based upon a projection operator that takes points near the data set onto a smooth surface. Figure 8.5 shows a cloud of points and the corresponding MLS surface produced by a variant of the PMLS surfaces, called VMLS, which were encoded in PointShop 3D (see Zwicker et al. [429] for further details).

Given a point $\mathbf{x}$ near $\mathcal{S}$, the projection procedure involves two steps:

- Computation of a hyperplane H in $\mathbb{R}^3$ that approximates the surface $\mathcal{S}$ *locally*.
- Computation of the projection of $\mathbf{x}$ on a local polynomial approximation of $\mathcal{S}$ over H.

Remarkably, both steps are performed using the MLS method. Of course, the degree of approximation depends on the degree of polynomials used in the second step. More specifically, the projection procedure can be detailed as follows:

- **The local approximating hyperplane**. The computation of the local approximating hyperplane (sometimes called stencil)

$$H = \{\mathbf{x} \in \mathbb{R}^3 : \mathbf{x} \cdot \mathbf{n} - D = 0\}, \tag{8.42}$$

here defined by a unit vector $\mathbf{n}$ and an offset $D \in \mathbb{R}$ (the symbol $\cdot$ stands for the standard inner product in $\mathbb{R}^3$), is accomplished by minimising the following weighted sum of squared distances

$$\sum_{i=1}^{N} (\mathbf{x}_i \cdot \mathbf{n} - D)^2 \, w(||\mathbf{x}_i - \mathbf{p}||). \tag{8.43}$$

where $\mathbf{p}$ is the orthogonal projection of $\mathbf{x}$ onto H, and w is a non-negative weight function that typically only depends on the Euclidean distance. Taking into account that the weights $w(||\mathbf{x}_i - \mathbf{p}||)$ decrease with the Euclidean distance $||\mathbf{x}_i - \mathbf{p}||$, we conclude that the resulting hyperplane H approximates the tangent hyperplane to $\mathcal{S}$ near $\mathbf{p}$. In the minimisation process of (8.42), we may find several local minima, so we choose the closest to $\mathbf{p}$, i.e. the one that minimises $|\mathbf{x} \cdot \mathbf{n} - D|$. Note that H defines a local orthonormal coordinate system with $\mathbf{p}$ at the origin, so that the coordinates of the sample points $\{\mathbf{x}_i\}_{i=1}^{N}$ of $\mathcal{S}$ in this coordinate system are (u_i, v_i, h_i), where (u_i, v_i) are the parameter values in H and $h_i = \mathbf{x}_i \cdot \mathbf{n} - D$ is the height of $\mathbf{x}_i$ over H.

- **The local approximating polynomial P.** This is the first part of the second step. Let $\{\mathbf{p}_i\}_{i=1}^{N}$ be the orthogonal projections of the sample points $\{\mathbf{x}_i\}_{i=1}^{N}$ onto H, i.e. $\mathbf{p}_i = (u_i, v_i, 0)$. The 2-dimensional local approximation to $\mathcal{S}$ is given by a degree 2 polynomial P that minimises the weighted least squares error

$$\sum_{i=1}^{N}(P(\mathbf{p}_i) - h_i)^2 \, w(||\mathbf{x}_i - \mathbf{p}||). \tag{8.44}$$

- **The projection of x onto the local polynomial approximation of the surface.** This is the second part of the second step. Once determined the approximating polynomial P, we can determine the projection of $\mathbf{x}$ onto the MLS surface $\mathcal{S}$ as follows

$$\Pi(\mathbf{x}) = \mathbf{p} + P(\mathbf{0}) \cdot \mathbf{n}, \tag{8.45}$$

where $\Pi(\mathbf{x})$ denotes such a projection operator.

In short, Levin's two-step minimisation procedure first determines a local reference frame that approximates the surface, and then computes the polynomial function that approximates the surface locally. The reason behind the use of a local stencil H of data points rather than the whole set of data points in the domain lies in the fact that points near the fixed point have more influence on the value of the function than those points far away. Applying the stencil to several points, we are able to construct a global approximation to the whole surface by blending those local approximations.

However, Amenta and Kil [15] show that Levin's projection has undesirable behaviour near corners and edges. As Ochotta et al. also note in [307], the failures of Levin's two-step procedure near corners and edges occur regardless of the sampling density and neighbourhood size. In fact, the first minimisation step may fail to output the expected reference frame, and consequently the expected normal to the surface. This is illustrated in Figure 8.6; (a) shows the expected local hyperplane that approximates the surface, while (b) depicts a wrong local hyperplane. In the latter case, the MLS projection operator pinches and extends away from the points $\mathbf{x}_i$.

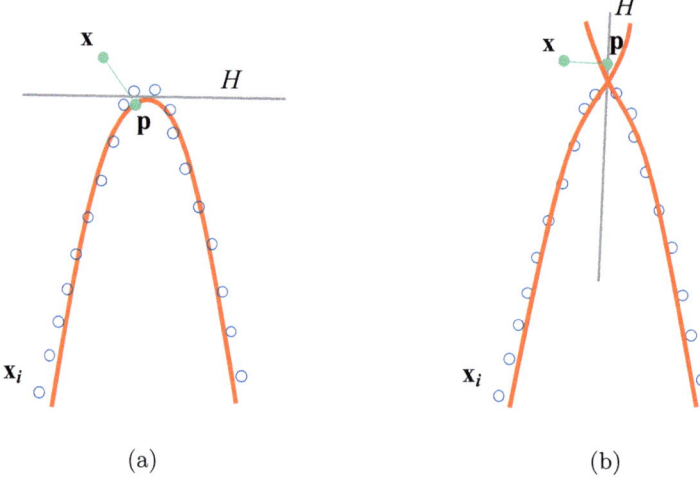

Fig. 8.6. The first step of Levin's procedure for MLS surfaces: (a) the expected local approximating hyperlane and (b) the wrong local approximating hyperlane H for a point **x** near a curve.

Intuitively, we may think of the sampling density of points as the source of Levin's first-step minimisation problem. In fact, from the differential geometry, we know that such approximating local plane exists for smooth surfaces, but the first step of Levin's MLS procedure is not finding it correctly. Ochotta et al. [307] also argue that this deficiency results from performing the minimisation in two different steps. Once determined a planar approximation to the surface at the first step, there is no way for the second step to correct such a local reference frame. Instead, they use a unified projection operator that guarantees that the projection is always orthogonal to f, not to the hyperplane H. A former solution to this problem is due to Fleishman et al. [139], who propose a robust projection scheme that does not fail near sharp features (say, corners and edges), but this is accomplished at expense of extra processing steps.

Note that Levin's MLS surface can be viewed as a point-set surface, which is defined as the set of stationary points of a projection operator. It seems that this surface definition was first used by Alexa et al. [5] in the context of point-based modelling and rendering. See [138] and [319] for other applications of the concept of point-set surfaces. Amenta and Kil [6] explicitly define a point-set surface as the set of local minima of an energy function along the directions provided by a vector field. Adamson and Alexa [1] give a simplified definition of an implicit surface which is tailored for efficient ray tracing, as well as sampling conditions that guarantee the manifold reconstruction of the surface.

8.4 RBF Implicit Surfaces

Radial basis functions (RBFs) were introduced by Hardy [176] and Duchon [120]. Hardy empirically discovered, in the field of cartography, that a linear combination of multiquadrics—now known as multiquadric RBFs—produce a single interpolant that fits all the given scattered point data. In mathematics, the seminal work on thin-plate splines and other RBFs is due to Duchon [121, 119, 120] and later Madych and Nelson [254, 255, 253].

Since then, RBFs have gained in popularity in various disciplines of science and engineering, namely computer graphics, computer aided design, neural networks, fluid mechanics and simulation, geodesy and cartography, and applied mathematics. One of the most influential contributions to the theory of RBFs is due to Micchelli [269], who established solid theoretical results on the invertibility of the RBF interpolation matrix. Micchelli's work focused on "Laplace transforms" related to Gaussian RBFs. Also, Franke [145] published a survey paper on several interpolating methods which helped to shed light on the essence of the theory of RBFs. A more comprehensive study of the RBFs is given by Buhmann [73].

In computer graphics, Nielson [300] introduced the use of RBFs to build up interpolants for 3D data where the interpolation centres do not form a regular grid, but the first surface reconstruction algorithms based on RBF interpolants are usually credited to Savchenko et al. [349], Carr et al. [78], and Turk and O'Brien [392]. Savchenko and colleagues proposed RBFs to interpolate implicit surfaces. Carr and colleagues used the thin-plate spline RBF to interpolate incomplete surfaces extracted from 3D medical graphics [78]. Turk and O'Brien proposed the variational implicit surface, which is essentially a RBF implicit surface built up from the interpolation of a cloud of points in a way similar to thin-plate interpolation [392]. This interpolation method provides the surface of minimal curvature that passes through the data points. Solving this minimisation problem is accomplished using variational interpolation; hence the variational implicit surfaces. For further details on variational interpolation, the reader is referred to Duchon [120].

8.4.1 RBF Interpolation

Let us then consider the problem of interpolating a multivariate function $f : \Omega \subset \mathbb{R}^3 \to \mathbb{R}$, from a set of sample values $\{f(\mathbf{x}_i)\}_{i=1}^N$ on a scattered point data set $\{\mathbf{x}_i\}_{i=1}^N \subset \Omega$. A popular method for interpolating multivariate scattered data is using RBFs. To reconstruct f efficiently, it suffices to approximate f locally by a real-valued function ϕ at each *centre* $\mathbf{x}_i$. This real-valued function is called *radial basis function* (RBF) and depends on the Euclidean distance $||\cdot||$ from each centre $\mathbf{x}_i$, so that $\phi(||\mathbf{x} - \mathbf{x}_i||) = \phi_i(\mathbf{x})$. That is, ϕ is radially symmetric.

The characteristic property of a radial basis function is that its response decreases (or increases) monotonically with distance from its centre.

8 Implicit Surface Fitting

In Figure 8.4, we can observe various RBFs. In the MLS formulation, these radial functions are used as weight functions. Here, they are used as basis functions of splines that approximate functions and interpolate data.

The RBF interpolation requires an appropriate choice of the radial basis functions. An important class of radial basis functions is the class of thin-plate functions. The thin-plate functions are the basis functions of thin-plate splines. Thin-plate functions are a special case of polyharmonic functions, which means that thin-plate splines are a particular case of polyharmonic splines. In fact, Duchon [121] shows that solving for thin-plate splines to interpolate sample points in 2D is equivalent to using the biharmonic radial basis function $\phi(d) = d^2 \log(d)$ to interpolate these points; in 3D, the thin-plate solution amounts to interpolating the sample points using the triharmonic radial radial function $\phi(d) = d^3$. Turk and O'Brien [392] just opted by using the triharmonic thin-plate RBF $\phi(||\mathbf{x} - \mathbf{x}_i||) = ||\mathbf{x} - \mathbf{x}_i||^3$, while Carr et al. [78] used the uniharmonic thin-plate RBF $\phi(||\mathbf{x} - \mathbf{x}_i||) = ||\mathbf{x} - \mathbf{x}_i||$. Chosen the radial basis function ϕ, we are able to build up approximations to the function f as follows

$$F(\mathbf{x}) = \sum_{i=1}^{N} w_i \phi(||\mathbf{x} - \mathbf{x}_i||) \tag{8.46}$$

where the interpolant $F(\mathbf{x})$ is a result of the sum of N radial basis functions, each of which is associated with a distinct centre $\mathbf{x}_i$, and weighted by an adequate coefficient w_i.

In order to guarantee the positive-definiteness—one of the conditions to ensure the uniqueness—of the solution, it is necessary to add a low-degree polynomial to the interpolant F in Equation (8.46). The result is the *general form* of an RBF interpolant:

$$F(\mathbf{x}) = P(\mathbf{x}) + \sum_{i=1}^{N} w_i \phi(||\mathbf{x} - \mathbf{x}_i||), \quad \mathbf{x} \in \mathbb{R}^n \tag{8.47}$$

where $P(\mathbf{x})$ is a polynomial of degree at most k. Duchon [120] shows that a linear polynomial P suffices to guarantee the smoothness of the interpolant. The biharmonic and triharmonic splines likely are the smoothest interpolants in the sense that they interpolate the scattered data and minimise certain energy functionals [77].

Surface reconstruction through RBF interpolation thus requires to approximate a real-valued function $f(\mathbf{x})$ by the RBF interpolant $F(\mathbf{x})$ given the scalar values f_i at the distinct points $\mathbf{x}_i$ ($i = 1, \ldots, N$). To compute the interpolant $F(\mathbf{x})$ in Equation (8.47), we have to first determine the weights w_i and the vector of unknown coefficients $\mathbf{c} = [c_1, \ldots, c_k]^T$ that give P in terms of its basis $\mathbf{b}(\mathbf{x}) = [b_1(\mathbf{x}), \ldots, b_k(\mathbf{x})]^T$. The weights w_i are such that $F(\mathbf{x})$ satisfies the interpolation conditions

$$F(\mathbf{x}_i) = f_i, \quad i = 1, \ldots, N. \tag{8.48}$$

This system of equations has more parameters than data, so we have to further impose the following orthogonality conditions on the weights w_i:

$$\sum_{i=1}^{N} w_i P(\mathbf{x}_i) = 0, \quad \text{for all polynomials } P \text{ of degree at most } k. \qquad (8.49)$$

Equations (8.48) and (8.49) may be then written in matrix form as

$$\begin{pmatrix} \mathbf{A} & \mathbf{B} \\ \mathbf{B}^T & \mathbf{O} \end{pmatrix} \begin{pmatrix} \mathbf{w} \\ \mathbf{c} \end{pmatrix} = \begin{pmatrix} \mathbf{f} \\ \mathbf{0} \end{pmatrix} \qquad (8.50)$$

where

$$\mathbf{A} = \begin{pmatrix} \phi_{11} & \phi_{12} & \cdots & \phi_{1N} \\ \phi_{21} & \phi_{22} & \cdots & \phi_{2N} \\ \vdots & \vdots & \ddots & \vdots \\ \phi_{N1} & \phi_{N2} & \cdots & \phi_{NN} \end{pmatrix}$$

with $\phi_{ij} = \phi(||\mathbf{x}_j - \mathbf{x}_i||)$ for $i, j = 1, \ldots, N$, and

$$\mathbf{B} = \begin{pmatrix} b_{11} & b_{12} & \cdots & b_{1k} \\ b_{21} & b_{22} & \cdots & b_{2k} \\ \vdots & \vdots & & \vdots \\ b_{N1} & b_{N2} & \cdots & b_{Nk} \end{pmatrix}$$

with $b_{ij} = b_j(\mathbf{x}_i)$, for $i = 1, \ldots, N$, $j = 1, \ldots, k$. The submatrix $\mathbf{O}$ is the $(k \times k)$ zero matrix.

Solving the system of Equations (8.50) determines the coefficients $\mathbf{c} = [c_1, \ldots, c_k]^T$ of the the polynomial P, the weights $\mathbf{w} = [w_1, \ldots, w_N]^T$, and hence $F(\mathbf{x})$. In fact, the system (8.50) is symmetric and positive semi-definite, which means that there is a unique solution for the w_i and c_j [160]. Moreover, the system can be directly solved using the symmetric LU decomposition if the system is limited to a few thousand constraints; for example, Turk and O'Brien [392] employed this LU technique to interpolate implicits together with a degree-one polynomial P.

The RBF interpolation allow us to reconstruct complex and smooth implicit surfaces with arbitrary topological shape from a set of constraint points, weights and a k-order polynomial. Consequently, it is straightforward to compute the normal vector and the curvature at a given point of the surface, as necessary in many geometric applications, and ensure C^2 continuity. Another nice property of RBFs is that they impose no restrictions on the scattered data points. In particular, RBFs do not require that these scattered data points lie in any type of regular grid or space decomposition, as usual in the marching cubes and its variants. Unfortunately, the interpolation process is very time-consuming when the number of data points goes above a few thousands, since it requires the global evaluation of all the basis functions.

8.4.2 Fast RBF Interpolation

Standard RBFs have the disadvantage that they only cope with small scattered data sets, i.e. data sets having up to two thousand points approximately. To overcome this problem, Carr et al. [76, 77] introduced a kind of fast RBFs to reconstruct surfaces from very large point data sets. Speeding up the RBF interpolation was made feasible thanks two mechanisms:

- Fast multipole method (FMM).
- RBF centre reduction.

The FMM was originally developed by Greengard and Rokhlin [168] in computational physics. This method allows the fast summation of potential fields (harmonic RBFs) as needed in particle simulations. More specifically, it allows the matrix-vector product of the form given by Equation (8.46) to be computed in $\mathcal{O}(n \log(n))$ or even $\mathcal{O}(n)$ operations instead of the $\mathcal{O}(n^2)$ operations performed on the direct product. In fact, the FMM was originally designed for the fast evaluation of harmonic RBFs in 2D and 3D, and later extended to higher-order polyharmonic RBFs by Beatson et al. [36, 37], just those used by Carr et al. [76] .

The fast RBF interpolation also uses the centre reduction technique to make possible the reconstruction of surfaces from very large point data sets. Unlike the standard RBFs, the fast RBFs do not use all the input scattered points $\mathbf{x}_i$ as RBF centres (or interpolation nodes). By using a significantly smaller point data subset of centres within a predefined fitting accuracy δ, one can reconstruct the surface with the same visual quality. This allows us then to remove the redundant detail or noise before going to the fitting stage. This can be done using a greedy algorithm as the one described by Carr et al. [76].

To render an RBF implicit surface, we have to bear in mind that it can be defined as the zero set of a single RBF fitted to the given scattered point data. Therefore, this surface can be directly displayed using a ray tracer, or indirectly using any implicit surface polygonisation algorithm described in previous chapters. For example, Carr et al. [76] use the marching tetrahedra to polygonise RBF implicit surfaces.

8.4.3 CS-RBF Interpolation

The main drawback of the RBFs, including the fast RBF, lies in their global nature, which results from the use of noncompactly supported basis functions. This means that even a small change on a single constraint point or centre affects the entire interpolated surface. On the contrary, compactly supported RBFs (CS-RBFs) follow the principle of locality in a way similar to the classical B-splines [408]. Changing the position of a given centre $\mathbf{x}_i$ causes only a local change of the interpolant and the corresponding surface.

8.4 RBF Implicit Surfaces

Therefore, compactly supported RBFs (CS-RBFs) allow for a better control since the influence of each radial function is local. This control is determined by the radius of the radial basis function at each centre. Wendland [408] introduced the following family of compactly supported radial basis functions to interpolate an implicit surface from scattered point data:

$$\phi = \begin{cases} (1-r)_+^p P(r) & \text{if } 0 \leq r < 1 \\ 0 & \text{if } r \geq 1 \end{cases} \quad (8.51)$$

where $(1-r)_+^p$ is the truncated power function, and $P(r)$ is a low degree polynomial. In particular, the radial basis functions in Table 8.1 are derived from (8.51).

Table 8.1. Wendland's compactly supported radial basis functions (CS-RBFs).

Dimension	Radial Basis Function (RBF)	Continuity
$d=1$	$(1-r)_+$	C^0
	$(1-r)_+^3(3r+1)$	C^2
	$(1-r)_+^5(8r^2+5r+1)$	C^4
$d=3$	$(1-r)_+^2$	C^0
	$(1-r)_+^4(4r+1)$	C^2
	$(1-r)_+^6(35r^2+18r+3)$	C^4
	$(1-r)_+^8(32r^3+25r^2+8r+1)$	C^6
$d=5$	$(1-r)_+^3$	C^0
	$(1-r)_+^5(5r+1)$	C^2
	$(1-r)_+^7(16r^2+7r+1)$	C^4

The radius of support of these functions is equal to 1, though any radius is allowed by scaling of the basis function. By summing up the contributions of these polynomial, positive-definite and compactly supported RBFs associated to the centres, we obtain a piecewise polynomial interpolant of minimal degree. As expressed in Table 8.1, the dimension and smoothness of the interpolant F depend on the dimension d and continuity C^k of the radial basis functions.

8.4.4 The CS-RBF Interpolation Algorithm

The CS-RBF interpolation algorithm to reconstruct implicit surface from scattered data sets is outlined in Algorithm 31 (cf. Morse et al. [282]) and illustrated in Figure 8.7.

We are here assuming that the number n of centres that is less than the number N of data points, i.e. we are using centre reduction. The steps 4–6 are the core of Algorithm 31 since they are also the steps we need to compute the standard RBF interpolant given by (8.47).

In the case of compactly supported RBFs, these three steps are locally performed for each centre as follows:

Algorithm 31 The CSRBF Implicit Surface Reconstruction

1: **procedure** CSRBF($\{\mathbf{x}_i\},\{\mathbf{f}_i\},P(\mathbf{x})$) ▷ $i \leftarrow 1, N$
2:     $F \leftarrow P(\mathbf{x})$
3:     **for** $k \leftarrow 1, n$ **do** ▷ $n \leq N$
4:         Build up the system of equations (8.50)
5:         Solve the system of equations (8.50)
6:         Evaluate the local interpolant F_k
7:         $F \leftarrow F + F_k$
8:     **end for**
9:     Ray-trace or polygonise the implicit surface F
10: **end procedure**

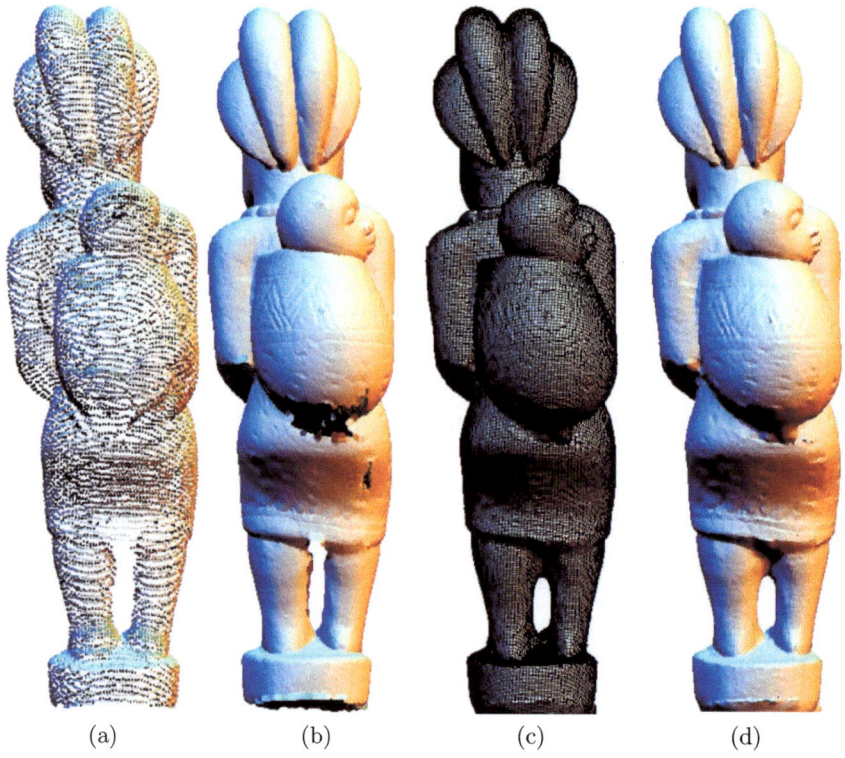

(a)　　　　　(b)　　　　　(c)　　　　　(d)

Fig. 8.7. Multilevel CS-RBF surface reconstruction of an African statue (woman and child): (a) a subset of the original point dataset; (b) the original point dataset with 220,318 points with some missing local regions of points; (c) the mesh of the reconstructed surface; (d) the flat-shaded reconstructed surface. (This African statue dataset is courtesy of Tamy Boubekeur's repository. The pictures were generated from the RBF3D testbed due to Ohtake et al. [312].)

- *Building up the system of equations.* Using radial basis functions of finite support has a significant impact on the sparsity of the resulting matrix of the system of equations (step 4), simply because $\phi(||\mathbf{x}_i - \mathbf{x}_k||) = 0$ for all $(\mathbf{x}_i, \mathbf{x}_k)$ farther apart than the radius of support. An immediate consequence is that only $O(n)$ storage is required. Besides, Morse et al. [282] use a k-d tree to find all points within the radius r of a given centre, which is performed in $O(\log n)$ time.
- *Solving the system of equations.* To solve the system of equations (step 5), Morse and colleagues use a direct (LU) sparse matrix solver [116]. The computational complexity of this LU solver seems to be $O(n^{1.5})$ at most [345].
- *Evaluating the local interpolant.* Finally, one evaluates the local interpolant F_k at each using (8.46), where N now denotes the number of data points within the radius of support.

A similar reconstruction algorithm based on CS-RBFs is due to Ohtake et al. [310, 312]. They use a multiresolution approach together with a simple RBF centre reduction to achieve high-quality approximations, even when the density of point data is not uniform or there is the need for repairing incomplete data, as illustrated in Figure 8.7.

8.5 MPU Implicit Surfaces

The partition-of-unity (PoU) approach has attracted considerable attention in recent years because of its local nature. The essence of this approach can be traced back to Shepard's blending method [363], which was originally thought of as a good means of building a global solution function by blending local solution functions using smooth, local weights that sum up to one everywhere in the domain. Thus, each local function has a limited domain of influence in the resulting global function. This is illustrated in Figure 8.8 for four positive real functions f_1, f_2, f_3, and f_4 that are defined locally about four points $\mathbf{x}_1$, $\mathbf{x}_2$, $\mathbf{x}_3$, and $\mathbf{x}_4$, respectively.

In general, we split a bounded domain $\Omega \in \mathbb{R}^d$ into a number of slightly overlapping subdomains Ω_i that cover Ω, that is $\Omega \subseteq \bigcup_i \Omega_i$. These subdomains are intervals in $\mathbb{R}$ (Figure 8.8), circles in $\mathbb{R}^2$ (Figure 8.9), and spheres in $\mathbb{R}^3$ that overlap near their boundaries.

Then, we build a partition of unity on the set $\{\Omega_i\}$ of subdomains. Such a partition of unity is nothing more than a collection of nonnegative functions ϕ_i with compact support $\mathrm{supp}(\phi_i) \subseteq \Omega_i$ that satisfy the condition $\sum_i \phi_i = 1$ in Ω.

For each subdomain $\{\Omega_i\}$, we form the set P_i of the data points inside $\{\Omega_i\}$, computing then a local function f_i that fits the points of P_i. The global fitting function F is then the result of a combination of the local functions f_i weighted by the partition functions ϕ_i as follows:

256   8 Implicit Surface Fitting

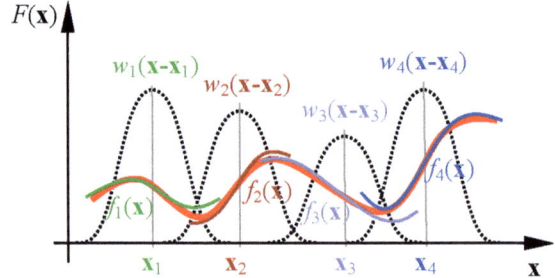

Fig. 8.8. Four local functions f_i ($i = 1, 2, 3, 4$) blended by weighting functions w_i centred at the points $\mathbf{x}_i$ on the real line $\mathbb{R}$. The graph of the resulting function is the red solid curve.

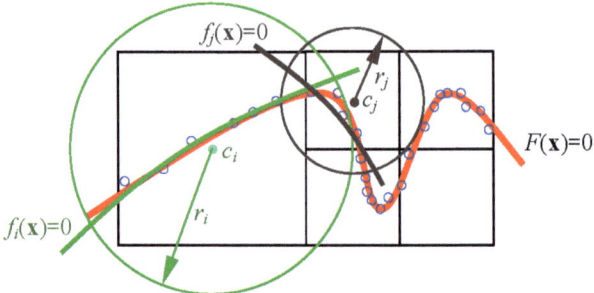

Fig. 8.9. Three local functions f_i ($i = 1, 2, 3$) blended by weighting functions w_i centred at the points $\mathbf{c}_i$ on the real line $\mathbb{R}$.

$$F(\mathbf{x}) = \sum_{i=1}^{n} \phi_i(\mathbf{x}) f_i(\mathbf{x}) \qquad (8.52)$$

For example, in Figure 8.8, the resulting function F in red is built from a combination of four local functions f_i, which are associated to four weight functions w_i. The condition $\sum_i \phi_i = 1$ is just obtained from the smooth weight functions w_i using the following normalisation formula:

$$\phi_i(\mathbf{x}) = \frac{w_i(\mathbf{x})}{\sum_j w_j(\mathbf{x})}. \qquad (8.53)$$

Equation (8.52) can be then re-written as

$$F(\mathbf{x}) = \frac{\sum_i w_i(\mathbf{x}) f_i(\mathbf{x})}{\sum_i w_i(\mathbf{x})} \qquad (8.54)$$

Starting from this framework, various hierarchical implicit surface reconstruction algorithms have been proposed in the literature. Possibly, the most

8.5 MPU Implicit Surfaces

known of these algorithm is the multilevel partition-of-unity (MPU) method introduced by Ohtake et al. [308]. This method uses an adaptive 2^n-tree subdivision of the domain which allows for manifold reconstruction even from large point datasets.

In 2D, reconstructing an implicit curve is carried out using an adaptive quadtree-based subdivision, as illustrated in Figure 8.9. In this case, the subdivision of a square into four smaller squares depends on the shape changes of the curve inside such a square.

In 3D (see Figure 8.10), the MPU method to reconstruct an implicit surface involves three key ingredients:

- An octree space partitioning that subdivides the bounding box domain into smaller boxes or cubes. This space subdivision adapts to local shape variations of the reconstructing surface. However, the subdomains are not those cubes. Instead, the subdomains are spheres that contain the cubes and match their centres in order to cover the domain.
- Piecewise quadratic functions f_i that fit the local shape of the surface, i.e. a piecewise quadratic function fits the data points in each box subdomain or octree cell. These piecewise quadratic functions are called the *local shape functions*.
- Weighting functions (partitions of unity) that blend those quadratic functions together.

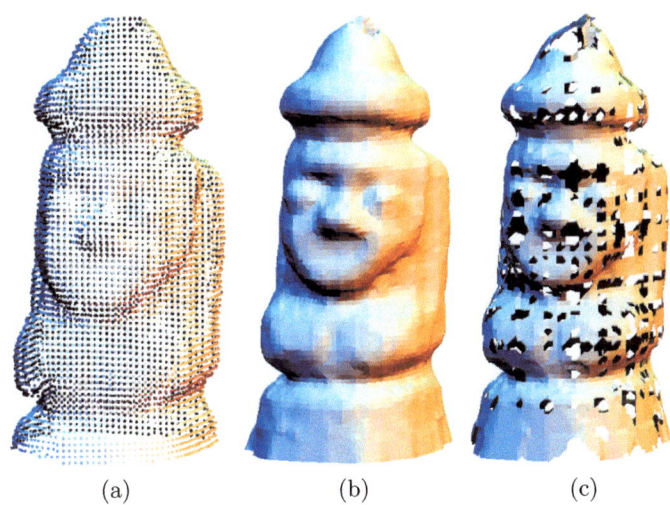

Fig. 8.10. Surface reconstruction of a Moai statue via MPU implicits: (a) dataset with 6053 points; (b) mesh of the reconstructed surface with support size of 0.5; (c) mesh of the reconstructed surface with support size of 0.4. (Moai's dataset is courtesy of Tamy Boubekeur's repository. The pictures were generated from the MPU testbed due to Ohtake et al. [308].)

8 Implicit Surface Fitting

This approach allows for other local fitting functions. This flexibility is reinforced by the fact that sharp features (e.g. ridges and corners) can accurately represented by extending the set of local fitting functions with supplementary shape functions. The previous work of Ohtake et al. [309] (and the references therein) influenced the inclusion of those sharp features in MPUs.

8.5.1 MPU Approximation

MPUs can be used to *approximate* or *interpolate* a point dataset. If we intend to approximate a point dataset, we should use the general trivariate quadratic polynomial $P(\mathbf{x})$ to form weight functions

$$\omega_i(\mathbf{x}) = P\left(\frac{3\,\|\mathbf{x} - \mathbf{c}_i\|}{2\,r_i}\right) \tag{8.55}$$

which are centred at $\mathbf{c}_i$ (i.e. centres of the octree cells) and have a spherical support of radius r_i.

The octree data structure is thus the core of the MPU-based implicit surface reconstruction algorithm. For each leaf octree cell containing sample points of the physical surface, we have to generate a local shape function f_i that approximates such a surface in the cell. Each approximant f_i is built using a least-squares fitting framework similar to MLS surfaces.

There are three possible approximants, depending on the number of data points inside the ball associated to a given cell, as well as the distribution of the normals at the data points. The possible approximants are the following:

- A *general quadric* in $\mathbb{R}^3$. This trivariate quadratic polynomial serves the purpose of approximating larger regions of the surface, in particular those unbounded regions and regions with two or more sheets inside a given octree cell.
- A *bivariate quadratic polynomial* in local coordinates. This polynomial is suited to approximate a single local smooth patch.
- A *piecewise quadric surface* that fits a sharp feature (i.e. a corner or an edge). The piecewise nature of this quadric makes it adequate for modelling regions with differential singularities like cusps and creases.

Let N be the number of points in a ball centred at $\mathbf{c}$, the centre of a given octree cell, and $\mathbf{n}$ a unit normal vector at $\mathbf{c}$. This unit normal vector is determined from the normalised weighted arithmetic mean of the normals assigned to data points inside the ball, with the weights taken from Equation (8.55). Let also θ be the maximal angle between $\mathbf{n}$ and those normals inside the ball. The choice of the local surface fit depends on the values of N and θ.

Local Fit of a General Quadric

The surface is locally approximated with a general quadric if the following condition is satisfied:

$$N > 2N_{\min} \quad \text{and} \quad \theta \geq \frac{\pi}{2} \tag{8.56}$$

where $N_{\min}$ takes on the empirical value 15. In this case, the local shape function is given by the general form of a quadric surface

$$f(\mathbf{x}) = Ax^2 + By^2 + Cz^2 + 2Dxy + 2Exz + 2Fyz + 2Gx + 2Hy + 2Iz + J \tag{8.57}$$

or, equivalently, in the matrix notation as

$$f(\mathbf{x}) = \mathbf{x}^T \mathbf{A} \mathbf{x} + \mathbf{b}^T \mathbf{x} + c \tag{8.58}$$

where

$$\mathbf{A} = \begin{pmatrix} A & D & E \\ D & B & F \\ E & F & C \end{pmatrix}, \quad \mathbf{b}^T = \begin{pmatrix} 2G \\ 2H \\ 2I \end{pmatrix} \quad \text{and} \quad c = J$$

The unknowns $A, B, C, D, E, F, G, H, I$, and J are computed by means of a minimisation procedure described by Ohtake et al. [308].

Local Fit of a Bivariate Quadratic Polynomial

Using a bivariate quadratic polynomial to fit the surface locally requires that the following condition be satisfied:

$$N > 2N_{\min} \quad \text{and} \quad \theta < \frac{\pi}{2} \tag{8.59}$$

Such a bivariate quadratic shape function is given by the second term of the following expression of f

$$f(\mathbf{x}) = w - (Au^2 + 2Buv + Cv^2 + Du + Ev + F) \tag{8.60}$$

where (u, v, w) are the coordinates of a local coordinate system with the origin at $\mathbf{c}$ and such that the normal vector $\mathbf{n}$ at $\mathbf{c}$ is orthogonal to the plane (u, v) along w, i.e. (u, v, w) are the coordinates of $\mathbf{x}$ in the new coordinate system. Once again, the unknown coefficients A, B, C, D, E, and F are computed by means of a minimisation procedure described in [308].

Local Approximation of Sharp Features

The following condition

$$N \leq 2N_{\min} \tag{8.61}$$

indicates that there may be a sharp edge or corner inside the ball B centred at $\mathbf{c}$. In this case, we consider a *piecewise* quadratic function instead of the quadratic functions (8.57) and (8.60) to fit the data points inside B. The automatic recognition of sharp edges and corners is carried out by a procedure proposed by Kobbelt et al. [213].

The MPU Approximation Algorithm

Algorithm 32 (jointly with Algorithm 33) describes the MPU approximation to reconstruct an implicit surface from a scattered point dataset with precision ϵ_0.

As said above, the algorithm uses an adaptive octree subdivision of the bounding box Ω into subsidiary cubes Ω_k. The algorithm also uses a K-d tree for sorting and collecting data points inside the sphere that encloses each leaf cube. If the sphere is empty (i.e. no data points), its associated interior cube will not be subdivided any further. Otherwise, one first estimates a local max-norm approximation error using the Taubin distance [383] given by

$$\epsilon = \max_{||\mathbf{x}_i - \mathbf{c}|| < R} \frac{||f(\mathbf{x}_i)||}{||\nabla f(\mathbf{x}_i)||} \tag{8.62}$$

Algorithm 32 The MPU Implicit Surface Approximation

1: **procedure** MPUAPPROXIMATION(Ω,**x**,ϵ_0)
2:     $\sum_w \leftarrow 0$ ▷ initialises denominator of (8.54)
3:     $\sum_{wf} \leftarrow 0$ ▷ initialises numerator of (8.54)
4:     $\Omega \leftarrow$ MPU(**x**,ϵ_0)   ▷ Algorithm 33
5:     $F \leftarrow \frac{\sum_{wf}}{\sum_w}$ ▷ formula (8.54)
6: **end procedure**

Algorithm 33 The MPU Implicit Surface Reconstruction

1: **procedure** MPU(**x**,ϵ_0)
2:     **if** $|\mathbf{x} - \mathbf{c}_i| > r_i$ **then**
3:        return;   ▷ excludes points beyond the radius of support r_i
4:     **end if**
5:     **if** f_i has not been generated yet **then**
6:        Create f_i ▷ local shape fit
7:        Compute ϵ_i ▷ Taubin's distance
8:     **end if**
9:     **if** $\epsilon_i > \epsilon_0$ **then**
10:        **if** no children created **then**
11:           Create children Ω_k ▷ octree subdivision
12:        **end if**
13:        **for** each child Ω_k **do**
14:           $\Omega_k \leftarrow$ MPU(**x**,ϵ_0)
15:        **end for**
16:     **else**
17:        Compute $\omega_i(\mathbf{x})$ ▷ formula (8.55)
18:        $\sum_w \leftarrow \sum_w + \omega_i(\mathbf{x})$ ▷ updates denominator of (8.54)
19:        $\sum_{wf} \leftarrow \sum_{wf} + \omega_i(\mathbf{x}) f_i(\mathbf{x})$ ▷ updates numerator of (8.54)
20:     **end if**
21: **end procedure**

If $\epsilon > \epsilon_0$ (Algorithm 33, step 9), where ϵ_0 is a user-predefined threshold, one subdivides the cube into eight smaller cubes (Algorithm 33, step 11), and the approximation is delegated down to its child cubes (Algorithm 33, steps 13–14). Otherwise, one computes the corresponding weight function $w_i(\mathbf{x})$ and updates the numerator and denominator (Algorithm 33, steps 17–18) of the global MPU function (Algorithm 32, step 5).

The computation of the local MPU function $f_i(\mathbf{x})$ is carried out by step 6 of Algorithm 33. Recall that the choice of this local shape function depends on the values of N and θ, say the number of points inside the ball enclosing a given leaf cube and the maximal angle between the normal at the centre of the ball and the normals at the points inside the ball.

8.5.2 MPU Interpolation

MPUs can be also used to interpolate the point dataset. In this case, we do not use the weights of Equation (8.55). Instead, we use the inverse-distance singular weights proposed by Franke and Nielson [146]:

$$w_i(\mathbf{x}) = \left[\frac{(r_i - ||\mathbf{x} - \mathbf{c}_i||)_+}{r_i ||\mathbf{x} - \mathbf{c}_i||}\right]^2 \tag{8.63}$$

where

$$(d)_+ = \begin{cases} d & \text{if } d > 0 \\ 0 & \text{otherwise.} \end{cases} \tag{8.64}$$

8.6 Final Remarks

Boissonnat [56] apparently first addressed the problem of surface reconstruction from scattered datasets in the 1980s. Another reference work on surface reconstruction from point cloud is due to Hoppe et al. [192]. Since then surface reconstruction has become an active research topic in computer graphics and visualisation. As seen in the beginning of the present chapter, three major approaches have been used to fit surfaces to point clouds, depending on the geometric representation for such surfaces: simplicial surface, parametric surface, and implicit surface.

Nevertheless, this chapter is mainly concerned with algorithms that fit implicit surfaces to such point clouds. Basically, there are two families of methods for fitting an implicit surface to a cloud of points: interpolation methods and approximation methods. Interpolation methods generate surfaces that exactly pass through the data points, while approximation methods produce surfaces that pass near the data points. Thus, approximation methods are the adequate choice to reconstruct surfaces from point datasets with noise.

Interpolation and approximation methods include finite elements [123], RBFs, MLS, and MPUs. The latter three take advantage over the former method because they do not require a consistent tessellation of the function domain, and are known as mesh-free methods. These mesh-free methods are also known as Galerkin methods in numerical analysis, computation and engineering.

In surface reconstruction, we also find polygonal interpolants. The reader is referred to Wachspress [403], who proposed rational polynomial interpolants for convex polygons. See also Sukumar and Malsch [381] for a recent review on the construction of polygonal interpolants. Unsurprisingly, polygonal interpolants is now an active research field in computer graphics and geometric modelling, in particular with respect to the construction of barycentric coordinates on irregular polygons [140, 193, 268, 380, 381, 382].

Part IV

Designing Complex Implicit Surface Models

Overview

Part IV introduces the idea of implicit surfaces to build and animate complex 3D models. There are several approaches and in this part we focus on constructing models from parts that are themselves built from geometric elements. In Chapter 9 various modelling techniques are described, and in the following chapters, we discuss how these techniques may be applied to static models, animated models and a simulation model using the common theme of implicit surfaces as the base modelling methodology. Describing natural phenomenae has long been a challenge in computer graphics, so we take examples from nature to illustrate the usefulness of these techniques.

Skeletal implicit surfaces have been variously called *metaballs* [306], *soft objects* [422] and *blobby modelling* [47]; however these terms are no longer generally applicable as the surfaces may be neither blobby nor soft. Such implicit surface models are constructed from combinations of skeletal elements, such as points, line segments, circles, etc. Each primitive generates a scalar field and an implicit surface is defined as an isosurface in the field, (see also Overview of Part I and Chapter 1, *level sets* of a function).

Designing with implicitly defined surfaces offers various advantages; many geometric operations, for example, are simplified. These include the standard set operations (union, intersection, difference, etc.) of constructive solid geometry (CSG), functional composition with other implicit functions, and inside/outside tests. As noted in Chapter 2, the relationships between the skeletal elements is most easily stored in a tree structure and in this chapter we will use the *BlobTree* described in [417].

As mentioned in earlier chapters, visualising the surfaces can be done either by direct ray tracing using an algorithm as described in [106] and [206] or by first converting to polygons [178, 422].

9
Skeletal Implicit Modelling Techniques

In this chapter we review skeletal implicit modelling techniques and introduce functions that have proved to be very useful in the construction of complex models.

9.1 Distance Fields and Skeletal Primitives

When a function f is applied to a point $\mathbf{p} \in \mathbb{R}^3$, the result is a scalar value $f(\mathbf{p}) \in \mathbb{R}$. A surface $\mathcal{S} \subset \mathbb{R}^3$ is given by

$$f(\mathbf{p}) = C \qquad (9.1)$$

where C is any scalar value in $\mathbb{R}$. Implicit surfaces can also be written as $f(\mathbf{p}) - C = 0$ where C is nonzero. The surface $\mathcal{S}$ is an *isosurface* of the *scalar field* produced by $f(\mathbf{p})$, and v is the *isovalue* that produces $\mathcal{S}$. In computer graphics, $\mathcal{S}$ is commonly known as an *implicit surface*.

Another type of implicit surface is the *distance field*, defined with respect to some geometric entity $\mathsf{T} \subset \mathbb{R}^3$, e.g. a point, curve, surface, or solid:

$$f_\mathsf{T}(\mathbf{p}) = \min_{\mathbf{q} \in \mathsf{T}} |\mathbf{q} - \mathbf{p}| \qquad (9.2)$$

Intuitively, $f_\mathsf{T}(\mathbf{p})$ is the shortest distance from $\mathbf{p}$ to T. Hence, when $\mathbf{p}$ lies on T, $f_\mathsf{T}(\mathbf{p}) = 0$.

One challenge when analysing implicit surfaces is visualising the underlying scalar fields. A common technique is to regularly sample f on a 2D planar slice through the field and map the values to grayscale, creating a *field image* (Figure 9.1(a)). Another useful visualisation can be created by applying a sin function to the values of f before mapping to grayscale. This creates a *contour diagram* (Figure 9.1(d)).

Nonzero isovalues can be used with distance fields to define *offset surfaces*, where $f_\mathsf{T}(\mathbf{p}) = C$ and $C > 0$. Here C is the distance from the offset surface

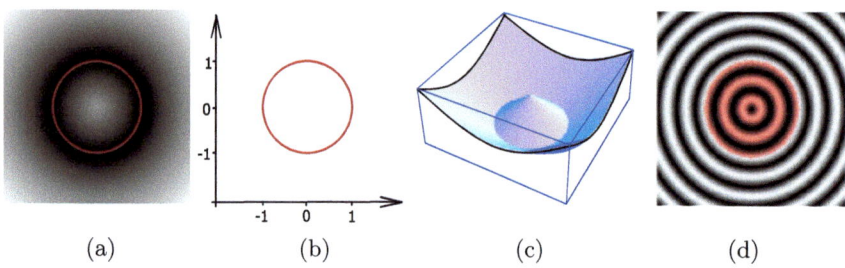

(a)　　　　　　(b)　　　　　　(c)　　　　　　(d)

Fig. 9.1. A 2D implicit circle defined by the distance field $f = \sqrt{x^2 + y^2} - 1$. A 2×2 region of the infinite distance field f is visualised in (a) by sampling f at each pixel and mapping the value to grayscale. The circle lies on the zero isocontour $f = 0$, highlighted in red in (a) and shown explicitly in (b). The field f is plotted as a standard height map in (c). In (d), a contour diagram is created by applying $(1 + \sin(k \cdot f))/2$ to the value at each pixel before mapping to grayscale. The area inside the zero isocontour, $f < 0$, is highlighted in red.

to T. Note that if T is a closed surface, then $f_\mathsf{T}(\mathbf{p}) = v$ defines two new surfaces—one "inside" the old surface, and another "outside." If T has no interior, as is the case for points, curves, open surfaces, and solids, then only one offset surface is defined. In this case, $f_\mathsf{T}(\mathbf{p}) = C$ is referred to as a *skeletal primitive*.

The idea of using implicit surfaces as a procedural method for building models by combining primitive parts was put forward in a paper by Ricci in the early seventies [341]. As argued by Bloomenthal [51], perhaps the most useful procedural method for defining f is that of skeletal implicit surfaces [51, 306, 422]. Originally introduced to the field of computer graphics by Blinn [47] for the visualisation of electron density fields, skeletal implicit surfaces are constructed from skeletal primitives, each defining their own potential function. The correspondence between skeletal elements and implicit primitives allows for intuitive model construction using skeletal elements, while maintaining the inherent advantages of implicit surfaces. This higher-level control reduces the degrees of freedom in the model but there are several advantages. For example, $f(\mathbf{p})$ can only have a single value at any point in space. That value is either on the surface, or it is not and thus, self-intersections are impossible. Self-intersections can be highly undesirable with other modelling methodologies such as polygon meshes.

Each component $f_i(\mathbf{p})$ of the implicit function may be split into a distance function $d_i(\mathbf{p})$ and a potential function $g_i(r)$, where r stands for the distance to the skeleton. Using $\circ$ as the operation of composition of functions, each component $f_i(\mathbf{p})$ can be then written as follows:

$$f_i(\mathbf{p}) = g_i(r) \circ d_i(\mathbf{p}).$$

Potential functions g_i are chosen so that the field values are maximum on the skeleton and generally fall off to zero at some chosen distance from the

9.1 Distance Fields and Skeletal Primitives 269

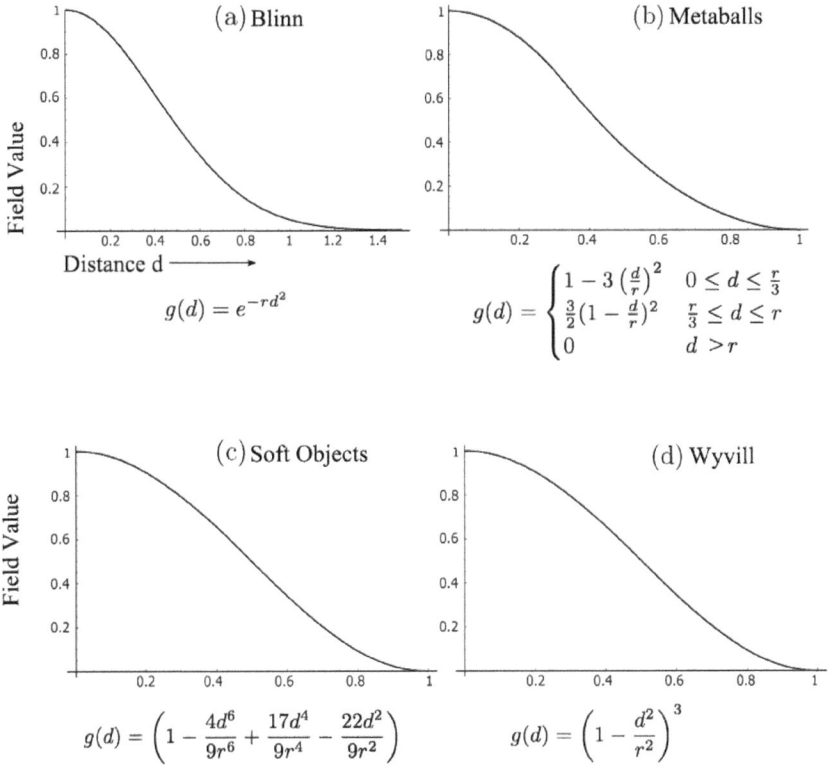

Fig. 9.2. Potential functions. (a) Blinn's Gaussian or "blobby" function, (b) Nishimura's "metaball" function, (c) Wyvill et al's "soft objects" function, and (d) the Wyvill function.

skeleton. One of the distinguishing features of the early work was to derive a number of different potential functions (see Figure 9.2). An implicit model is generated by combining the influences of k skeletal elements, the potential field contributed by each element i will be denoted by f_i. The contributions can be combined in several different ways and together they define a scalar field f.

In the simple case where the resulting surfaces are blended together, the global potential field $f(x, y, z)$ of an object, the implicit function, may be defined as:

$$f(\mathbf{p}) = \sum_{i=0}^{k-1} f_i(\mathbf{p})$$

The surface of the object may be derived from the implicit function $f(\mathbf{p})$ as the points of space whose value equals a threshold denoted by C. The isosurface, $f(\mathbf{p}) = C$ of a skeletal point and a radius of influence is a sphere. In turn, the isosurface of two skeletal points with the same radius of influence

Fig. 9.3. Two skeletal points placed in close proximity.

is is shown in Figure 9.3, where the field at any point **p** is calculated as in Equation (9.1).

9.2 The *BlobTree*

The *BlobTree* was introduced in [418] and proposed a tree structure that extended the CSG tree (see [338]) to include various blending operations. Various other similar systems were also published at that time. To understand the different approaches return to the definition of the implicit function.

For a given potential function f, the volume $\mathcal{V}$, and corresponding implicit surface $\mathcal{S}$, are defined by the following equations:

$$\mathcal{V} = \{\mathbf{p} \in \mathbb{R}^3 | f(\mathbf{p}) \leq 0\} \tag{9.3}$$
$$\mathcal{S} = \{\mathbf{p} \in \mathbb{R}^3 | f(\mathbf{p}) = 0\} \tag{9.4}$$

Originally introduced to the field of computer graphics by Ricci [341], two distinct classes of potential function have emerged. The first class, here termed f zero functions (f_Z) for clarity, assumes implicit primitives are defined as in Equations (9.3) and (9.4). In this case the potential functions defined by the primitives typically vary over the whole of space. The second class, here termed f_C functions, employ a modified version of Equations (9.3) and (9.4) as follows:

$$\mathcal{V} = \{\mathbf{p} \in \mathbb{R}^3 | f(\mathbf{p}) \geq C\} \tag{9.5}$$
$$\mathcal{S} = \{\mathbf{p} \in \mathbb{R}^3 | f(\mathbf{p}) = C\} \tag{9.6}$$

where $C > 0$, and is a user defined value (typically 0.5). Note that for visualisation, if the solver expects an f_Z function, an f_C function is simply modified by scaling by negative one and adding C.

The salient point in this classification lies in the fact that many common modelling operations with implicit surfaces employ separate mathematical operations dependent on whether f_Z functions or f_C functions are being used. For example, f_C functions support summation blending as presented in Section 9.5, which if applied to f_Z functions results in a shrinking of the input surfaces. In some cases, f_Z functions offer functional composition operations with superior properties in the resulting potential values [28, 317]. In contrast, the main advantage of f_C functions is that primitives are easily defined such that $f(\mathbf{p}) = 0$ outside a boundary, thus allowing the use of standard optimisation techniques such as bounding volumes or spatial subdivision. An alternative classification of potential functions focuses on bounded versus unbounded fields [28]. This approach is useful when considering the optimisation of function evaluations; however, beyond considerations of optimisation it is not usefully applied when examining the properties of functional composition operations. For example, many unbounded forms of f_C functions, such as those defined by Blinn [47], are easily incorporated into systems employing bounded f_C functions.

9.3 Functional Composition Using f_Z Functions

f_Z functions often define implicit primitives such that the distance from the surface $\mathcal{S}$ to a given point $\mathbf{p}$ is given as $d = f(\mathbf{p})$. Skeletal primitives are thus easily employed by defining f as:

$$f = f_{\text{distance}}(\mathbf{p}) - r \qquad (9.7)$$

where r is the desired distance from the skeletal element to the surface $\mathcal{S}$ and f_{distance} is a distance function as previously defined [29].

Pasko et al. [317] introduced the theory of *R-functions* to computer graphics. Their method uses R-functions to perform binary set-theoretic operations on two implicit primitives with potential functions f_1 and f_2, here denoted as $\cup_R$ for union, $\cap_R$ for intersection and $\setminus_R$ for difference. The methods they present work under the condition of reversing the inequality from Equation (9.3), such that the volume V is defined by the following inequality $f(\mathbf{p}) \geq 0$. A variety of forms are presented, the most useful in practice being:

$$\cup_R f = f_1 + f_2 + \sqrt{f_1^2 + f_2^2} \qquad (9.8)$$

$$\cap_R f = f_1 + f_2 - \sqrt{f_1^2 + f_2^2} \qquad (9.9)$$

$$\setminus_R f = f_1 - f_2 - \sqrt{f_1^2 + f_2^2} \qquad (9.10)$$

These functions have C^1 discontinuities only where $f_1 = f_2 = 0$. Alternative forms exist which can ensure C^m continuity for a given value of m if needed, thus providing some advantages over Ricci's [341] max-min method (see Section 9.5.2).

Extensions to this formulation allow for soft transitions and blending by the addition of a displacement function $d(f_1, f_2)$, where d has a maximal absolute value at $d(0,0)$ and asymptotically approaches a zero value with increasing values of the arguments. The following has been presented as a useful form of d for blending purposes [317]:

$$d(f_1, f_2) = \frac{a_0}{1 + \left(\frac{f_1}{a_1}\right)^2 + \left(\frac{f_2}{a_2}\right)^2} \qquad (9.11)$$

where a_0, a_1 and a_2 are user defined constants controlling the shape of the blend. The displacement function is added directly to the values of Equations (9.8) to (9.10), for example a blending union operation is defined as:

$$f = f_1 + f_2 + \sqrt{f_1^2 + f_2^2} + \frac{a_0}{1 + \left(\frac{f_1}{a_1}\right)^2 + \left(\frac{f_2}{a_2}\right)^2} \qquad (9.12)$$

Pasko et al. subsequently introduced *bounded blending* to provide localised control of blends using R-functions [318]. Two methods were proposed, the first replaced the displacement function from Equation (9.11) with a function providing a local area of influence. This local area of influence allowed for blends with similar properties to those defined for f_C functions employing field functions with varying hardness factors. It also has the advantage of being repeatedly applied in a hierarchical system. The second, and more interesting approach was the use of a bounding solid to control the locality and extent of blending. Unlike controlled blending, bounded blending allows two implicit primitives to be blended in one region and non-blended in another region simultaneously.

A more robust approach to controlling the shape of blends has been defined by Barthe et al. [28, 30]. They defined a method using free-form curves modified in the Euclidean user space to control the shape of the blend between two implicit primitives. This method represents a unified and intuitive approach to a wide array of modelling operations, and provides superior properties in the resulting potential field over all other known methods; however, it does not allow for the truly localised blending control offered by bounded blending.

In the *BlobTree* system models are defined by expressions which combine implicit primitives and the operators ∪ (union), ∩ (intersection), − (difference), + (blend), ⋄ (super-elliptic blend), and w (warp). The *BlobTree* is not only the data structure, built from these expressions but also a way of visualising the structure of the models. The operators listed above are binary with the exception of warp which is a unary operator. In fact it is more efficient to use n-ary rather than binary operators. The *BlobTree* incorporates affine transformations as nodes so that it is also a scene graph.

9.4 Combining Implicit Surfaces

Given a set of k implicit primitives with implicit surface functions f_i, where $i \in [0, k-1]$, any method which combines them to produce a final potential function f may be used as a modelling operation. Typically, only certain operations are considered useful, those that combine primitives in an intuitive fashion while maintaining desirable properties in the resulting potential function. Figure 9.4 shows the results of the operators discussed. Three implicit point primitives have been combined in various ways. The top row shows a

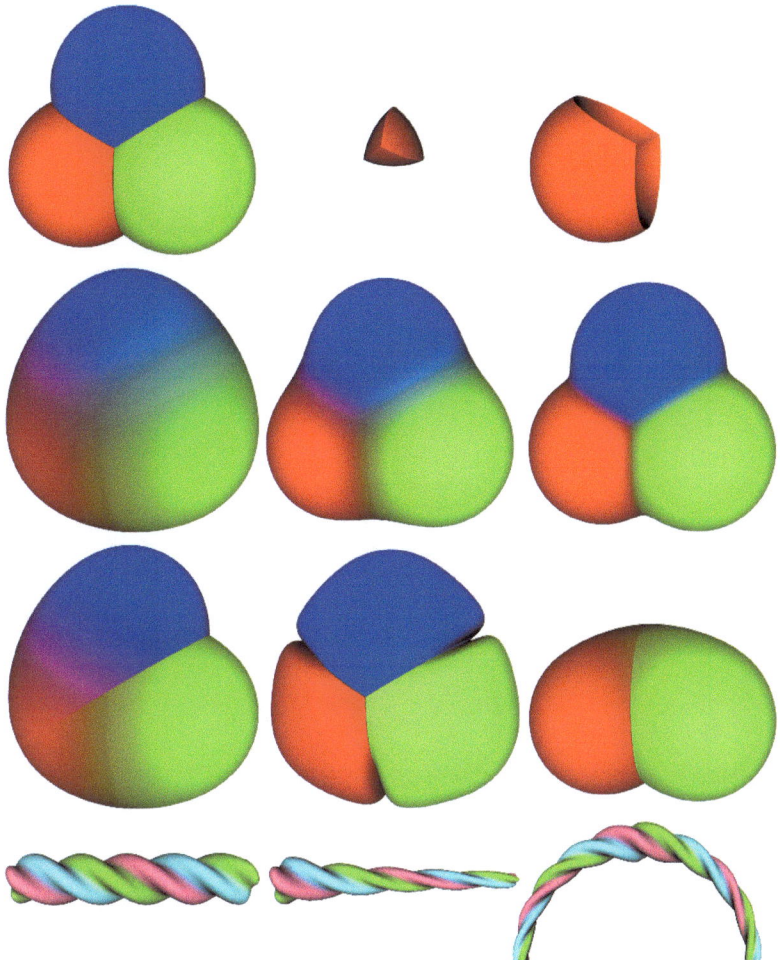

Fig. 9.4. Implicit modelling operators, identified from left to right. Top row: union, intersection, difference. Second row: super-elliptic blending with n=1, n=3, n=10. Third row: controlled blending, precise contact modelling, bounded blending. Bottom row: three blended line primitives twisted (left), twisted and tapered (centre), and twisted, tapered and bent (right).

union of the resulting spheres, the intersection of the spheres, i.e. only the volume that is common to all three remains after the operation. In the third image the green and blue spheres have been subtracted from the red sphere. On the second row we see the results of super-elliptic blending. The blend operator is described in Equation (9.14), see Section 9.5. The third row shows the results of using the precise contact modelling operator described in Section 9.5.3, and the bottom row the results of applying warp operations to three implicit cylinders, see Section 9.6.

9.5 Blending Operations

One of the main advantages of modelling with skeletal implicit surfaces is the ease with which primitives are blended together using *summation* as follows:

$$f = \sum_{i=0}^{k-1} f_i \tag{9.13}$$

This is the most compact and efficient blending operation which can be applied to implicit surfaces as illustrated in Figure 9.3.

The *super-elliptic blending* allows the modeller to control the amount of blending to achieve a large range of blends [341]. Given the same set of k implicit primitives, and a blending power $n \in [1, +\infty)$, super-elliptic blending is defined as:

$$f = \left(\sum_{i=0}^{k-1} (f_i)^n\right)^{\frac{1}{n}} \tag{9.14}$$

Here, Equation (9.13) can be seen to be a special case of Equation 9.14 with $n = 1$. Moreover,

$$\lim_{n \to +\infty} \left(\sum_{i=0}^{k-1} (f_i)^n\right)^{\frac{1}{n}} = \max_{i=0}^{k-1} (f_i) \tag{9.15}$$

Thus, as n varies from 1 to $+\infty$, it creates a set of blends varying between summation blending and a union of the implicit primitives as defined by Equation (9.17).

The choice of potential function can now be seen to be important, as it directly affects the shape of blended surfaces constructed using Equations (9.13) and (9.14), as well as impacting the overall efficiency of surface visualisation. One choice for such potential function is [420]:

$$g(d) = \left(1 - \frac{d^2}{r^2}\right)^3 \tag{9.16}$$

This function, shown in Figure 9.2(d), satisfies the desired conditions while providing intuitive properties when combining primitives and being efficient for evaluation purposes. Many alternative field functions have been proposed [46, 204, 306, 422]. One of the main advantages in many of these formulations is the inclusion of a hardness factor. A hardness factor affects the blending properties of primitives to control the smoothness of resulting blends. An efficient field function with a hardness factor was introduced by Blanc and Schlick [46], who also review previous approaches. By incorporating a hardness factor into a primitive's field function, desired blending properties can be associated with individual primitives, rather than with the blends between them; however, this approach is of limited use in hierarchical modelling where a series of blends with varying degrees of hardness may be desired.

One property arising from blends defined by Equation (9.13), in conjunction with the field function from Equation (9.16), is the occurrence of bulging when working with an articulated skeleton. This can be seen in Figure 9.5. Bloomenthal examined this topic in detail [51], and determined that convolution is the best solution to this problem. Succinctly, the field function is defined by convolving a filter over the signal defined by the skeleton. Completely bulge-free surfaces result when blending two line segments with coincident endpoints, and bulge-free blends in branching situations can be obtained through the use of polygonal skeletal elements under certain conditions.

The main drawbacks of convolution are its implementation complexity and increased computational complexity; however, if perfectly bulge-free blends are required this is currently the preferred method. Convolution may properly be viewed as another field function, and skeletal implicit primitives based on convolution may be used in the same way as primitives defined using other field functions.

9.5.1 Hierarchical Blending Graphs

Bloomenthal points out in [54], that by simply using the summation operator for blending, unwanted blending can occur. Figure 9.6 shows the problem. The cartoon creature consists of a number of point primitives. Blending is required between some of them but not all. The problem can be partially solved by using a group that contains objects which do not blend (see [74]) but for full control over a chain of primitives where blending is required between arbitrary pairs controlled blending via a blending graph is required as originally defined by Guy et al. [171], and subsequently implemented as a blending node in the *BlobTree* [417]. A fundamental limitation with previous implementations of blending graphs for implicit surfaces has been the flat nature of the data

Fig. 9.5. Bulge when skeletal implicit primitives (two line segments with coincident endpoints) are blended using summation.

Fig. 9.6. Unwanted blending between the leg and the body (left) and result of using a hierarchical blending graph (right).

structures employed. They have not supported hierarchical compositions of blending graphs, requiring instead that all nodes involved in the controlled blending be defined in a single graph. A simple example of the problem is shown in Figure 9.7. On the left is the *BlobTree childCBNode* constructed using controlled blending as follows:

```
childCBNode = controlledBlendGraph()
childCBNode.addBlendGroup( redSphere, blueSphere )
childCBNode.addBlendGroup( redSphere, greenSphere )
```

Each blend group added to the graph defines a blend between the inputs, thus the red and blue spheres are blended and the red and green spheres are blended. The blue and green spheres do not blend as they are not explicitly defined to do so by the blend graph. The centre image in Figure 9.7 showing *parentCBNode*, defined by the following, illustrates the problem:

```
parentCBNode = controlledBlendGraph()
parentCBNode.addBlendGroup( childCBNode, yellowSphere )
```

When blending the yellow sphere to *childCBNode*, it must blend globally, demonstrated by the yellow sphere blending with both the green and blue spheres defined within *childCBNode*. This is due to the blending graph employed by *childCBNode* being a purely internal structure. If the user wishes the yellow sphere to blend only with some of the elements within *childCBNode*, they must rewrite their model definition to include all four original spheres within one controlled blending graph node. Although this simple example is easily rewritten, in practical situations this can be highly problematic. Requiring all input nodes to be children of a single controlled blending node makes hierarchical model construction at best extremely difficult, and generally not possible.

To solve this, blending graphs are redefined such that they have internal and external blend groups. Internal blend groups define blends among the

Fig. 9.7. Three examples using controlled blending. See Section 9.5.1 for the corresponding functions defining the three models. Left: childCBNode. Centre: parentCBNode using childCBNode as input. Right: parentCBNode using hierachical-CBNode as input.

various children of the controlled blending node. External blend groups define those parts of the field which will be used for blending in an hierarchical controlled blending environment. For example, *childCBNode* may be redefined in a hierarchical fashion as *HCB* (hierarchical controlled blending) such that only the green sphere is available for external blending as follows:

```
HCB = hierarchicalControlledBlendGraph()
HCB.addInternalBlendGroup( redSphere, blueSphere )
HCB.addInternalBlendGroup( redSphere, greenSphere )
HCB.addExternalBlendGroup( greenSphere )
```

The right image in Figure 9.7 shows the result when *parentCBNode* is redefined to use *HCB* as follows:

```
parentCBNode = hierarchicalControlledBlendGraph()
parentCBNode.addInternalBlendGroup(HCB,yellowSphere)
```

The yellow sphere now blends only with the green sphere, as this was the only one nominated for external blending by *HCB*. This approach scales well, for example *parentCBNode* may in turn define its own external blending groups.

9.5.2 Constructive Solid Geometry

As previously stated, implicit surfaces are inherently useful for *solid modelling* operations. Ricci introduced a *constructive geometry* for defining complex shapes from operations such as union, intersection, difference and blend upon primitives [341]. The surface was considered as the boundary between the half spaces $f(\mathbf{p}) < 1$, defining the inside, and $f(\mathbf{p}) > 1$ defining the outside. This initial approach to solid modelling evolved into *constructive solid geometry* or CSG [338]. CSG is typically evaluated bottom up according to a binary tree, with low-degree polynomial primitives as the leaf nodes, and internal nodes representing Boolean set operations. These methods are readily adapted for use in implicit modelling, and in the case of skeletal implicit surfaces the Boolean set operations union $\cup_{\max}$, intersection $\cap_{\min}$ and difference $\setminus_{\minmax}$ are defined as [417]:

$$\cup_{\max} f = \max_{i=0}^{k-1} (f_i) \tag{9.17}$$

$$\cap_{\min} f = \min_{i=0}^{k-1} (f_i) \tag{9.18}$$

$$\setminus_{\minmax} f = \min\left(f_0, 2C - \max_{j=1}^{k-1}(f_j)\right) \tag{9.19}$$

As pointed out by Pasko et al. [317], these operations introduce C^1 discontinuities in the resulting potential field, and are thus problematic when used in hierarchical blending situations. In particular, subsequent blending

278   9 Skeletal Implicit Modelling Techniques

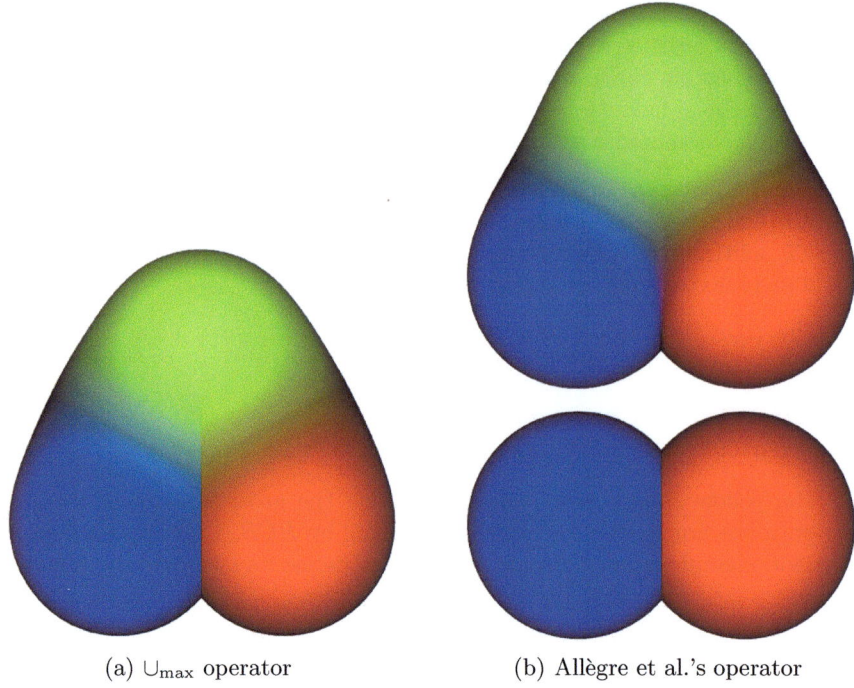

(a) $\cup_{max}$ operator   (b) Allègre et al.'s operator

Fig. 9.8. Blending between a point (green), and the result of a union operation between two points (red and blue). In Figure 9.8(a) an undesirable artifact is observed due to a discontinuity in the field resulting from the underlying $\cup_{max}$ operator (Equation (9.17)). In Figure 9.8(b) this discontinuity has been removed using an alternative formulation due to Allègre et al. [6].

operations which overlap the C^1 discontinuity result in crease like artifacts, shown in Figure 9.8(a). These discontinuities are also problematic for some $2D$ parametrisation methods for implicit surfaces [387].

One attempt to solve this problem was presented by Allègre et al. [6], and is based on the theory of R-functions [317]. Their method conserves the sharp transition expected from CSG operations at the join of the two input surfaces, and produces a C^1 continuous field everywhere else. The result is that subsequent blending operations do not contain any crease like artifacts, as shown in Figure 9.8(b). Nevertheless, their approach is of limited use as the union operator they described effectively compresses the blending radius of the resultant potential field. The result is that after combining four or more primitives with Allègre et al.'s operator, there are parts of the surface where subsequent blending operations will not produce any noticeable blending. In an hierarchical modelling environment this can quickly lead to degenerate models with no inherent blending properties. The intersection operator they describe has the inverse effect, resulting in too much blending.

A better solution to this problem was presented by Barthe et al. [31], based on their earlier methods employed for CSG operations on f_Z functions [28]. In contrast to the method of Allègre et al., they maintain a good transition of field values outside the surface for later summation blending operations. The key problem with their approach is its overwhelming complexity as they succinctly point out [31]:

> "We point out that it is obvious that our operators are computationally expensive ..." (page 138)

Their solution depends on solving a highly complex set of polynomials, and all computations must be carried out using complex numbers, even where the results are real.

9.5.3 Precise Contact Modelling

Precise contact modelling, or PCM, is a method of deforming implicit surface primitives in contact situations, while maintaining a precise contact surface with C^1 continuity [74].

PCM is implemented by the inclusion of a deforming function $D(\mathbf{p})$ as follows. For each pair of objects, collision is first detected using a bounding box test. Once it is established that there is likely a collision PCM is applied. A local, geometric deformation term, D_i, is computed and added to f_i. The volume of the colliding objects is divided into an interpenetration region and a deformation region. The result of applying D_i is that the interpenetration region is compressed so that contact is maintained without interpenetration occurring (see Figure 9.9). The effect of D_i is attenuated to zero within the propagation region so that the volume outside of the two regions is not deformed.

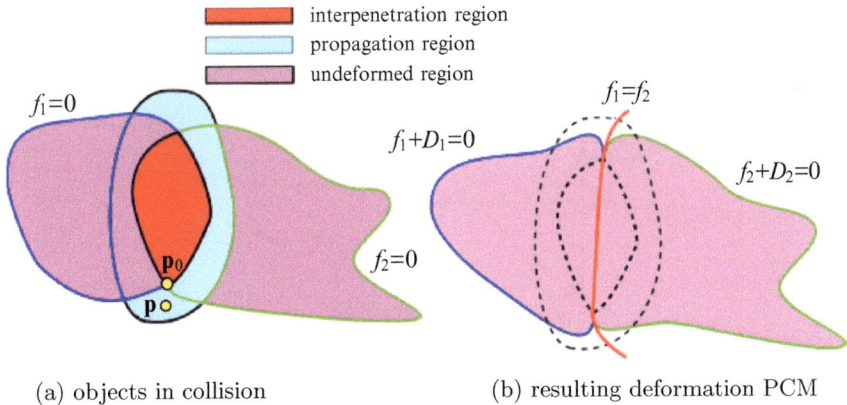

Fig. 9.9. A 2D slice through objects in collision showing the various regions and PCM deformation.

Deformation in the Interpenetration Region

Given two skeletal elements generating fields $f_1(\mathbf{p})$ and $f_2(\mathbf{p})$, the surface around each one is calculated as:

$$f_1(\mathbf{p}) + D_1(\mathbf{p}) = 0 \qquad (9.20)$$
$$f_2(\mathbf{p}) + D_2(\mathbf{p}) = 0 \qquad (9.21)$$

These equations have a common solution or contact surface (red line in Figure 9.9) in the interpenetration region. For some $\mathbf{p}$ in the interpenetration region, we then have:

$$D_1(\mathbf{p}) = -f_1(\mathbf{p}) \qquad (9.22)$$
$$D_2(\mathbf{p}) = -f_2(\mathbf{p}) \qquad (9.23)$$

Thus, the contact surface corresponds to those points in the interpenetration region for which $f_1(\mathbf{p}) = f_2(\mathbf{p})$. Intuitively, the deeper within object 1, object 2 penetrates, the higher the implicit value of object 1 and thus the more object 2 will be compressed.

Deformation in the Propagation Region

The function D_i is defined to produce a smooth junction at the boundary of the interpenetration region, in other words where $D_i = 0$ but its derivative is greater than zero. From here to the boundary of the propagation region, D_i is used to attenuate the propagation to zero. The nearest point on the interpenetration region boundary, $\mathbf{p_0}$ is found by following the gradient.

Figure 9.10 illustrates the behaviour of the deformation in the propagation region. Within the propagation region $D_i(\mathbf{p}) = h_i(r)$, where $\mathbf{p}$ is the point whose implicit value is being calculated and $r = \|\mathbf{p} - \mathbf{p_0}\|$. The radius w_i in Figure 9.10 is set by the user and defines the size of the propagation region, so that no deformation occurs beyond this region (see also Subsection 9.5.4).

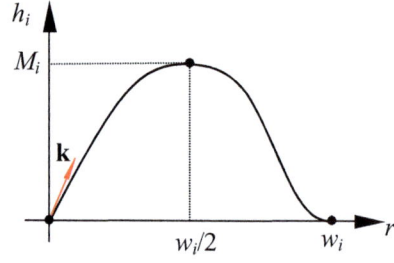

Fig. 9.10. The function $h_i(r)$ is the value of the deformation function D_i in the propagation region.

The equation for h_i is formed in two parts by two cubic polynomials that are designed to join at $r = \frac{w_i}{2}$ where the slope is zero. More specifically,

$$h_i(r) = \begin{cases} cr^3 + dr^2 + kr & \text{if } r \in [0, \frac{w_i}{2}] \\ \frac{4a_0}{w^3}(r-w)^2 \frac{(4r-w)}{w^3} & \text{if } r \in [\frac{w_i}{2}, w_i] \end{cases}$$

where $w = w_i$, $a_0 = M_i$ (the maximum value of h_i), $c = \frac{4(wk-4a_0)}{w^3}$, and $d = \frac{4(3a_0-wk)}{w^2}$.

It is desirable that we have C^1 continuity, going from interpenetration to the propagation region. Thus $h_i'(0) = k$ in Figure 9.10, the directional derivative of D_i at the junction (marked as $\mathbf{p}_0$ in Figure 9.9). As indicated in Equation (9.22), $D_i = -f_i$ in the interpenetration region, thus:

$$k = \|\nabla f_i(\mathbf{p}_0)\|.$$

In addition to the size w_i of the propagation region, we have to control how much the objects inflates in the propagation region. This inflation control is done with reference to the parameter α_i, which takes on a value provided by the user. This factor α relates the maximum value M_i of h_i and the current minimum value $D_{i,\min}$ of D_i, which is negative in the interpenetration region; hence $M_i = -\alpha_i D_{i,\min}$. Thus, an object will be compressed in the interpenetration region will inflate in the propagation region.

PCM is only an approximation to a properly deformed surface, but is an attractive algorithm due to its simplicity. More recently, it has been applied to implicit surfaces of an articulated skeleton in a branching situation, where distant parts of the skeleton could be defined not only to avoid blending, but to deform each other using PCM [18].

$$f = f_j + D_j, \text{ where } f_j = \max_{i=0}^{k-1}(f_i) \tag{9.24}$$

The reader is referred to [74, 108] for a more complete description of the method.

9.5.4 Generalised Bounded Blending

Pasko et al. introduced the idea of bounded blending [318], in which two functional solids defined by potential functions f_1 and f_2, could be locally blended using R-functions based on a third bounding potential function f_b. This method provides control over local blending between two solids; however, Pasko et al.'s approach has two fundamental limitations. The first problem is the reliance on R-functions as the base blending operator, which not only restricts the user to this specific set of modelling operations, but also restricts the application of this technique to pairwise blends. The second shortcoming of their method is that it does not allow interpolation between two arbitrary blending operations.

Generalised bounded blending (GBB) extends this simple concept, opening up a plethora of new modelling possibilities for arbitrary implicit surfaces. Although inspired by the work of Pasko et al., the underlying mathematical formulation is distinct. The essence of GBB is the use of an interpolation operation to interpolate between distinct blending operations applied to the same set of k implicit primitives (with potential functions f_i where $i = 0, ..., k-1$). The interpolation is controlled by a separate implicit primitive with potential function f_b, which bounds the region of interest. Defining the two blending operations as f_{o1} and f_{o2}, the field defined by GBB is given as follows:

$$f = f_b \cdot f_{o1}(f_0, ..., f_{k-1}) + (1.0 - f_b) \cdot f_{o2}(f_0, ..., f_{k-1}) \qquad (9.25)$$

In principle, any potential function can be used to define the bounding region defined by f_b. In practice, the following constraints are necessary to produce intuitive results:

- $f_b \in [0, 1]$
- $\nabla f_b = 0$ when $f_b = 0$
- $\nabla f_b = 0$ when $f_b = 1$

Good results can be obtained by making f_b a sigmoid function, which decreases smoothly from 1 to 0 over $[0, 1]$. In this research f_b was defined as a distance function computed by $f_b(p) = ((1 - r)^2)^3$ where r is the distance from a skeletal primitive. Nevertheless, any arbitrary field can be used for f_b within Equation (9.25), including non-smooth fields such as those produced by CSG operations although the results can not be useful.

Six different blending operations which may be applied between two or more primitives in the *BlobTree* are shown in Figure 9.11. In this case only two

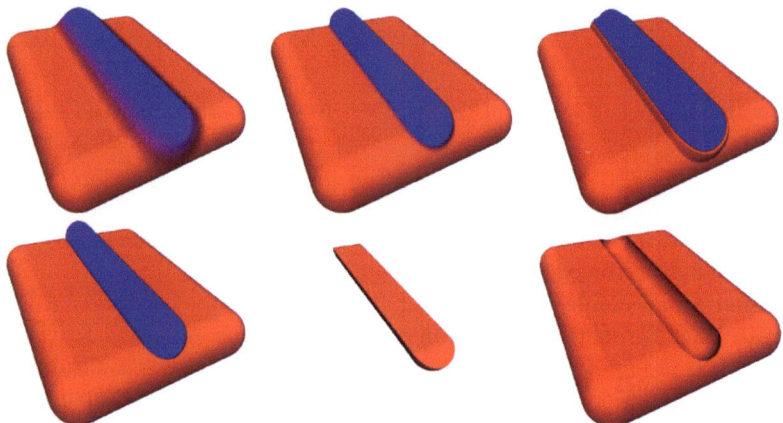

Fig. 9.11. Six methods used to blend two implicit primitives (a skeletal rectangle primitive, red, and a skeletal line primitive, blue) labelled from left to right. Top row: summation blend, super elliptic blend with n = 10, precise contact blending. Bottom row: union, intersection, difference.

9.5 Blending Operations   283

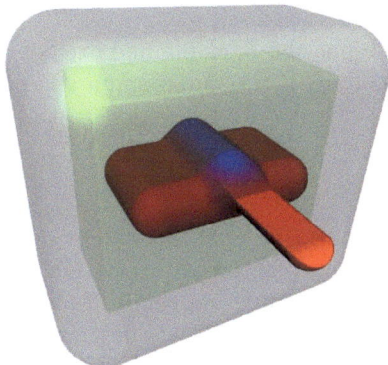

Fig. 9.12. Application of generalised bounded blending (GBB) to interpolate between a blend and intersection of two implicit primitives. The bounding solid, with potential function f_b, is defined by a block primitive and is visualised by two transparent surfaces. Within the inner surface $f_b = 1$, outside the outer surface $f_b = 0$, and in between $f_b \in (0, 1)$.

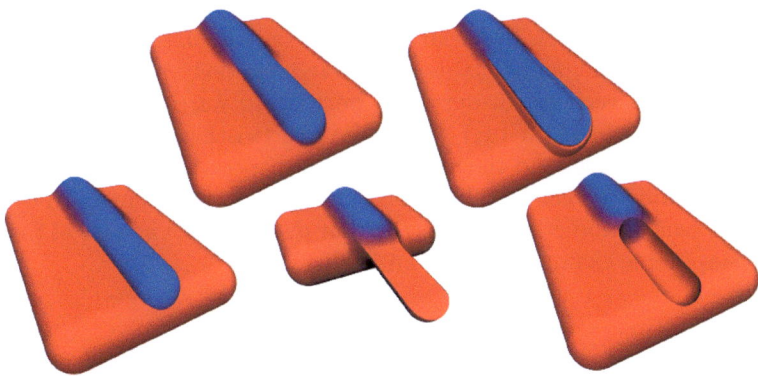

Fig. 9.13. Five examples of GBB. In each case an interpolation is made from the summation blend operation shown in Figure 9.11, to each of the other blend operations shown in Figure 9.11 using the bounding solid shown in Figure 9.12. Labelled from left to right. Top row: super-elliptic blend with $n = 10$, precise contact blending. Bottom row: union, intersection, difference.

input primitives are used for clarity, whereas in practice all of these operations are n-ary. The application of GBB is illustrated in Figure 9.12, which shows an interpolation between the summation blend and intersection operations (Figure 9.11). The bounding solid, with potential function f_b, is defined by a block primitive and is visualised by two transparent surfaces. Within the inner surface $f_b = 1$, outside the outer surface $f_b = 0$, and in between $f_b \in (0, 1)$. Where f_b changes from 0 to 1, the surface is interpolating between the intersection and summation blend operations. The final surface is shown in Figure 9.13

without the bounding solid visualised, along with 4 other examples showing interpolation between summation blending and each of: super-elliptic blending, precise contact modelling, union, intersection and difference.

9.6 Deformations

Modelling operations may also be applied to individual implicit primitives. *Spatial warping* is one method in this category. Examples of spatial warps include the Barr warps: twist, taper, and bend [27]. Warping is easily applied to implicit surfaces by providing a single warping function $w(\mathbf{p})$ [417]. Given an arbitrary choice for w, and a single implicit primitive with implicit surface function f_0, f is defined as follows:

$$\mathbf{p}' = w(\mathbf{p}) \qquad (9.26)$$
$$f(\mathbf{p}) = f_0(\mathbf{p}') \qquad (9.27)$$

This differs from other modelling operations in that the operator is applied to the input (the position) before it is provided to the implicit surface function, rather than operating on the returned value. Barr applies the warp function to wireframe models, and thus uses the warp function w_i to change the coordinates of the vertices. In our case, we wish to warp space, thus we use the inverse warp function $(w_i)^{-1}$. The inverse twisting operation is a twist with a negative angle, and the inverse tapering operation is a taper with the inverse shrinking coefficient. The inverse of bend cannot be produced by modifying the bend parameters (see Barr [27] for details of the inverse of the bend operation). Some examples are shown in Figure 9.4.

9.7 BlobTree Traversal

In the earlier sections we have reviewed some of the operations that have been incorporated into the *BlobTree* and we are now in a position to understand the tree traversal using the Ricci operators for CSG.

The BlobTree is a binary tree similar to CSG tree. Each tree node is either a primitive or an operation. More specifically, leaf nodes store primitives, whereas other nodes accommodate operations (e.g. warp, blend, union, intersection and difference).

Given a node $\mathcal{N}$ and a point $\mathbf{p}$, the recursive function, $f(\mathcal{N}, \mathbf{p})$ returns the appropriate value, which obviously depends on the primitive or operation stored in the node. So,

- If $\mathcal{N}$ is a *primitive*,
$$f(\mathcal{N}, \mathbf{p}) = f(\mathbf{p})$$
returns the field value at $\mathbf{p}$. This is the basic requirement of a rendering algorithm such as [396, 419].

- If $\mathcal{N}$ is a *warp*,
$$f(\mathcal{N}, \mathbf{p}) = f(L(\mathcal{N}), w(\mathbf{p})).$$

- If $\mathcal{N}$ is a *blend*,
$$f(\mathcal{N}, \mathbf{p}) = f(L(\mathcal{N}), \mathbf{p}) + f(R(\mathcal{N}), \mathbf{p}).$$

- If $\mathcal{N}$ is an *union*,
$$f(\mathcal{N}, \mathbf{p}) = \max(f(L(\mathcal{N}), \mathbf{p}), f(R(\mathcal{N}), \mathbf{p})).$$

- If $\mathcal{N}$ is an *intersection*,
$$f(\mathcal{N}, \mathbf{p}) = \min(f(L(\mathcal{N}), \mathbf{p}), f(R(\mathcal{N}), \mathbf{p})).$$

- If $\mathcal{N}$ is a *difference*,
$$f(\mathcal{N}, \mathbf{p}) = \min(f(L(\mathcal{N}), \mathbf{p}), -f(R(\mathcal{N}), \mathbf{p})),$$

where $L(\mathcal{N})$ and $R(\mathcal{N})$ denote the left and right sub-trees of the node $\mathcal{N}$.

The operations above are only a few of the operations in the *BlobTree*. The *BlobTree* is also a scene graph, so that it includes nodes describing geometric transformations as well as many nodes introduced for such operations as animation and texturing. There are now many implementations of the *BlobTree* and research software can be downloaded over the internet.

9.8 Final Remarks

In this chapter we have reviewed some of the techniques that make modelling with skeletal implicit primitives an intuitive process, and that contribute towards the design of complex of models. Models can be represented by the *BlobTree*, in which each node represents one of these techniques. Complex models are visualised by traversing the *BlobTree* and finding a value attached to each point. To render such a model, an isosurface is found. The *BlobTree* nodes represent different types of blending, space deformation, CSG and also affine transformations. Other types of nodes may be found in the literature, for example, for texturing [353] and animation [150].

In the following chapters we take a look at some applications of the techniques discussed, and how they may be used to design complex models.

10
Natural Phenomenae-I: Static Modelling

Implicit modelling as an underlying metaphor provides a large number of techniques that facilitate building of complex models. The *BlobTree* provides tools that make use of blending, CSG, deformation, precise contact modelling and other procedural techniques. Figure 10.1 shows a sea anemone model that was built using the *BlobTree*. The spines were placed procedurally using spiral phyllotaxis and blended to the base. The base of the anemone deforms to fit the rock using precise contact modelling. In the following sections we explore methods for describing complex models from the natural world using the implicit methodology.

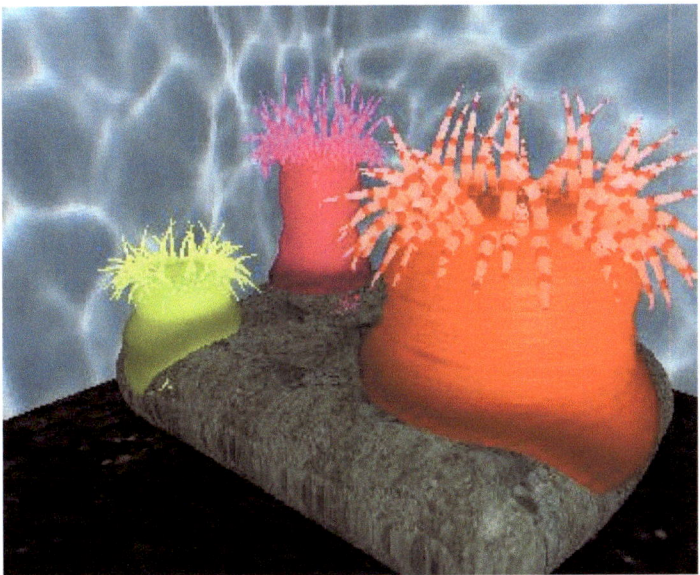

Fig. 10.1. Anemone illustrates the use of PCM to "fit" the rock.

288   10 Natural Phenomenae-I: Static Modelling

10.1 *Murex Cabritii* Shell

The seemingly simple mathematical character of shells, which yield a great variety of beautiful shapes, has attracted much attention from computer modellers. Two motivations for such work are to synthesise realistic images that can be incorporated into computer-generated scenes, and to gain a better understanding of the mechanism of shell formation [144, 266]. Two open problems in the modelling of shells are finding a good method to represent thin spines, and to capture the thickness of the shell walls [144]. In this chapter both of the above problems are addressed using the *BlobTree*. A model of *Murex cabritii* is described which includes large spines, shell walls of non-zero thickness, and allows different textures to be applied to different parts of the shell, while blending textures automatically where these parts join. A preliminary version of this work was published in [152].

10.2 Shell Geometry

As reviewed in [144, 266], the surface of a shell without protrusions may be defined by sweeping a closed generating curve C in the shape of the aperture of the shell along a logarithmic helico-spiral S. The scale of the generating curve increases in geometric progression as the angle of rotation around the shell's axis increases arithmetically.

The helico-spiral is conveniently described in a cylindrical coordinate system (Figure 10.2). The radius R (distance of a point P on the helico-spiral

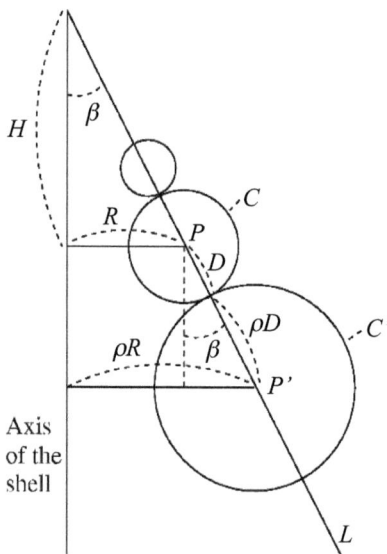

Fig. 10.2. One-half of a longitudinal cross-section of a turbinate shell.

from the shell axis) is an exponential function of the angle of revolution θ around the axis:

$$R(\theta) = R_0 \rho^{(\frac{\theta}{360°})}; \; R_0 > 0, \; \rho > 1, \; \theta \geq 0 \tag{10.1}$$

where R_0 is the initial radius and ρ is the ratio of the radii corresponding to a rotation of 360°. The vertical displacement H of point P increases in proportion to the radius:

$$H(\theta) = R(\theta) \cot \beta, \; \beta > 0 \tag{10.2}$$

where β is the angle between the axis of the spiral and a line L passing through successive whorls of the helico-spiral (Figure 10.2). A whorl is defined as a single volution of a spiral shell, or one turn about the axis.

The size of the generating curve C at point P can be determined under the assumption that C is a circle of radius D lying in the plane including the shell axis and the point P, and that the circles in consecutive whorls are tangential to each other. From Figure 10.2 we then obtain:

$$D(\theta) = \frac{R(\theta)}{\sin \beta} \left(\frac{\rho-1}{\rho+1} \right) \tag{10.3}$$

In the case of noncircular generating curves, Equation (10.3) remains useful as an approximate indicator of the curve size.

10.3 Murex Cabritii

To model *Murex cabritii* requires a description of the parts of the shell. The model is derived from observations made from Figure 10.3, and from a written description of the shell found in [336] page 507, which lists the following features:

Fig. 10.3. *Murex cabritii.*

- A smallish, oval aperture in a strongly convex body whorl.
- A long slender canal below the main body whorl, narrowly open, with three axial rows of four to five spines.
- Each whorl has three varices (ridges) which bear several sharp curving spines.
- Beaded axial riblets (or small bumps) are present between varices.

For the remainder of this chapter a whorl is redefined as a volution of the shell beginning at one varix, and ending after three varices have been formed. From Figure 10.3 it has been estimated that a whorl corresponds to a rotation of $\theta_{\text{whorl}} = 348°$ about the axis of the shell, thus the angle between successive varices θ_{varix} is equal to $116°$. This redefinition is employed as model construction is more usefully guided by angles at which varices occur than arbitrary intervals of $360°$.

The model presented in this chapter is constructed with $w_{\text{count}} = 7$ whorls, each having $v_{\text{count}} = 3$ varices, thus a total of $w_{\text{count}} v_{\text{count}} = 21$ varices are in the final model. Five to six spines (as observed in Figure 10.3) are modelled in the axial rows rather than four to five as described above. The bumps occur periodically both parallel and perpendicular to the helico-spiral, and five sets of bumps are added along the helico-spiral between each pair of varices. The y-axis in the standard coordinate system is defined as the axis of rotation of the shell. The following parameters are used to define the helico-spiral for the model:

$$\begin{aligned} \beta &= 22.5° \\ \rho &= 1.3 \\ R_0 &= 0.2 \\ D(\theta) &= \frac{R(\theta)}{\sin \beta} \frac{\rho-1}{\rho+1} = 0.341 R(\theta) \end{aligned} \quad (10.4)$$

10.4 Modelling *Murex Cabritii*

Procedural techniques are used to construct a *BlobTree* defining the model of *Murex cabritii*. To implement the procedures introduced in this chapter, the Python interface similar to that outlined in [388] was used. To describe the construction of the *BlobTree*, the following notation is introduced. Arbitrary *BlobTree* models are described using the symbol B, with specific instances denoted using appropriate subscripts. For example, skeletal implicit primitives, which are the basic building blocks from which models are constructed, are denoted as follows:

$$\begin{aligned} B_{\text{point}} &\to \text{Skeletal point primitive} \\ B_{\text{line}} &\to \text{Skeletal line primitive} \end{aligned} \quad (10.5)$$

Models are defined by expressions which combine *BlobTrees* using a mixture of basic operators ∪ (union), ∩ (intersection), − (difference), + (blend),

10.4 Modelling *Murex Cabritii*

$\oplus_n$ (super-elliptic blend) and functional composition operators f_{control} (controlled blend), f_{transl} (translate), f_{scale} (scale), f_{rot} (rotate), f_{twist} (twist warp), f_{taper} (taper warp), f_{bend} (bend warp), f_{textG} (gradient interpolated 2D texture mapping), and f_{textF} (field interpolated 2D texture mapping). Additionally $\sum$ and $\bigoplus_n$ are used to represent the blend and super-elliptic blend, respectively, of multiple *BlobTrees* using limit style notation. These operators all correspond to the well-defined implicit surface modelling operations introduced in Chapter 9.

At the lowest level, these operators act on one or more primitives. As a valid *BlobTree* results from each operation, which may be passed as input to other operators, hierarchical models are easily constructed.

The functional composition operators differ from the basic operators in that they require additional parameters to the input *BlobTrees*. Notationally this is defined as $f_{\text{operator}}(p_1, p_2, \ldots, p_n)(B_1, B_2, \ldots, B_m)$ for an arbitrary operator with n parameters and m input *BlobTrees* as follows:

$f_{\text{transl}}(x,y,z)(B)$	$\rightarrow$ translate by (x,y,z)
$f_{\text{scale}}(x,y,z)(B)$	$\rightarrow$ scale by (x,y,z)
$f_{\text{rot}}(\theta, axis)(B)$	$\rightarrow$ rotate by θ about the given axis using the right-hand rule
$f_{\text{transf}}(m)(B)$	$\rightarrow$ transform by matrix m
$f_{\text{taper}}(n)(B)$	$\rightarrow$ taper by n along the positive y-axis
$f_{\text{bend}}(\theta, d)(B)$	$\rightarrow$ bend by θ degrees about the z-axis over a distance of d units
$f_{\text{control}}(b_1, \ldots, b_n)(B_1, \ldots, B_m)$	$\rightarrow$ Controlled blend of m *BlobTrees* where each b_i defines a blend group and $b_i \subseteq \{1, \ldots, m\}$
$f_{\text{textG}}(t)(B)$	$\rightarrow$ apply texture t using gradient interpolated texture mapping
$f_{\text{textF}}(t)(B)$	$\rightarrow$ apply texture t using field interpolated texture mapping

For clarity, the numerical parameters to the above functional composition operators will at times be omitted in the following discussion.

Construction of the *BlobTree* defining a *Murex cabritii* shell is discussed next. Section 10.4.1 describes building the main body whorl of the shell. Creation of the varices is discussed in Section 10.4.2, followed by the addition of bumps in Section 10.4.3 and the spines on the lower canal in Section 10.4.4. Creating the aperture is described in Section 10.4.5 and the application of 2D textures is discussed in Section 10.5.

10.4.1 Main Body Whorl

The formulas in Section 10.2 determine position (Equations (10.1) and (10.2)) and size (Equation (10.3)) of a generating curve along a helico-spiral, such that if successive curves are placed along the helico-spiral and connected in a polygonal mesh, an approximation of the surface of the shell is obtained. For example, Fowler et al. [144] used piecewise Bézier curves to construct generating curves, which were applied to model a great variety of shells.

A similar method is used to create the implicit model. A generating implicit surface B_g is first defined (for example using a skeletal implicit point primitive). The placement of an instance of B_g on the helico-spiral at any angle θ is then performed in three steps:

1. Scale by $D(\theta)$—Equation (10.3).
2. Translate by $(R(\theta), H(\theta), 0)$—Equations (10.1) and (10.2).
3. Rotate by θ about the y-axis.

that is, the function

$$P(B, \theta) = f_{\text{rot}}(\theta, a_y) \cdot f_{\text{transl}}(R(\theta), H(\theta), 0) \cdot f_{\text{scale}}(D(\theta), D(\theta), D(\theta)) \cdot B \quad (10.6)$$

transforms an arbitrary *BlobTree* B as described above. To construct the whorl, instances of B_g are placed at fixed intervals of θ_g along the helico-spiral using Equation (10.6). The value assigned to θ_g must be chosen with care. If θ_g is too large, then a smooth blend along the helico-spiral will not be realised. In contrast, if θ_g is too small, then the tight overlap of the many instances of B_g will lead to poor blending properties when adding detail to the shell.

To incorporate controlled blending, each whorl is modelled in three whorl sections, which are contained between successive varices along the whorl, and thus correspond to a rotation of θ_{varix} about the axis of the shell. Each whorl section is created by placing five instances of B_g on the helico-spiral such that $\theta_g = \theta_{\text{varix}}/5$. The *BlobTree* for a whorl section $B_{\text{whorl}_v}^w$ which immediately precedes varix v on whorl w is given by:

$$B_{\text{whorl}_v}^w = \sum_{i=1}^{5} P(B_g, (3w + v - 1)\theta_{\text{varix}} + \theta_g i) \quad (10.7)$$

Figure 10.4 shows a whorl section composed of five point primitives placed along a helico-spiral, as the radius of the field defined by each primitive is increased, the resulting blended surface tends toward a shell whorl with a circular aperture.

To avoid unwanted blending between consecutive whorls, controlled blending (see Section 9.5.1) is applied to create the main shell body B_{body} using the following procedure:

$$L_{\text{whorlsections}} = (B_{\text{whorl}_i^j}), \text{with } i = 1, \ldots, v_{\text{count}}, j = 1, \ldots, w_{\text{count}}$$
$$L_{\text{blendpairs}} = \{(j, j+1) : j \in \{1, 2, \ldots, w_{\text{count}} v_{\text{count}} - 1\}\} \quad (10.8)$$
$$B_{\text{body}} = f_{\text{control}}(L_{\text{blendpairs}})(L_{\text{whorlsections}})$$

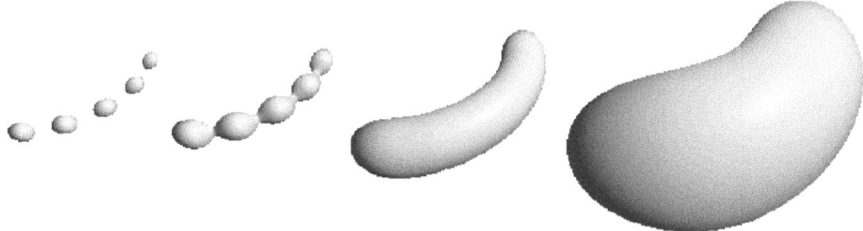

Fig. 10.4. Five point primitives placed on a helico-spiral. As the size of the field produced by each primitive increases, the resulting surface forms part of the main body whorl of a shell.

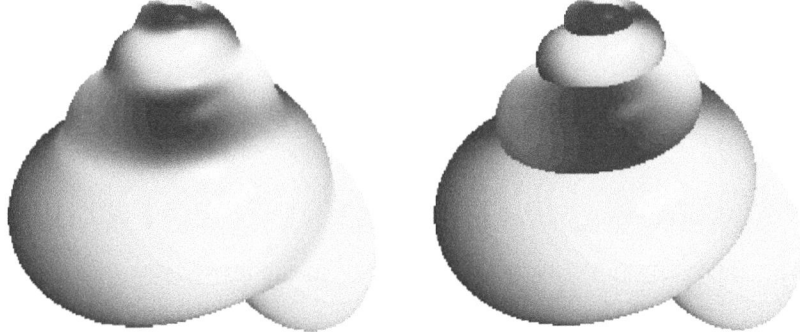

Fig. 10.5. Each whorl of a shell is composed of three sections (shown in Figure 10.4). On the left all sections blend with all other sections, on the right controlled blending constrains each section to blend only with its two neighbours along the helico-spiral.

Each whorl section is blended with its two immediate neighbours, and not with any other whorl sections. The resulting surface is thus smooth along the helico-spiral, while adjacent whorls do not blend together. A comparison of the results obtained with and without controlled blending is shown in Figure 10.5, where four whorls and a total of twelve whorl sections were modelled using a point primitive as the generating surface. The whorl sections are assigned dark and light colours in an alternating pattern so that each whorl section is easily distinguished.

To incorporate the long slender canal below the main body whorl, a tapered line primitive, subsequently bent with a bend operator, was placed below and blended to a point primitive as follows:

$$B_g = f_{\text{transf}}(B_{\text{point}}) + f_{\text{transf}}\left(f_{\text{bend}}\left(f_{\text{taper}}(f_{\text{transf}}(B_{\text{line}})) \right) \right) \quad (10.9)$$

The generating surface and the resulting whorl it defines are shown in Figure 10.6.

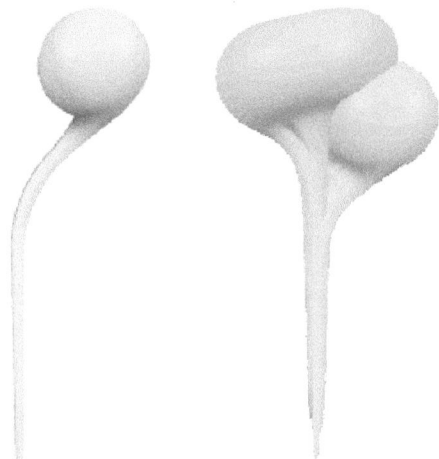

Fig. 10.6. On the left is the generating surface used for the model of *murex cabritii*, on the right is the whorl this surface defines.

10.4.2 Constructing Varices

Varices are the spiny ridges extending out from the main body whorl at even intervals of θ_{varix} around the axis of the shell. The varix is modelled primarily as a series of curving spines of varying size. The relative size and location of spines for varix v on whorl w is determined on a per-whorl basis, (for values see [153]). Individual spines are modelled using tapered line primitives which are bent by 30° over four units of length. Thus, spines shorter than 4 units are bent less than 30°, and spines longer than 4 units are not bent over their whole length. All spines are modelled with the same thickness. By applying taper such that the amount of taper is inversely proportional to the length of each individual spine, a uniform thickness at the tip is achieved, regardless of individual spine length. Construction of the ith spine $B_{\text{spine}_i}^{w}$ for a varix on whorl w using this method.

The left hand image in Figure 10.7 shows the resulting series of spines for $w = 7$ blended to a whorl section. The result does not accurately reflect the form observed in Figure 10.3, as the spines in the varix of *Murex cabritii* are not free-standing, but are blended together in a ridge. A circle primitive (which defined a toroid implicit surface) is added to connect the spines to each other near the shell surface. The effect of this operation is seen in the centre image of Figure 10.7. A new problem now emerges in that the base of the spines are obscured by the toroid ridge. To make the spines stand out from the ridge, a suitable scale is introduced:

$$B_{\text{spine}_i}^{w} = f_{\text{bend}}(30,4) \cdot f_{\text{rot}}(-90, a_z) \cdot f_{\text{taperZ}}\left(\frac{8}{\delta_i^w}\right) \cdot \\ \cdot f_{\text{taperX}}\left(\frac{4}{\delta_i^w}\right) \cdot f_{\text{scale}}(1, \delta_i^w, 3) \cdot B_{\text{line}} \quad (10.10)$$

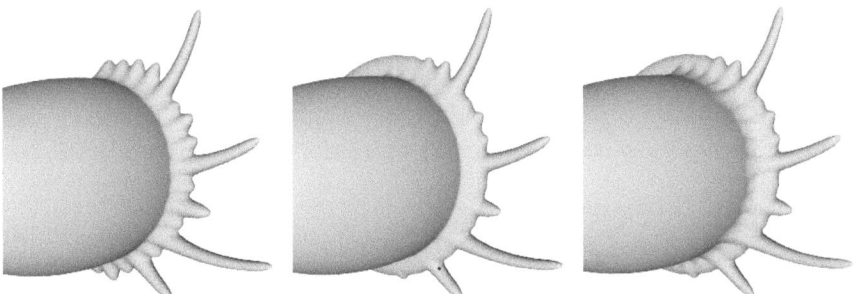

Fig. 10.7. Creation of a varix. Left: bent tapered line primitives are placed as curved spines. Centre: a circle primitive is used to create a toroid ridge blending the spines together. Right: spines are modified as and super-elliptic blending is employed.

where a_z stands for the z-axis and δ_i^w the relative size of curving spines at each of 3 varices per whorl in the model of *murex cabritii*. In this case, the spines are scaled by a factor of 3 in the z axis, and subsequently tapered by an increased amount in the z-axis. The resulting spines are much wider in the z-axis at their base, but gradually revert toward a circular aperture along their length. When the spines are positioned along the helico-spiral, the z-axis in their local coordinate system is transformed to be parallel to the helico-spiral, thus the base of the spines are lengthened along the helico-spiral. Finally, to avoid an overly smooth blending of the spines both with each other, and with the toroid ridge, super-elliptic blending (Equation (9.14)) was used with a blend factor of 3 to blend all of the components of the varix together. The base *BlobTree* for a varix $B_{\text{varix}}{}^w$ on whorl w is thus defined as:

$$B_{\text{varix}}{}^w = f_{\text{transf}}(B_{\text{circle}}) \oplus_3 \bigoplus_{i=1}^{s_{\text{count}}} {}_3 f_{\text{rot}}(\alpha_i, a_z) \left(f_{\text{transl}}(r_a, 0, 0)(B_{\text{spine}_i}{}^w)\right) \tag{10.11}$$

using $B_{\text{spine}_i}{}^w$ from Equation (10.10). The vth varix $B_{\text{varix}_v}{}^w$ on whorl w is defined by using Equation (10.6) with $B_{\text{varix}}{}^w$ as follows:

$$B_{\text{varix}_v}{}^w = P\left(B_{\text{varix}}{}^w, (3w+v)\theta_{\text{varix}}\right) \tag{10.12}$$

The right image of Figure 10.7 shows the final varix blended with the final whorl section using Equation (10.12), that is the result of $B_{\text{varix}_{v_{\text{count}}}}{}^{w_{\text{count}}} + B_{\text{whorl}_{v_{\text{count}}}}{}^{w_{\text{count}}}$.

10.4.3 Constructing Bumps

Individual bumps B_{bump} were modelled using single point primitives scaled by $(s_x, s_y, s_z) = (1.4, 1.0, 1.5)$ as follows:

$$B_{\text{bump}} = f_{\text{scale}}(s_x, s_y, s_z)(B_{\text{point}}) \tag{10.13}$$

Five sets of bumps were placed at regular intervals along the helico-spiral between each successive set of varices. Definition of a single set of bumps $B_{\text{bumpset}}{}^w$ for whorl w is done in a similar fashion to the placement of spines on a varix, utilising a set of empirically determined values. One bump was placed for every two spines present in the varices for the given whorl as follows:

$$B_{\text{bumpset}}{}^w = \sum_{i=1}^{s_{\text{count}}/2} f_{\text{rot}}(\alpha_{2i}, a_z)(f_{\text{transl}}(r_a, 0, 0)(B_{\text{bump}})) \qquad (10.14)$$

To place five sets of bumps $B_{\text{bumpsection}}{}^w_v$ before a given varix v on whorl w, where $(3w + v)\theta_{\text{varix}}$ is the angle of the varix along the helico-spiral, Equation (10.6) is used as follows:

$$B_{\text{bumpsection}}{}^w_v = \sum_{i=1}^{5} P\left(B_{\text{bumpset}}{}^w, \left(3w + v - 1 + \frac{i}{6}\right)\theta_{\text{varix}}\right) \qquad (10.15)$$

The left image of Figure 10.8 shows the final bump section and final whorl section blended together using Equation (10.15), that is the result of $B_{\text{bumpsection}}{}^{w_{\text{count}}}_{v_{\text{count}}} + B_{\text{whorl}}{}^{w_{\text{count}}}_{v_{\text{count}}}$. The bumps are placed as desired; however, the overlap in blending regions causes undesirable amounts of blending between adjacent bumps. The first step to solving this problem is to employ super-elliptic blending again. For the *Murex cabritii* model $n = 3$ has been found to work well.

A localised method is used to provide additional relief for the more tightly packed bumps on the top and bottom of the whorl. Bumps are scaled based on their rotation from the horizontal plane again as defined by empirically determined values for α_i. To produce a more organic feel, the sizes of individual bumps were further modified using the function normal(μ, σ), which returns a pseudo-random number with a normal distribution, where μ is the mean and σ is the standard deviation. The default scale values (s_x, s_y, s_z) are thus replaced by (s'_x, s'_y, s'_z) defined as follows:

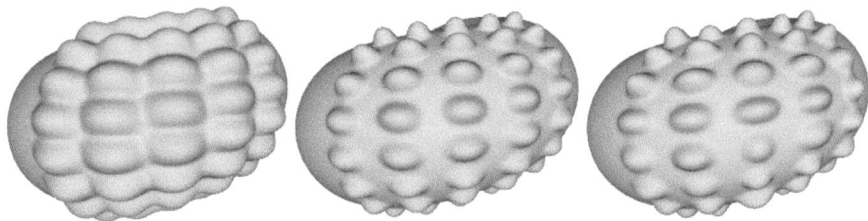

Fig. 10.8. Creation of bumps. Left: bumps arranged and blended. Centre: non-uniform scaling and super-elliptic blending $\oplus_3$ are applied. Right: bumps are randomly scaled, and rotated such that their long axis is aligned locally with the helico-spiral.

$$s'_x = \text{normal}\left(s_x, \tfrac{s_x}{10}\right)$$
$$s'_y = \text{normal}\left(s_y, \tfrac{s_x}{10}\right) \quad (10.16)$$
$$s'_z = \text{normal}\left(s_z, \tfrac{s_x}{10}\right)$$

An additional problem arises because the bumps are scaled non-uniformly, so they are not locally aligned with the helico-spiral. This effect can be observed in both the left and centre images from Figure 10.8. To counteract this effect, each individual bump is rotated about the x-axis before they are positioned, such that their longer axis is locally parallel to the helico-spiral.

10.4.4 Constructing Axial Rows of Spines

One row of axial spines protrudes from the lower canal below each varix on the last whorl of the shell. Individual axial spines B_{axspine} are modelled using tapered line primitives, in a similar fashion to the curving spines in the varices from Equation (10.10). The relative sizes and number of axial spines are determined separately for each row. The resulting spines, blended with the lower whorl, are shown in the left image of Figure 10.9.

As with the bumps, to produce a more organic looking object random variation is introduced. Each spine is randomly bent by 3° to 9°, one to three times, using a corresponding number of bend operators. The result is shown in the right image of Figure 10.9. For details of the spine placement, the reader is referred to Galbraith [153].

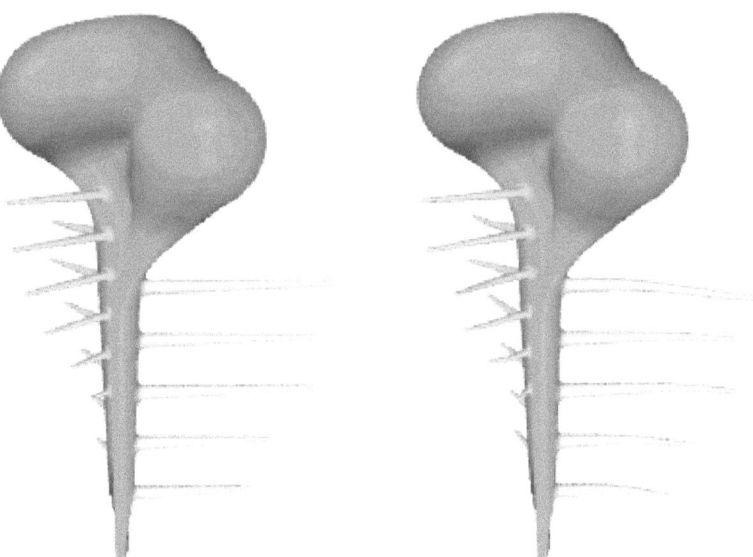

Fig. 10.9. Axial rows of 5-6 spines. Left: spines are straight. Right: each spine is randomly bent 3° to 9° 1-3 times.

10.4.5 Construction of the Aperture

To create the aperture, a solid model, defined as B_{aperture}, is constructed in the shape of the aperture. A CSG difference operation is then used to remove this material from the main body of the shell, thus creating an opening. B_{aperture} is modelled using the same technique as that described for the main body whorl.

A generating surface given by

$$B_{g_{\text{aperture}}} = f_{\text{transf}}(B_{\text{point}}) + f_{\text{transf}}(B_{\text{cone}}) + \\ + f_{\text{transf}}\left(f_{\text{bend}}\left(f_{\text{taper}}\left(f_{\text{transf}}(B_{\text{line}})\right)\right)\right) \quad (10.17)$$

is created, which is slightly smaller in each dimension orthogonal to the helico-spiral, than B_g. Equation (10.17) defines $B_{g_{\text{aperture}}}$, which is formed in a similar fashion to B_g. A point primitive slightly smaller than that of the main whorl's, is blended to a tapered and subsequently bent line primitive, also slightly smaller than that of B_g. To this is also added an inverted cone primitive which extends the inner edge of $B_{g_{\text{aperture}}}$ to the edge of the previous whorl. This is done to ensure that no material is left between the outer shell wall and that of the previous whorl. From Equation (10.17), and similar to Equation (10.7), the corresponding whorl section B_{apwhorl} is defined as follows:

$$B_{\text{apwhorl}} = \sum_{i=1}^{7} P(B_{g_{\text{aperture}}}, (w_{\text{count}} v_{\text{count}} - 1)\theta_{\text{varix}} + \theta_g . i) \quad (10.18)$$

Equation (10.18) describes a whorl equivalent to $B_{\text{whorl}}{}_{v_{\text{count}}}^{w_{\text{count}}}$ plus two additional instances of $B_{g_{\text{aperture}}}$. This implies that the aperture will extend θ_{varix} degrees into the shell from the opening. The two additional instances of $B_{g_{\text{aperture}}}$ ensure that B_{apwhorl} extends beyond the termination of $B_{\text{whorl}}{}_{v_{\text{count}}}^{w_{\text{count}}}$, which in turn ensures that the aperture makes a clean opening. $B_{g_{\text{aperture}}}$ and B_{apwhorl} are shown in the left and centre images of Figure 10.10. As defined, B_{apwhorl} is not suitable for creating the opening in the shell, because B_{apwhorl} overlaps significantly with the previous whorl. Before it is used to create the opening, one more modelling operation is performed, shown in the right image of Figure 10.10. A difference operation is used to remove the previous whorl from B_{apwhorl}, thus defining B_{aperture}, which creates an opening in the shell.

$$B_{\text{aperture}} = B_{\text{apwhorl}} - \sum_{v=1}^{v_{\text{count}}} B_{\text{whorl}}{}_{v}^{w_{\text{count}}-1} \quad (10.19)$$

The summation term defines the second to last whorl using Equation (10.7). The right image in Figure 10.10 shows the resulting model. The aperture in the main shell body is then created as:

$$B_{\text{bodywithaperture}} = B_{\text{body}} - B_{\text{aperture}} \quad (10.20)$$

10.4 Modelling *Murex Cabritii*   299

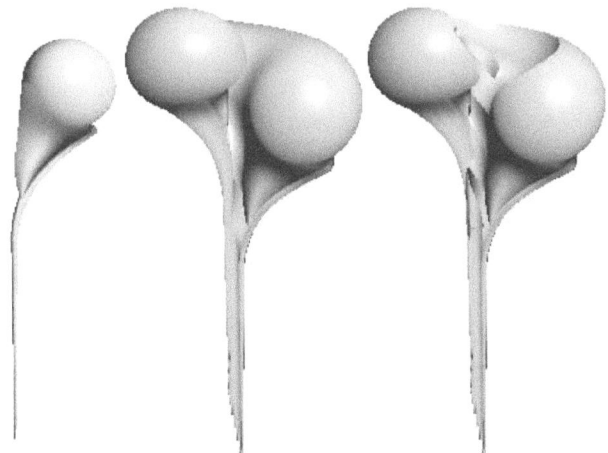

Fig. 10.10. Creating an aperture. Left: the generating surface B_{g_aperture}. Centre: the resulting whorl from Equation (10.18). Right: the final aperture after difference is applied to remove the previous whorl.

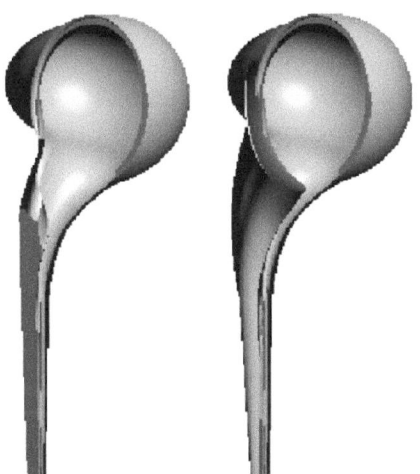

Fig. 10.11. Creating the aperture. Left: the opening which is carved out by Equation (10.20). Right: the final shape of the opening after adding in $B_\text{insidewall}$ as in Equation (10.21).

where B_body is the complete shell without the opening from Equation (10.9).

The left image in Figure 10.11 shows the result when subtracting B_aperture from the last whorl section alone. This figure illustrates that the opening is present only in the last third of the last whorl; however, this is a sufficient size of opening for any view position set outside the shell to give the impression that the shell is hollow. Only if the view position is set inside the aperture is the solid nature of the model revealed.

A more serious problem arises from the observation that the aperture should be oval (as described in Section 10.3). To generate the desired oval aperture, $B_{\text{bodywithaperture}}$ is revised to include an inside wall:

$$B_{\text{bodywithaperture}} = (B_{\text{body}} - B_{\text{aperture}}) \cup B_{\text{insidewall}} \qquad (10.21)$$

$B_{\text{bodywithaperture}}$ with and without the inside wall is shown in Figure 10.11.

10.5 Texturing the Shell

The final step in producing a photorealistic model of *Murex cabritii* is the application of four 2D textures, using two separate texturing methods. The textures used are shown in Figure 10.12, and were created using standard paint programs.

Each whorl section in the main body whorl is textured with the texture shown in Figure 10.12(a) using gradient interpolated texture mapping f_{textG} [387]. This method allows a single texture to be applied to an arbitrary *BlobTree*. As whorl sections are blended to each other, the resulting textures on each whorl section are blended together. By placing the varices directly over these regions, discontinuities in the resulting texture blends are concealed.

Note that since the original work was done on texturing the shell, several new techniques have been designed to ease the task of texturing any point set object. Work is proceeding on this field but the following paper is of interest, [353].

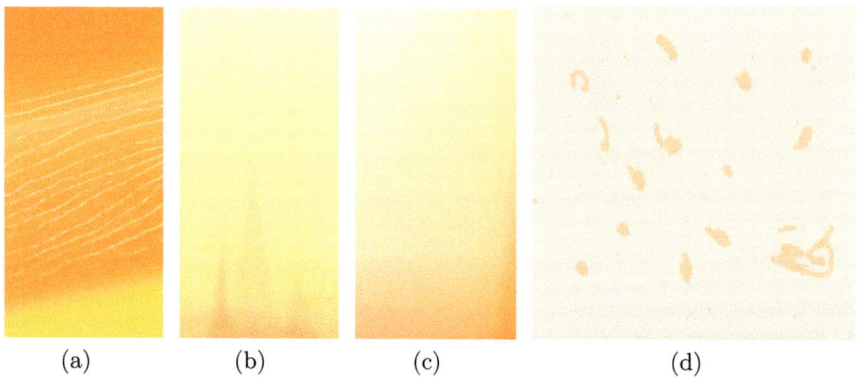

Fig. 10.12. Textures and their corresponding uses in the model of *Murex cabritii*: (a) main body whorl; (b) spines in varices; (c) axial rows of spines; (d) bumps on main whorl.

10.6 Final Model of *Murex Cabritii*

It is now possible to redefine the main geometry of the shell B_{body} (originally defined in Equation (10.9)), including textures, as follows:

$$B_{\text{vbwhorl}_v^w} = f_{\text{textG}}(B_{\text{whorl}_v^w}) + f_{\text{textF}}(B_{\text{varix}_v^w}) + f_{\text{textF}}(B_{\text{bumpsection}_v^w})$$
$$L_{\text{whorlsections}} = (B_{\text{whorl}_i^j}), \text{with } i = 1, \ldots, v_{\text{count}}, \; j = 1, \ldots, w_{\text{count}}$$
$$L_{\text{blendpairs}} = \{(j, j+1) : j \in \{1, 2, \ldots, w_{\text{count}} v_{\text{count}} - 1\}\}$$
$$B_{\text{body}} = f_{\text{control}}(L_{\text{blendpairs}})(L_{\text{whorlsections}})$$
(10.22)

The key change between Equations (10.9) and 10.22, is the use of $B_{\text{vbwhorl}_v^w}$ in place of $B_{\text{whorl}_v^w}$. $B_{\text{vbwhorl}_v^w}$ incorporates texture maps, and blends a whorl section with its corresponding varix and bump section. Starting with this formulation for B_{body}, the *BlobTree* model of *Murex cabritii* B_{murex} is defined as follows:

$$B_{\text{murex}} = (B_{\text{body}} - B_{\text{aperture}}) \cup B_{\text{insidewall}} + f_{\text{textF}}(B_{\text{axspinerows}}) \quad (10.23)$$

10.7 Shell Results

A comparison of a photograph of *Murex cabritii*, with the resulting model of *Murex cabritii* defined by Equation (10.23), is shown in Figure 10.13. The following areas of the model remain open to improvement: the opening was modelled by observing the opening on similar shells (*Murex troschel*); the position and number of spines and bumps were based on a single view of the shell, the number and placement of these features was arbitrary and suddenly change from one whorl to another; the textures were created in a paint program and pasted on to give a good approximation only; the varices do not extend to the lower canal. A major extension of the model would be to use reaction diffusion techniques [144] to place spines and bumps on the shell.

10.8 Final Remarks

The description of the model of *Murex cabritii* presented in this chapter illustrates how the *BlobTree* may be used to construct models of complex phenomena, based solely on simple geometric primitives and a small set of implicit surface modelling operations. This demonstrates concretely that not only implicit surfaces are a valid choice for modelling natural forms, but in addition that they can create models for which other methods such as L-systems fail. Specifically, large protrusions on a shell surface have been modelled simply by switching from a parametric to an implicit definition of the shell form.

Fig. 10.13. (a) *Murex cabritii*; (b) model of *Murex cabritii*.

11
Natural Phenomenae-II: Animation

In this chapter we develop the idea that implicit modelling is useful for animation and continue with the theme that examples from nature present a challenge for any modelling methodology. Recently animations of the accretion of ice and speleothem formations were produced using a version of the *BlobTree* modified for voxel-based simulation (see Figure 11.1). In this chapter we explore the construction and animation of a growing poplar tree using the techniques of implicit modelling.

11.1 Animation: Growing *Populus Deltoides*

In this chapter, a method is described for producing a photorealistic model of a growing tree based on the methods presented in [150]. Branching structures with smoothly blending junctions are a key feature of many natural phenomena, for example herbaceous plants, trees, coral, shells, icicles, speleothems, and animals. Several methods have been applied to model this phenomenon; however, previous work in the area of tree modelling [48, 201, 251, 257] fails to model branching structures that are not universally smoothly blending. Common features of trees such as the branch bark ridge shown in Figure 11.2(a), may combine both smooth and nonsmooth components in a single branching point. Another feature of trees, the bud-scale scar shown in Figure 11.2(b) and (c), may vary from a nonsmooth to a smooth blend over time. We demonstrate how these features may be modelled.

We also model the mechanism used to define the architecture of a growing tree during many growing seasons. A common approach to modelling developmental sequences of plant growth is to simulate the temporal development of plant structures. In these models, the development and final structure of a plant model emerges from the developmental rules. In contrast we describe the developmental growth of a tree using an extension of the *global-to-local methodology* [151, 290, 331]. The advantage of this approach is that complex

304   11 Natural Phenomenae-II: Animation

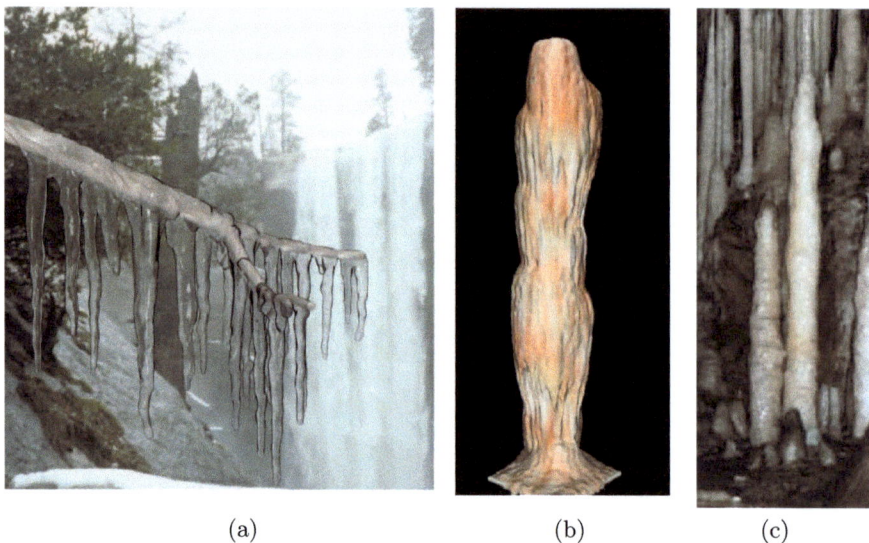

Fig. 11.1. (a) Implicit ice against a real backdrop; (b) implicit speleothem; (c) real speleothem.

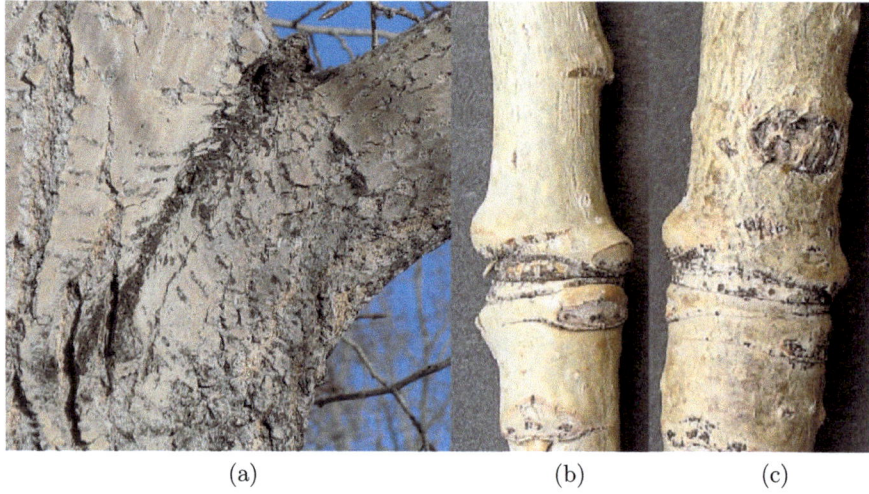

Fig. 11.2. Photographs of poplar trees showing: (a) the branch bark ridge; (b) bud-scale scars at age two years; (c) four years.

developmental sequences may be defined without knowledge of the underlying biological processes. Instead, the method allows specification of the model directly based solely on observed phenomena. Using these techniques, realistic visualisations of growing trees may be achieved.

11.2 Visualisation of Tree Features

Trees differ from herbaceous plants in that their structure grows over many years. One year's growth arising from a single bud is defined as a shoot. The elementary portion of the axis of a branch is an internode and a metamer is an internode which bears lateral organs (leaves, internodes, buds, fruits). A shoot typically consists of a sequence of metamers, each with associated leaf and lateral bud, terminated by an apical bud. The reader is referred to [414] for a thorough description of tree development and form, and to [195] for a description of *populus deltoides* (see Figure 11.3).

Fig. 11.3. *Populus deltoides* (Eastern cottonwood) at 27 years.

11.2.1 Modelling Branches with the *BlobTree*

The structure of a tree branch may be modelled using a skeletal line segment to represent each internode and blending the result. As pointed out by Bloomenthal [51], to achieve perfectly bulge-free blending both along a branch and at branching points the only method currently available within the implicit surface paradigm is convolution based on polygonal skeletons. As perfectly bulge-free blends are considered inappropriate for visualising many tree branching structures, convolution surfaces are not employed. Bloomenthal also suggested that bulging in ramiform structures using convolution surfaces based on line segments may be reduced by offsetting the skeletal elements [51]. In this section it will be demonstrated that the same approach can be used with skeletal implicit surfaces to achieve a sufficient degree of bulge-free blending, while maintaining smooth transitions at branching points.

To model branches of tapering thickness, individual internodes are represented by skeletal cone primitives, shown in Figure 11.4(a). The radius of the cone is set to $r_{c_b} = r_b - r_t$ at the base and $r_{c_t} = 0$ at the top, where r_b defines the base radius of the internode and r_t defines the top radius ($r_t \leq r_b$). The distance d_i from the cone to the visualised isosurface is $d_i = r_t$. Branches are modelled by summing the fields defined by cone primitives representing successive internodes.

To mitigate unwanted bulging (shown in Figure 11.4(b)), the length of the skeletal primitive representing the basal internode A is reduced by:

$$\delta_A = 1.75 r_{B_t} \quad (11.1)$$

where r_{B_t} is the top radius of internode B, as illustrated in Figure 11.5. The value 1.75 used in Equation (11.1) was determined empirically, and assumes that the underlying field function f_{field} for both skeletal primitives is defined as in Equation (9.16). Figure 11.5 shows the resulting isosurfaces when skeletal line primitives A and B are blended. On the left, A and B are connected at the branching point P, resulting in a large bulge. In contrast, the right image shows the result of reducing the length of A by δ_A.

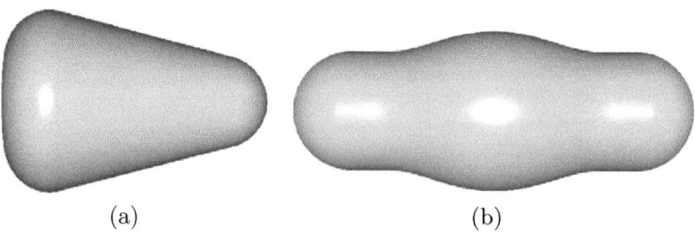

(a) (b)

Fig. 11.4. (a) A *BlobTree* cone primitive; (b) bulge when two line segments with coincident end points are blended.

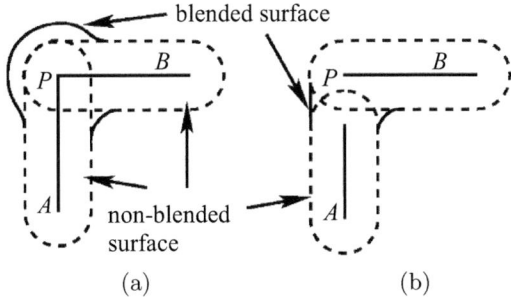

Fig. 11.5. (a) Two skeletal primitives A and B intersect at branching point P, resulting in bulging; (b) the length of A is shortened to reduce bulging.

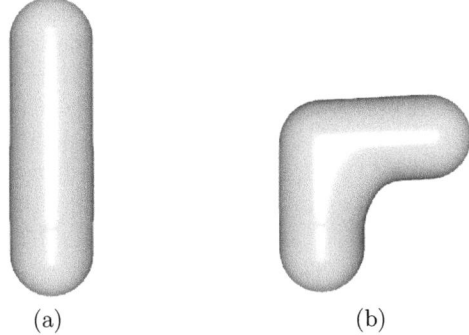

Fig. 11.6. Two cone primitives with their end points offset are blended using summation: (a) 180° separation; (b) 90° separation.

This method does not completely remove the bulging between internodes, but makes it practically unnoticeable, while maintaining a large blending radius in the interior angle between adjoining internodes, shown in Figure 11.6. The scale factor of 1.75 used in Equation (11.1) could be varied by up to 3% without affecting the bulge appreciably.

Internodes in a tree may be fairly short, which introduces an additional problem. For example, in *populus deltoides* they typically vary from near 0 cm to 5 cm. As the tree grows, internodes near the base thicken considerably, so that the value of δ_A from Equation (11.1) quickly exceeds the length of the internode for which it was computed. To overcome this, for each step in the animation, the complete set of internodes representing the branch are first computed as a series of connected line segments. Subsequently this set of line segments is locally decimated based on the desired thickness of the branch, until all remaining line segments meet the condition that their length exceeds the local value of δ_A. The line simplification starts at the base of the branch, which will have the thickest internode, and proceeds to the tip of the branch, where no simplification is typically required as the branch is at its thinnest.

11.2.2 Modelling the Branch Bark Ridge and Bud-scale Scars

Generalised bounded blending (GBB, see Section 9.5.4) is applied to model the bark ridge ridges and bud-scale scars in trees by interpolating between summation blending and precise contact modelling (PCM, see Section 9.5.3). This is appropriate, as in nature the shape is a consequence of the collision between bark volumes, and PCM was designed to model collision deformations. It can also be observed that the collision volume is bounded and there is a smooth change from the deformed part of the volume to the smoothly blended part.

To model branching points, the lateral branch is first shortened at its base and subsequently offset from the main branch using a variation of Equation (11.1). To create the branch bark ridge, GBB is applied as in Figure 11.7, with the base branch shown in red, and the lateral branch in blue. To capture the blackened colour of the branch bark ridge, the texture is set to black where both the bounding field and the deformation due to PCM are maximal. As either the value of the bounding field or of the deformation decreases, the surface attributes are interpolated between those defined by the bounding field, and those defined for the branches as shown in Figure 11.8. On the left the same situation is shown as in Figure 11.7, except that where the field is modified by the PCM blend, the surface attributes are defined by the bounding field, illustrated by the green ridge. On the right we see the application of texture maps and a black colour defined by the bounding field to produce visually realistic results.

To model bud-scale scars, the bounding field f_b is initially configured such that PCM is applied uniformly. This is used to combine both lateral and apical shoots with their basal branch. The effect can be seen in the left-hand image in Figure 11.9. The ring around the main branch denotes a bud-scale scar formed by the apical bud, and separates two years growth on a single branch.

Fig. 11.7. Application of GBB to blend a lateral branch (shown in blue) to its base branch (shown in red) by interpolating between PCM (left) and summation blending (centre). The bounding field is visualised by two transparent surfaces, the inner one bounding a region where $f_b = 1$ and the outer one representing the zero surface $f_b = 0$.

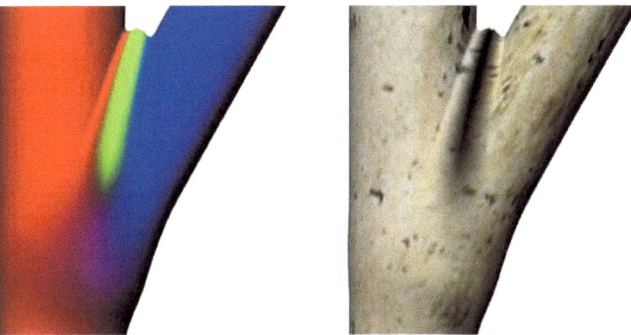

Fig. 11.8. Modifying the texture in the region of PCM deformation. Example image (left) and branch bark ridge model (right).

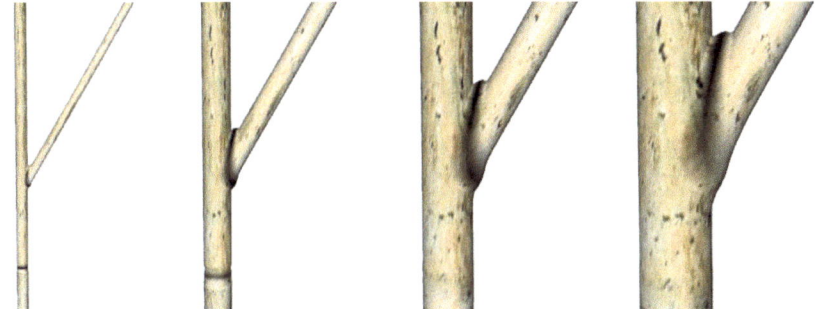

Fig. 11.9. Evolution of bud-scale scars and branch bark ridge over time. Left-right: 1 year, 4 years, 7 years, 10 years.

The other ring on the lateral branch represents a bud-scale scar formed by a lateral bud.

To model the variation of bud-scale scars and the branch bark ridge over time, the field due to f_b is modified over time. For bud-scale scars formed by apical buds, f_b shrinks slowly such that the bud-scale scars slowly disappear. For bud-scale scars formed by lateral branches, f_b is slowly moved such that the bud-scale scar becomes a branch bark ridge. Results for a simple branching situation during ten years are shown in Figure 11.9.

11.3 Global-to-Local Modelling of a Growing Tree

The *global-to-local methodology* was originally applied to model continuous developmental sequences of growth for two types of lilac inflorescence by Galbraith et al. [151]. It was subsequently extended to model a wide variety of static plant structures [331], and to animate continuous developmental sequences of herbaceous plant growth [290].

The extension of the global-to-local methodology to model growing trees is based on the following observation. Branches of every order can be considered as a series of growth increments produced annually as shoots. The difference in branch length from one year to the next determines the length of that year's shoot.

From this observation, a two-step approach is developed. In the first step, the global-to-local methodology is used to define the tree's branching structure in terms of shoots. As in the previous work, self-similarity is used to apply the same set of functions recursively to model all orders of branches.

In the second stage, individual shoots are decomposed into metamers. This decomposition is variable throughout the structure of the tree, thus creating branching structures without self-similarity. Subsequently, the growth of shoots in each year is defined using the methods developed previously for herbaceous plants [290].

The branching structure is first described for branches of order 0 (the trunk) and 1. It is then extended to higher order branches. The following notation is used throughout this section; $y \in [0, y_{\max}]$ denotes the current year after germination, where $y_{\max}$ is the maximum age of the tree. Length of a branch in year y is denoted as l_y, and the following inequality is enforced:

$$l_y + l_{\min} \leq l_{y+1} \tag{11.2}$$

where $l_{\min}$ is a user-defined minimum length of shoot. For a shoot that grows during year y, the shoot length l_s is defined by $l_s = l_y$ if $y = 0$, and $l_s = l_y - l_{y-1}$ otherwise.

Let us denote x_s as the global position of a shoot along its branch, measured as the distance from the base of the branch to the shoot's position, and $x_{s_y} \in [0, 1]$ the relative position of a shoot along its branch in year y, where $x_{s_y} = x_s/l_y$. The relative position of a shoot along its branch during year $y_{\max}$ is denoted by $x_{s_{\max}} \in [0, 1]$, where $x_{s_{\max}} = x_s/l_{\max}$. Local parameters of tree components are determined interactively by the user with graphically defined functions G over the domain $[0, 1]$, defined using an interactive function editor. Unless stated otherwise, these functions are defined in terms of x_{s_y}, $x_{s_{\max}}$ or time $t_y = y/y_{\max}$. For example, the height of the trunk by year is defined by G_h as follows $l_y = G_h(t_y)$.

11.3.1 Crown Shape

The first step is to define the length of each branch for each year. The length of the trunk, or branch of order 0, is treated as a special case, and is determined by G_h as above. The length of lateral branches is determined by the use of two functions $G_{s1}(x_{s_y})$ defining the desired silhouette of the tree when it is young, and $G_{s2}(x_{s_y})$ when it is mature, where x_{s_y} is determined by the position of the branch's parent shoot (the shoot that the branch grows from). The length of a branch during each year is then determined as follows:

11.3 Global-to-Local Modelling of a Growing Tree   311

$$l_{d_y} = l_{p_y} \cdot \left((1 - t_y) \cdot G_{s1}(x_{s_y}) + t_y \cdot G_{s2}(x_{s_y}) \right) \quad (11.3)$$

$$l_y = \max(l_{d_y}, l_{y-1} + l_{\min}) \quad (11.4)$$

where l_{d_y} is the determined length before enforcing Equation (11.2). Figure 11.10 shows the result where the young tree is narrow, and the mature tree has a well-rounded crown, as in *Populus deltoides*. An alternate form is shown in Figure 11.11, defining a tree which grows more like an evergreen.

To ensure a smooth progression of the overall crown shape, the input to functions G_{s1} and G_{s2} is defined in terms of x_{s_y}, rather than $x_{s_{\max}}$. As x_{s_y} varies for a given branch by year, the introduction of high frequency variations in these functions will cause erratic growth patterns for individual branches year by year. Therefore, it is important that G_{s1} and G_{s2} are defined as smooth curves. A significant side effect of this formulation is that the appearance of branches on a specific shoot cannot be controlled using G_{s1} and G_{s2}.

To provide a finer level of control, a scale factor s_s unique to each shoot is introduced. s_s is defined by $G_{ss}(x_{s_{\max}})$ for each shoot as follows:

$$s_s = G_{ss}(x_{s_{\max}}) \quad (11.5)$$

The scale factor is used in several ways in the modelling process. Initially it provides a direct method for controlling the length of branches by scaling the result of Equation (11.4) by s_s. Figure 11.12 shows an exaggerated example where a discontinuous function is used to create a repeating pattern of short, medium and long branches.

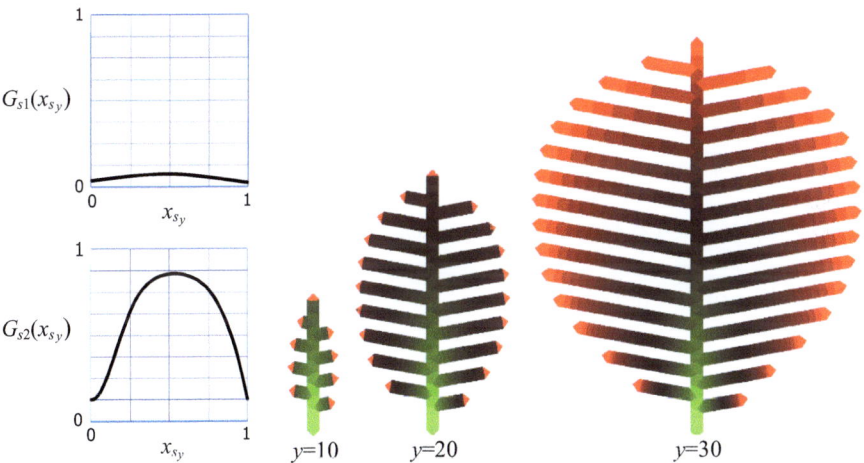

Fig. 11.10. Length of branches is determined by interpolation between two silhouette curves based on the age of the tree y. The silhouette is defined by $G_{s1}(x_{s_y})$ when $y = 0$, and $G_{s2}(x_{s_y})$ when $y = 30$. x_{s_y} defines position of the branch along the trunk.

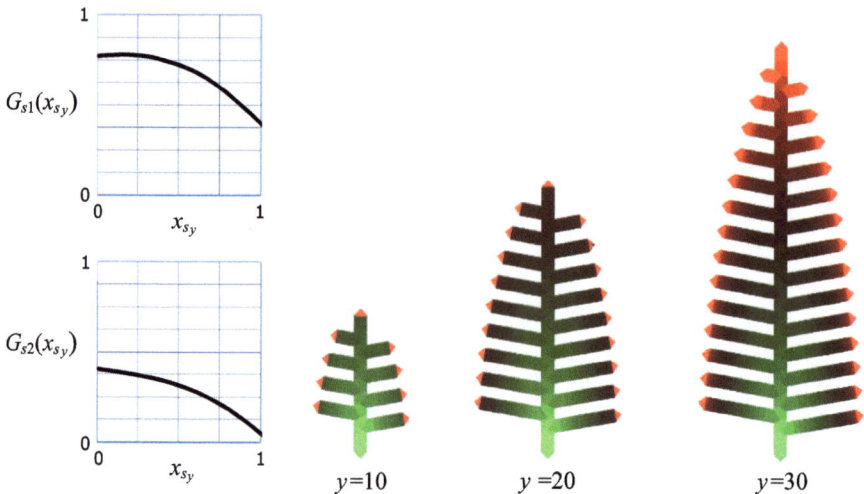

Fig. 11.11. Length of branches is determined by interpolation between two silhouette curves based on the age of the tree y. The silhouette is defined by $G_{s1}(x_{s_y})$ when $y = 0$, and $G_{s2}(x_{s_y})$ when $y = 30$ x_{s_y} defines position of the branch along the trunk.

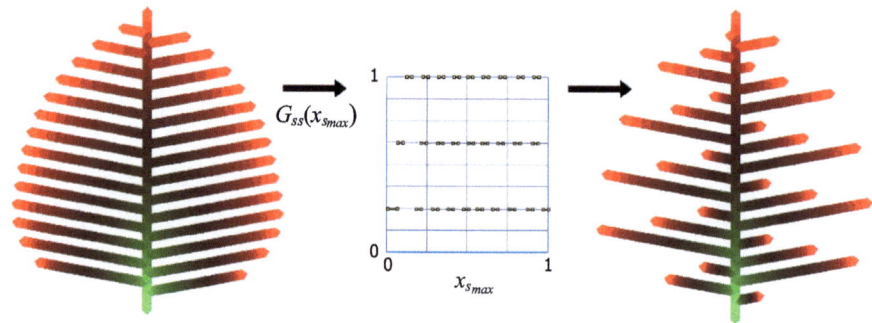

Fig. 11.12. Modifying branch length with the scale factor s_s.

11.3.2 Shoot Structure

Once a shoot's position and size within the branching structure have been determined, it is decomposed into a sequence of m metamers defined by G_m as follows:

$$m = G_m(q) \tag{11.6}$$

where $q = l_s/l_{s_{\max}}$ and $l_{s_{\max}}$ is the maximum length of a shoot. The lengths of internodes l_i in the shoot are assumed to be equal for each metamer where $l_i = l_s/m$. The position of lateral organs relative to their parent shoot is then given by $x_l \in [0, 1]$, where $x_{l_j} = j/m$ and $j \in [1, m]$.

11.3 Global-to-Local Modelling of a Growing Tree

During the year following a shoot's growth, new shoots grow from the apical bud and from some of the lateral buds. The lateral bud closest to the tip of the shoot produces the longest lateral branch, and the length of lateral branches decreases with increasing distance from the tip. As defined in Section 11.3.1, all of the lateral branches growing from a common shoot will have the same lengths.

To achieve the desired variation of lateral branch lengths along the shoot, each lateral branch is assigned a unique scale factor s_b using the function $G_{sb}(x_l)$ as follows:

$$s_b = G_{sb}(x_l) \cdot s_s \qquad (11.7)$$

where s_s is the scale factor of the branch's parent shoot. The final length of lateral branches is defined by scaling the result of Equation (11.4) by s_b, rather than by s_s as defined in Section 11.3.1. Defining G_{ss} such that lateral branches near the top of the shoot have higher scale values, and lateral branches near the base of the shoot have lower scale values, provides the desired arrangement of branch lengths. Additionally, lateral branches will not be created if s_b is below a user-defined threshold $s_{\min}$ allowing for some lateral buds to remain dormant as in real trees.

Lateral branches are arranged in a spiral phylotactic arrangement, where the lateral branch of each successive metamer is placed at an angle of 137.5° from the previous one in the plane perpendicular to the axis of the branch. Unnatural repetitive patterns can arise in the final structure if the longest branch from one year's shoot is nearly coplanar with the longest branch from the previous year's shoot.

To avoid such structures, the metamer count m for a given shoot is restricted such that the two longest lateral branches on the current shoot are placed at an angle at least 40 degrees away from the longest branch of the previous year's shoot.

11.3.3 Other Functions

A number of other functions have been determined to account for various visual properties. Full details are given in [153]. These are briefly described below:

- *Higher order branches.* These are modelled by recursive application of the functions employed to model branches of order 1. To account for variation in branch orders, some modifications are made. Lengths of higher order branches is defined as with order 1 branches, but are scaled based on the order of branching.
- *Avoidance of self-similar branching structures.* The key to avoiding the construction of self-similar branching structures lies in the variable decomposition of shoots into metamers. By defining $G_{ss}(x_{s_{\max}})$ (see Equation (11.5)) to vary continuously, considerable variation is achieved within the branching structure.

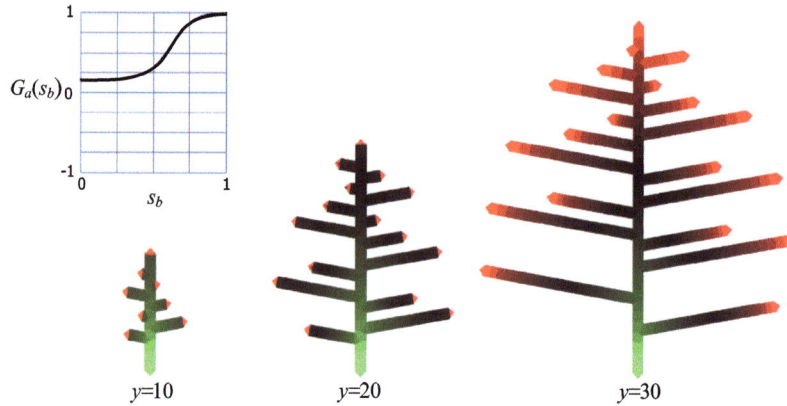

Fig. 11.13. Modelling loss of branches using the scale factor s_b, determined as in Figure 11.12. Due to the correlation between branch length and s_b, shorter branches are only observed near the top of the tree where they are younger.

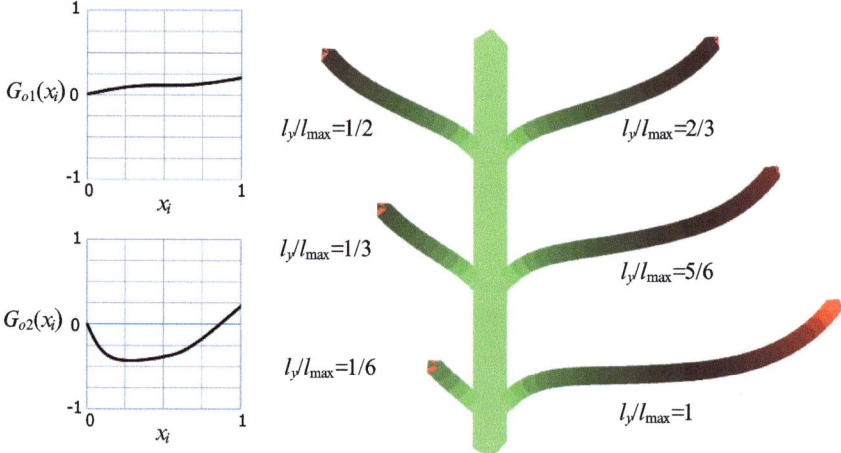

Fig. 11.14. Two functions, G_{o1} for branches of length 0, and G_{o2} for branches of length $l_{\max}$, define localised orientation of branch segments relative to the heading of the branch. Localised orientation of branch segments is defined by interpolating between the two functions based on current branch length.

- *Incorporating loss of branches.* As a tree grows, it sheds branches which that are no longer productive. Typically these are shorter branches in the interior of the canopy, where less light penetrates. The function designed for this introduces a correlation between branch length and branch lifespan. The result is that as the tree grows, shorter branches will tend to be lost sooner than long branches of the same order as illustrated in Figure 11.13.

- *Branch shape.* To produce more lifelike branching shapes, a method is introduced to define localised orientation for branch segments in terms of the relative position of an internode. This is done by interpolating between local orientations for a theoretical branch of length 0, and local orientations for a branch of maximum length. For each internode along a branch, a desired heading and radius is specified during construction of the branch. The result is shown in Figure 11.14 for six branches of varying length.

11.4 Results

Three animations illustrating various aspects of a growing poplar (*populus deltoides* in Latin) model were produced. Figure 11.9 demonstrates the effectiveness of the proposed method in modelling the branch bark ridge and bud-scale scars over time. Figure 11.15 shows frames illustrating the growth of a shoot during one year. Note the visible bud-scale scar at the base of the shoot where the bud scales are lost. Finally, the growth of *populus deltoides* during 27 years, including a succession of foliage in each year is shown in Figure 11.16. Figure 11.17 shows a series of images of real *populus deltoides* trees at different stages of growth. The shape and distribution of the branches is matched fairly well to the generated images shown in Figure 11.18. The tree at 27 years with a full canopy of leaves is shown in Figure 11.3.

Fig. 11.15. Growth of a shoot over 1 year.

Fig. 11.16. Foliage succession during one year.

Fig. 11.17. Photographs of *Populus deltoides* (Eastern cottonwood).

Fig. 11.18. Implicit surface models of *Populus deltoides* (Eastern cottonwood) at several developmental stages.

11.5 Final Remarks

In this chapter new methods for the creation of photo realistic animations of growing trees were presented. The main points of this work are summarised below:

11.5 Final Remarks

- A combination of blending, precise contact modelling, and generalised bounded blending was used to model bud-scale scars and branch bark ridges, and their evolution over time.
- A method was presented for producing continuous animations of growing trees using an extension of the global-to-local methodology.
- An animation of a growing *Populus deltoides* was defined using the above techniques.
- The use of the proposed *BlobTree* modelling and animation system to animate complex natural phenomena was demonstrated.

The main limitation of the current work continues to be the slow visualisation times. In cases where the proposed method for modelling branching points is not required, the algorithm for defining the architecture of a growing tree may easily be reused in any system supporting procedural modelling of plants. Examples of such systems include *L-systems* [330] and *xfrog* [109], whose use of surface models offers significant speed benefits. Conversely, it is possible to use alternate approaches to model the architecture of the tree, such as rule based models of tree growth using L-systems [294], as input to create a *BlobTree* model [149].

References

1. A. Adamson and M. Alexa. Approximating and intersecting surfaces from points. In *Proceedings of the 1st Eurographics/ACM SIGGRAPH Symposium on Geometry Processing*, pages 230–239, Aire-la-Ville, Switzerland, 2003. Eurographics Association.
2. S. Akkouche and E. Galin. Adaptive implicit surface polygonization using marching triangles. *Computer Graphics Forum*, 20(2):67–80, 2001.
3. G. Albertelli and R. Crawfis. Efficient subdivision of finite-element data sets into consistent tetrahedra. In *Proceedings of the IEEE Conference on Visualization'97*, pages 213–219. IEEE Computer Society Press, 1997.
4. G. Alefeld and J. Herzberger. *Introduction to Interval Computations*. Academic Press, New York, 1983.
5. M. Alexa, J. Behr, D. Cohen-Or, S. Fleishman, D. Levin, and C. Silva. Computing and rendering point set surfaces. *IEEE Transactions on Visualization and Computer Graphics*, 9(1):3–15, January–March 2003.
6. R. Allègre, A. Barbier, S. Akkouche, and E. Galin. A hybrid shape representation for free-form modeling. In *Shape Modeling International 2004*, pages 7–18, 2004.
7. E. Allgower and K. Georg. Simplicial and continuation methods for approximating fixed points and solutions to systems of equations. *SIAM Review*, 22(1):28–85, 1980.
8. E. Allgower and K. Georg. *Introduction to Numerical Continuation Methods*. SIAM's Classics in Applied Mathematics. SIAM Press, 2003.
9. E. Allgower and S. Gnutzmann. An algorithm for piecewise linear approximation of implicitly defined two-dimensional surfaces. *SIAM Journal on Numerical Analysis*, 24(2):452–469, 1987.
10. E. Allgower and S. Gnutzmann. Simplicial pivoting for mesh generation of implicitly defined surfaces. *Computer Aided Geometric Design*, 8(4):305–325, 1991.
11. E. Allgower and P. Schmidt. An algorithm for piecewise linear approximation of implicitly defined manifold. *SIAM Journal on Numerical Analysis*, 22(2):322–346, 1985.

12. N. Amenta, M. Bern, and M. Kamvysselis. A new Voronoi-based surface reconstruction algorithm. *ACM SIGGRAPH Computer Graphics*, 32(5):415–421, 1998.
13. N. Amenta, S. Choi, T. Dey, and N. Leekha. A simple algorithm for homeomorphic surface reconstruction. *International Journal of Computational Geometry and Applications*, 12(1–2):125–141, 2002.
14. N. Amenta, S. Choi, and R. Kolluri. The power crust. In *Proceedings of the 6th ACM Symposium on Solid Modeling and Applications*, pages 249–266, Ann Arbor, Michigan, USA, 2001. ACM Press.
15. N. Amenta and Y. Kil. The domain of a point set surface. In M. Alexa and S. Rusinkiewicz, editors, *Proceedings of Eurographics Symposium on Point-Based Graphics*, pages 139–147. ETH Zurich, Switzerland, Eurographics Association, June 2-4 2004.
16. B. Anderson, J. Jackson, and M. Sitharam. Descartes' rule of signs revisited. *American Mathematical Montlhy*, 105(5):447–451, 1998.
17. M. Andrade, J. Comba, and J. Stolfi. Affine arithmetic. In *Proceedings of Interval 94*. St. Petersburg, Russia, March 1994. Available at: www-graphics.stanford.edu/~comba/papers/aa-93-12-petersburg-paper.ps.gz.
18. A. Angelidis, P. Jepp, and M. Cani. Implicit modelling with skeleton curves: controlled blending in contact situations. In *Proceedings of the International Conference on Shape Modeling and Applications (SMI 2002)*, pages 137–144. IEEE Computer Society, May 2002.
19. B. Araújo. Curvature-dependent polygonization of implicit surfaces. Master's thesis, Instituto Superior Técnico, Universidade Técnica de Lisboa, Portugal, February 2008.
20. B. Araújo and J. Jorge. Adaptive polygonization of implicit surfaces. *Computers & Graphics*, 29(5):686–696, 2005.
21. C. Armstrong, A. Bowyer, S. Cameron, J. Corney, G. Jared, R. Martin, A. Middleditch, M. Sabin, and J. Salmon. *Djinn: A Geometric Interface for Solid Modelling: Specification and Report*. Information Geometers Ltd., Winchester, England, December 2000.
22. A. Atieg and G. Watson. Use of l_p norms in fitting curves and surfaces to data. *Australian and New Zealand Industrial and Applied Mathematics Journal*, 45:C187–C200, 2004.
23. D. Avis and B. Bhattacharya. Algorithms for computing d-dimensional voronoi diagrams and their duals. In F. Preparata, editor, *Computational Geometry*, volume 1 of *Advances in Computing Research*, pages 159–180. JAI Press, London, England, 1983.
24. R. Balsys and K. Suffern. Visualisation of implicit surfaces. *Computers & Graphics*, 25(1):89–107, February 2001.
25. R. Balsys and K. Suffern. Adaptive polygonisation of non-manifold implicit surfaces. In *Proceedings of the International Conference on Computer Graphics, Imaging and Vision: New Trends (CGIV'05)*, pages 257–263. IEEE Computer Society Press, 2005.
26. C. Barber, D. Dobkin, and H. Huhdanpaa. The quickhull algorithm for convex hulls. *ACM Transactions on Mathematical Software*, 22(4):469–483, December 1996.

27. A. H. Barr. Global and local deformations of solid primitives. In *Computer Graphics (SIGGRAPH 84 Conference Proceedings)*, volume 18, pages 21–30. ACM Press, 1984.
28. L. Barthe, N. A. Dodgson, M. A. Sabin, B. Wyvill, and V. Gaildrat. Two-dimensional potential fields for advanced implicit modeling operators. *Computer Graphics Forum*, 22(1):23–33, 2003.
29. L. Barthe, V. Gaildrat, and R. Caubet. Combining implicit surfaces with soft blending in a CSG tree. In *CSG Conference Series*, pages 17–31, Apr 1998.
30. L. Barthe, V. Gaildrat, and R. Caubet. Extrusion of 1D implicit profiles: theory and first application. *International Journal of Shape Modeling*, 7(2):179–198, 2001.
31. L. Barthe, B. Wyvill, and E. de Groot. Controllable binary CSG operators for soft objects. *International Journal of Shape Modeling*, 10(2):135–154, 2004.
32. S. Basu, R. Pollack, and M.-F. Roy. *Algorithms in Real Algebraic Geometry*. Springer-Verlag, New York, 2003.
33. B. Baumgart. Winged edge polyhedron representation. Stanford Artificial Intelligence Project, MEMO AIM-179 STAN-CS-320, Computer Science Department, Stanford University, 1972.
34. P. Baxa, V. Skala, and R. Moucek. Error estimation for isosurfaces. In *Proceedings of 6th International Conference on Computational Graphics and Visualization Techniques (COMPUGRAPHICS'97)*, pages 202–211, Vilamoura, Algarve, Portugal, 1997.
35. P. Baxandall. *Vector Calculus*. Oxford Applied Mathematics and Computing Science Series. Clarendon Press, Oxford, England, Oxford, UK, 1986.
36. R. Beatson, J. Cherrie, and D. Ragozin. Fast evaluation of radial basis functions: methods for four-dimensional polyharmonic splines. *SIAM Journal on Mathematical Analysis*, 32(6):1272–1310, 2001.
37. R. Beatson, M. Powell, and A. Tan. Fast evaluation of polyharmonic splines in three dimensions. *IMA Journal of Numerical Analysis*, 27(3):427–450, 2006.
38. T. Becker and V. Weispfenning. *Gröbner Bases*. Springer-Verlag, New York, 1993.
39. T. Belytschko, Y. Krongauz, D. Organ, M. Fleming, and P. Krysl. Meshless methods: an overview and recent developments. *Computer Methods in Applied Mechanics and Engineering*, 139(1–4):3–47, December 1996.
40. J. Berchtold, I. Voiculescu, and A. Bowyer. Interval arithmetic applied to multivariate Bernstein form polynomials. Technical Report 31/98 (`http://people.bath.ac.uk/ensab/G_mod/Bernstein/tr_31_98.html`), University of Bath, England, 1998.
41. Jakob Berchtold. *The Bernstein Basis in Set-theoretic Geometric Modelling*. PhD thesis, University of Bath, England, 2000.
42. F. Bernardini, C. Bajaj, J. Chen, and D. Schikore. Automatic reconstruction of 3D CAD models from digital scans. *International Journal of Computational Geometry and Applications*, 9(4–5):327–369, August-October 1999.
43. F. Bernardini, J. Mittleman, H. Rushmeier, C. Silva, and G. Taubin. The ball-pivoting algorithm for surface reconstruction. *IEEE Transactions on Visualization and Computer Graphics*, 5(4):349–359, October-December 1999.
44. F. Bigdeli. *Triangulations of Convex Polytopes and d-Cubes*. PhD thesis, University of Kentucky, Lexington, USA, 1991.
45. A. Blake and M. Isard. *Active Contours*. Springer-Verlag, Berlin, 1998.

46. C. Blanc and C. Schlick. Extended field functions for soft objects. In *Implicit Surfaces '95*, pages 21–32, 1995.
47. J. Blinn. A generalization of algebraic surface drawing. *ACM Transactions on Graphics*, 1(3):235–256, July 1982.
48. J. Bloomenthal. Modeling the mighty maple. In *Computer Graphics (SIGGRAPH 85 Conference Proceedings)*, volume 19, pages 305–311. ACM Press, 1985.
49. J. Bloomenthal. Polygonization of implicit surfaces. *Computer-Aided Geometric Design*, 5(4):341–355, November 1988.
50. J. Bloomenthal. An implicit surface polygonizer. In P. Heckbert, editor, *Graphics Gems IV*, Academic Press Graphics Gems Series, pages 324–349. Academic Press, 1994.
51. J. Bloomenthal. *Skeletal Design of Natural Forms*. PhD thesis, University of Calgary, 1995.
52. J. Bloomenthal. Surface tiling. In J. Bloomenthal, J. Blinn, M.-P. Cani-Gascuel, A. Rockwood, B. Wyvill, and G. Wyvill, editors, *Introduction to Implicit Surfaces*, The Morgan Kaufmann Series in Computer Graphics and Geometric Modeling, pages 127–165. Morgan Kaufmann Publishers, Inc., 1997.
53. J. Bloomenthal and K. Ferguson. Polygonization of nonmanifold implicit surfaces. *ACM SIGGRAPH Computer Graphics*, 29:309–316, August 1995. (SIGGRAPH'95).
54. J. Bloomenthal and B. Wyvill. Interactive Techniques for Implicit Modeling. *Computer Graphics*, 24(2):109–116, 1990.
55. G. Bodnár and J. Schicho. A computer program for the resolution of singularities. In H. Hauser, J. Lipman, F. Oort, and A. Quirós, editors, *Resolution of Singularities, A Research Textbook in Tribute to Oscar Zariski,*, volume 181 of *Progress in Mathematics*, pages 231–238. Birkhäuser, 2000.
56. J. Boissonnat. Geometric structures for three-dimensional shape representation. *ACM Transactions on Graphics*, 3(4):266–286, 1984.
57. J. Boissonnat and F. Cazals. Smooth surface reconstruction via natural neighbor interpolation of distance functions. In *Proceedings of the 16th ACM Symposium on Computational Geometry*, pages 223–232. ACM Press, 2000.
58. W. Boothby. *An Introduction to Differentiable Manifolds and Riemannian Geometry*. Number 63 in Pure and Applied Mathematics. Academic Press, Inc., New York, 1975.
59. P. Borodin, G. Zachmann, and R. Klein. Consistent normal orientation for polygonal meshes. In *Proceedings of Computer Graphics International (CGI'04)*, pages 18–25. IEEE Computer Society Press, 2004.
60. A. Bowyer. Computing Dirichlet tessellations. *The Computer Journal*, 24(2):162–166, 1981.
61. A. Bowyer. *sVLIs set-theoretic kernel modeller*. Information Geometers Ltd. and http://people.bath.ac.uk/ensab/G_mod/Svlis, 1995.
62. A. Bowyer, J. Berchtold, D. Eisenthal, I. Voiculescu, and K. Wise. Interval methods in geometric modelling. In *Geometric Modelling and Processing 2000*. IEEE Computer Society Press, April 2000.
63. A. Bowyer, R. Martin, H. Shou, and I. Voiculescu. Affine intervals in a CSG geometric modeller. In J. Winkler and M. Niranjan, editors, *Uncertainty in Geometric Computations*, pages 1–14. Kluwer Academic Publishers, 2001.

64. E. Brisson. Representing geometric structures in d dimensions: topology and order. *Discrete & Computational Geometry*, 9(4):387–426, 1993.
65. T. Brocker and K. Janich. *Introduction to Differential Topology*. Cambridge University Press, Cambridge, England, 1982.
66. K. Brodlie. *Mathematical Methods in Computer Graphics and Design*. Institute of Mathematics and Its Applications Conference Series. Academic Press, 1980.
67. I. Bronstein and K. Semendjajew. *Taschenbuch der Mathematik*. B. G. Teubner Verlagsgesellschaft, 1991.
68. J. Bruce and P. Giblin. *Curves and Singularities*. Cambridge University Press, Cambridge, England, 1984.
69. P. Brunet and I. Navazo. Solid representation and operation using extended octrees. *ACM Transactions on Graphics*, 9(2):170–197, 1990.
70. B. Buchberger. *On Finding a Vector Space Basis of the Residue Class Ring Modulo a Zero Dimensional Polynomial Ideal* (in German). PhD thesis, Universität Innsbruck, Austria, 1965.
71. K. Bühler. Fast and reliable plotting of implicit curves. In J. Winkler and M. Niranjan, editors, *Uncertainty in Geometric Computations*, pages 15–28. Kluwer Academic Publishers, 2002.
72. K. Bühler and W. Barth. A new intersection algorithm for parametric surfaces based on linear interval estimations. In W. Krämer and J. W. von Gudenberg, editors, *Scientific Computing, Validated Numerics, Interval Methods*. Kluwer Academic Publishers, 2001.
73. M. Buhmann. *Radial Basis Functions*. Cambridge University Press, Cambridge, England, 2003.
74. M. P. Cani. An implicit formulation for precise contact modeling between flexible solids. In *SIGGRAPH '93*, pages 313–320. ACM Press, Aug 1993.
75. H. Carr, T. Möller, and J. Snoeyink. Artifacts caused by simplicial subdivision. *IEEE Transactions on Visualization and Computer Graphics*, 12(2):231–242, March/April 2006.
76. J. Carr, R. Beatson, J. Cherrie, T. Mitchell, W. Fright, B. McCallum, and T. Evans. Reconstruction and representation of 3D objects with radial basis functions. *ACM SIGGRAPH Computer Graphics*, 35(3):67–76, August 2001.
77. J. Carr, R. Beatson, B. McCallum, W. Fright, T. McLennan, and T. Mitchell. Smooth surface reconstruction from noisy range data. In M Adcock, I. Gwilt, and Y.T. Lee, editors, *Proceedings of the 1st International Conference on Computer Graphics and Interactive Techniques in Australasia and South-East Asia (GRAPHITE'03), February 11–14, Melbourne, Australia*, pages 119–126. ACM Press, 2003.
78. J. Carr, W. Fright, and R. Beatson. Surface interpolation with radial basis functions for medical imaging. *IEEE Transactions on Medical Imaging*, 16(1):96–107, February 1997.
79. M. Cermak and V. Skala. Adaptive edge spinning algorithm for polygonization of implicit surfaces. In *Proceedings of the Computer Graphics International (CGI'04)*, pages 36–43. IEEE Computer Society Press, 2004.
80. M. Cermak and V. Skala. Edge spinning algorithm for implicit surfaces. *Applied Numerical Mathematics*, 49(3–4):331–342, June 2004.
81. S. Chan and E. Purisima. A new tetrahedral tessellation scheme for isosurface generation. *Computers & Graphics*, 22(1):83–90, January 1998.

82. R. Chandler. A tracking algorithm for implicitly defined curves. *IEEE Computer Graphics and Applications*, 8(2):83–89, March 1988.
83. S. Chapra and R. Canale. *Numerical Methods for Engineers* (5th Edition). McGraw-Hill, New York, 2006.
84. E. Chernyaev. Marching cubes 33: construction of topologically correct isosurfaces. Technical Report CN/95-17, European Organization for Nuclear Research (CERN), 1995. Available at: http://wwwinfo.cern.ch/asdoc/psdir/mc.ps.gz.
85. P. Cignoni, F. Ganovelli, C. Montani, and R. Scopigno. Reconstruction of topologically correct and adaptive trilinear surfaces. *Computers and Graphics*, 24(3):399–418, June 2000.
86. H. Cline, W. Lorensen, and S. Ludke. Two algorithms for the three-dimensional reconstruction of tomograms. *Medical Physics*, 15(3):320–327, May 1988.
87. E. Cohen, R. Riesenfield, and G. Elber. *Geometric Modeling with Splines: An Introduction*. A K Peters, Ltd., 2001.
88. D. Cohen-Steiner and F. Da. A greedy Delaunay-based surface reconstruction algorithm. *The Visual Computer*, 20(1):4–16, April 2004.
89. G. Collins and A.Akritas. Polynomial real root isolation using Descartes' rule of signs. In *Proceedings of the 3rd ACM Symposium on Symbolic and Algebraic Computations, Yorktown Heights, New York*, pages 272–275, 1976.
90. J. Comba and J. Stolfi. Affine arithmetic and its applications to computer graphics. In *Actas do VI Simpósio Brasileiro de Computação Gráfica e Processamento de Imagens (SIBGRAPI'93)*, pages 9–18, 1993.
91. D. Cox, J. Little, and D. O'Shea. *Ideals, Varieties, and Algorithms: an Introduction to Computational Algebraic Geometry and Commutative Algebra*. Undergraduate Texts in Mathematics. Springer-Verlag, New York, 1992.
92. H. Coxeter. Discrete groups generated by reflections. *The Annals of Mathematics*, 35(3):588–621, July 1934.
93. H. Coxeter. *Introduction to Geometry*. John Wiley & Sons, Inc., New York, USA, 1961.
94. B. Curless and M. Levoy. A volumetric method for building complex models from range images. *ACM SIGGRAPH Computer Graphics*, 30(3):303–312, August 1996.
95. J. Davenport, Y. Siret, and E. Tournier. *Computer Algebra — Systems and Algorithms for Algebraic Computation*. Academic Press, 1988.
96. M. Dayhoff. A contour-map program for x-ray crystallography. *Communications of the ACM*, 6(10):620–622, 1963.
97. M. de Berg, M. van Kreveld, M. Overmars, and O. Schwarzkopf. *Computational Geometry: Algorithms and Applications*. Springer-Verlag, New York, 2000.
98. C. de Boor. *A Practical Guide to Splines*. Number 27 in Applied Mathematical Sciences. Springer-Verlag, New York (revised edition), 2001.
99. A. de Cusatis Jr., L. de Figueiredo, and M. Gattass. Affine arithmetic and its applications to computer graphics. In *Actas do XII Simpósio Brasileiro de Computação Gráfica e Processamento de Imagens (SIBGRAPI'99)*, pages 65–71, 1999.
100. L. de Figueiredo. Surface intersection using affine arithmetic. In R. Bartels and W. Davis, editors, *Proceedings of the Conference on Graphics Interface'96*, pages 168–175. Canadian Information Processing Society, 1996.

101. L. de Figueiredo, J. Gomes, D. Terzopoulos, and L. Velho. Physically based methods for polygonization of implicit surfaces. In R. Booth and A. Fournier, editors, *Proceedings of the Conference on Graphics Interface (GI'92)*, pages 250–257. Morgan Kaufmann Publishers, Inc., 1992.
102. L. de Figueiredo and J. Stolfi. Adaptive enumeration of implicit surfaces with affine arithmetic. *Computer Graphics Forum*, 15(5):287–296, 1996.
103. L. de Figueiredo, J. Stolfi, and L. Velho. Approximating parametric curves with strip trees using affine arithmetic. *Computer Graphics Forum*, 22(2):171–179, 2003.
104. L. de Floriani and A. Hui. Data structures for simplicial complexes: an analysis and a comparison. In M. Desbrun and H. Pottmann, editors, *Proceedings of the 3rd Eurographics Symposium on Geometry Processing (SGP'05)*, pages 119–128. Eurographics Association, 2005.
105. E. de Forrest. On some methods of interpolation applicable to the graduation of irregular series. Annual Report of the Board of Regents of the Smithsonian Institute for 1871, Smithsonian Institute, 1873.
106. E. de Groot and B. Wyvill. Rayskip: Faster ray tracing of implicit surface animations. In *Proceedings of the 3rd International Conference on Computer Graphics and Interactive Techniques in Australasia and Southeast Asia (GRAPHITE'05)*, pages 31–36. ACM Press, 2005.
107. J. de Loera. *Triangulations of Polytopes and Computational Algebra*. PhD thesis, Cornell University, Ithaca, USA, 1995.
108. M. Desbrun and M. P. Cani-Gascuel. Active implicit surface for animation. *Graphics Interface '98*, pages 143–150, Jun 1998.
109. O. Deussen and B. Lintermann. *Digital Design of Nature: Computer-Generated Plants and Organics*. Springer-Verlag, 2005.
110. T. Dey, J. Giesen, N. Leekha, and R. Wenger. Detecting boundaries for surface reconstruction using co-cones. *International Journal of Computer Graphics and CAD/CAM*, 16:141–159, 2001.
111. T. Dey and S. Goswami. Tight cocone: a water-tight surface reconstruction. In *Proceedings of the 8th ACM Symposium on Solid Modeling and Applications*, pages 127–134, Seattle, Washington, USA, 2003. ACM Press.
112. S. Dineen. *Multivariate Calculus and Geometry*. Undergraduate Mathematics Series. Springer-Verlag, London, 1998.
113. D. Dobkin and M. Laszlo. Primitives for the manipulation of three-dimensional subdivisions. *Algorithmica*, 4(1):3–32, 1989.
114. D. Dobkin, S. Levy, W. Thurston, and A. Wilks. Contour tracing by piecewise linear approximations. *ACM Transactions on Graphics*, 9(4):389–423, 1990.
115. A. Doi and A. Koide. An efficient method of triangulating equi-valued surface by using tetrahedral cells. *IEICE Transactions*, E74(1):214–224, January 1991.
116. J. Dongarra, J. Croz, S. Hammarling, and I. Duff. A set of level 3 basic linear algebra subprograms. *ACM Transactions on Mathematical Software*, 16(1):1–17, March 1990.
117. Z. Du, V. Sharma, and C. Yap. Amortized bound for root isolation via sturm sequences. In D. Wang and L. Zhi, editors, *Proceedings of the International Workshop on Symbolic-Numeric Computing, Beijing, China*, 2005.
118. C. Duarte and J. Oden. H-p clouds: an h-p meshless method. *Numerical Methods for Partial Differential Equations*, 12(6):673–705, November 1996.

119. J. Duchon. Interpolation des fonctions de deux variables suivante le principe de la flexion des plaques minces. *RAIRO Modélisation Mathématique et Analyse Numérique*, 10(12):5–12, 1976.
120. J. Duchon. Splines minimizing rotation-invariant semi-norms in Sobolev spaces. In *Constructive Theory of Functions of Several Variables*, volume 571 of *Lecture Notes in Mathematics*, pages 85–100. Springer-Verlag, Berlin, 1977.
121. J. Duchon. Sur l'erreur d'interpolation des fonctions de plusieurs variables par les d^m splines. *RAIRO Modélisation Mathématique et Analyse Numérique*, 12(4):325–334, 1978.
122. M. Dürst. Letters: additional reference to marching cubes. *ACM SIGGRAPH Computer Graphics*, 22(4):72–73, 1988.
123. N. Dyn and S. Rippa. Data-dependent triangulations for scattered data interpolation and finite element approximation. *Applied Numerical Mathematics*, 12(1–3):89–105, May 1993.
124. B. Eaves. Properly labeled simplexes. In G. Dantzig and B. Eaves, editors, *Studies in Optimization*, volume 10 of *Studies in Mathematics*, pages 71–93. Mathematical Association of America, 1974.
125. M. Eck, T. DeRose, T. Duchamp, H. Hoppe, M. Lounsbery, and W. Stuetzle. Multiresolution analysis of arbitrary meshes. *ACM SIGGRAPH Computer Graphics*, 29:173–182, August 1995.
126. H. Edelsbrunner. *Algorithms in Combinatorial Geometry*, volume 10 of *EATCS Monographs on Theoretical Computer Science*. Springer-Verlag, Berlin, 1987.
127. H. Edelsbrunner and E. Mücke. Three-dimensional alpha shapes. *ACM Transactions on Graphics*, 13(1):43–72, 1994.
128. C. Edwards. *Advanced Calculus of Several Variables*. Academic Press, New York, 1973.
129. A. Eigenwillig, V. Sharma, and C. Yap. Almost tight recursion tree bounds for the Descartes method. In B. Trager, editor, *Proceedings of the 2006 ACM Symposium on Symbolic and Algebraic Computation*, pages 71–78. ACM Press, 2006.
130. B. Falcidieno and O. Ratto. Two-manifold cell decomposition of r-sets. *Computer Graphics Forum*, 11(3):391–404, 1992.
131. C. Fang, T. Chen, and R. Rutenbar. Floating-point error analysis based on affine arithmetic. In *Proceedings of the IEEE International Conference on Acoustics, Speech, and Signal Processing (ICASSP'03)*, volume 2, pages 561–564. IEEE Press, April 2003.
132. G. Farin. *Curves and Surfaces for CAGD: a Pratical Guide*. Academic Press, (4th edition), San Diego, 1997.
133. R. Farouki and V. Rajan. On the numerical condition of polynomials in Bernstein form. *Computer-Aided Geometric Design*, 4:191–216, 1987.
134. R. Farouki and V. Rajan. Algorithms for polynomials in Bernstein form. *Computer-Aided Geometric Design*, 5:1–26, 1988.
135. G. Fasshauer. Approximate moving least-squares approximation with compactly supported weights. In M. Griebel and M. Schweitzer, editors, *Meshfree Methods for Partial Differential Equations*, volume 26 of *Lecture Notes in Computer Science and Engineering*, pages 105–116. Springer-Verlag, Berlin, 2003.
136. G. Fasshauer. *Mesh-free Approximation Methods with MATLAB*, volume 6 of *Interdisciplinary Mathematical Sciences*. World Scientific Publishers, Singapore, 2007.

137. R. Finkel and J. Bentley. Quad trees: A data structure for retrieval on composite keys. *Acta Informatica*, 4(1):1–9, 1974.
138. S. Fleishman, M. Alexa, D. Cohen-Or, and C. Silva. Progressive point set surfaces. *ACM Transactions on Graphics*, 22(4):997–1011, October 2003.
139. S. Fleishman, D. Cohen-Or, and C. Silva. Robust moving least-squares fitting with sharp features. *ACM Transactions on Graphics*, 24(3):544–552, July 2005.
140. M. Floater. Mean value coordinates. *Computer-Aided Geometric Design*, 20(1):19–27, 2003.
141. M. Floater and K. Hormann. Surface parameterization: a tutorial and survey. In N. Dodgson, M. Floater, and M. Sabin, editors, *Advances in Multiresolution for Geometric Modelling*, Mathematics and Visualization, pages 157–186. Springer-Verlag, Heidelberg, 2005.
142. M. Floater and A. Iske. Multistep scattered data interpolation using compactly supported radial basis functions. *Journal of Computational and Applied Mathematics*, 73(1–2):65–78, October 1996.
143. S. Fortune. Voronoi diagrams and Delaunay triangulations. In D.-Z. Du and F. Hwang, editors, *Computing in Euclidean Geometry*, volume 1 of *Lecture Notes Series on Computing*, pages 193–230. World Scientific, 1992.
144. D. R. Fowler, H. Meinhardt, and P. Prusinkiewicz. Modeling seashells. *Computer Graphics*, 26:379–387, 1992.
145. R. Franke. Scattered data interpolation: tests of some methods. *Mathematics of Computation*, 38(157):181–200, 1982.
146. R. Franke and G. Nielson. Smooth interpolation of large sets of scattered data. *International Journal of Numerical Methods in Engineering*, 15(11):1681–1704, 1980.
147. M. Frontini and E. Sormani. Some variants of Newton's method with third-order convergence. *Applied Mathematics and Computation*, 140(2-3):419–426, August 2003.
148. H. Fuchs, Z. Kedem, and B. Naylor. On visible surface generation by a priori tree structures. In *Proceedings of the 7th Annual Conference on Computer Graphics and Interactive Techniques (SIGGRAPH'80)*, pages 124–133. ACM Press, 1980.
149. C. Galbraith, P. MacMurchy, and B. Wyvill. Blobtree trees. In *Computer Graphics International*, pages 78–85, 2004.
150. C. Galbraith, L. Mündermann, and B. Wyvill. Implicit visualization and inverse modeling of growing trees. *Computer Graphics Forum*, 23(3):351–360, 2004.
151. C. Galbraith, P. Prusinkiewicz, and C. Davidson. Goal oriented animation of plant development. In *10th Western Computer Graphics Symposium*, pages 19–32, University of Calgary, 1999. Department of Computer Science.
152. C. Galbraith, P. Prusinkiewicz, and B. Wyvill. Modeling murex cabritii seashell with a structured implicit surface modeler. *Visual Computer*, 18(2):70–80, 2002.
153. Callum Galbraith. *Modeling Natural Phenomena with Implicit Surfaces*. PhD thesis, Department of Computer Science, University of Calgary, 2005.
154. I. Gargantini and H. Atkinson. Ray tracing an octree: numerical evaluation of the first intersection. *Computer Graphics Forum*, 12(4):199–210, 1993.

155. J. Garloff. Convergent bounds for the range of multivariate polynomials. In *Interval Mathematics*, volume 212 of *Lecture Notes in Computer Science*, pages 37–56. Springer-Verlag, New York, 1985.
156. M. Garrity. Ray tracing irregular volume data. *ACM SIGGRAPH Computer Graphics*, 24(5):35–40, November 1990.
157. C. Gauss. *Theoria Motus Corporum Coelestium in Sectionibus Conicis Solem Ambientium*. Perthes & Besser, Hamburg, Germany, 1809. English translation by C. Harris, reprinted 1963, Dover, New York.
158. A. Geisow. *Surface Interrogations*. PhD thesis, University of East Anglia, England, 1983.
159. C. Gibson. *Singular points of smooth mappings*. Number 25 in Research Notes in Mathematics. Pitman Publishing Limited, London, 1979.
160. G. Golub and C. van Loan. *Matrix Computations*. Johns Hopkins Studies in Mathematical Sciences. The John Hopkins University Press, Baltimore, 1996.
161. A. Gomes. A concise b-rep data structure for stratified subanalytic objects. In L. Kobbelt, P. Schroder, and H. Hoppe, editors, *Proceedings of the Eurographics/ACM SIGGRAPH Symposium on Geometry Processing*, pages 83–93. Eurographics Association, 2003.
162. A. Gomes, A. Middleditch, and C. Reade. A mathematical model for boundary representations of n-dimensional geometric objects. In W. Bronsvoort and D. Anderson, editors, *Fifth Symposium on Solid Modeling and Applications*, pages 270–277. ACM Press, 1999.
163. L. Gonzalez-Vega and I. Necula. Efficient topology determination of implicitly defined algebraic plane curves. *Computer-Aided Geometric Design*, 19(9):719–743, 2002.
164. L. Gonzalez-Vega and G. Trujillo. Multivariate Sturm-Habicht sequences: real root counting on n-rectangles and triangles. *Revista Matemática de la Universitat Complutense de Madrid*, 10:119–130, 1997.
165. M. Gopi and S. Krishnan. A fast and efficient projection-based approach for surface reconstruction. In L. Gonçalves and S. Musse, editors, *Proceedings of the 15th Brazilian Symposium on Computer Graphics and Image Processing (SIBGRAPI'02)*, pages 179–186. IEEE Computer Society Press, 2002.
166. D. Grabiner. Descartes' rule of signs: another construction. *American Mathematical Montlhy*, 1067(9):854–855, 1999.
167. J. Gram. Über entwicklung reeler functionen in reihen mitelst der methode der kleinsten quadrate. *Journal für die reine und angewandte Mathematik*, 94:41–73, 1883.
168. L. Greengard and V. Rokhlin. A fast algorithm for particle simulations. *Journal of Computational Physics*, 73(2):325–348, December 1987.
169. A. Guéziec and R. Hummel. Exploiting triangulated surface extraction using tetrahedral decomposition. *IEEE Transactions on Visualization and Computer Graphics*, 1(4):328–342, December 1995.
170. L. Guibas and J. Stolfi. Primitives for the manipulation of general subdivisions and the computation of voronoi diagrams. *ACM Transactions on Graphics*, 4(2):74–123, 1985.
171. A. Guy and B. Wyvill. Controlled blending for implicit surfaces. In *Implicit Surfaces '95*, pages 107–112, 1995.
172. M. Hall and J. Warren. Adaptive polygonization of implicitly defined surfaces. *IEEE Computer Graphics & Applications*, 10(6):33–42, November 1990.

173. E. Halley. A new, exact and easy method of finding the roots of equations generally, and that without any previous reduction. *Philosophical Transactions of the Royal Society of London*, 18:136–145, 1694.
174. E. Hansen. A multidimensional interval Newton method. *Reliable Computing*, 12(4):253–272, 2006.
175. E. Hansen and R. Greenberg. An interval Newton method. *Applied Mathematics and Computation*, 12(2-3):89–98, 1983.
176. R. Hardy. Multiquadric equations of topography and other irregular surfaces. *Journal of Geophysical Research*, 76(8):1905–1915, 1971.
177. J. Hart. Sphere-tracing: a geometric method for the antialiased ray tracing of implicit surfaces. *The Visual Computer*, 12(10):527–545, 1997.
178. J. C. Hart and B. Baker. Implicit modeling of tree surfaces. In *Proceedings of Implicit Surfaces '96*, pages 143–152, Oct 1996.
179. E. Hartmann. A marching method for the triangulation of surfaces. *The Visual Computer*, 14(3):95–108, 1998.
180. B. Heap. Algorithms for production of contour maps over an irregular triangular mesh. Research Report NAC 10, National Physical Laboratory, February 1972.
181. W. Heidrich, P. Slusallek, and H.-P. Seidel. Sampling procedural shaders using affine arithmetic. *ACM Transactions on Graphics*, 17(3):158–176, 1998.
182. S. Helgason. *Differential Geometry, Lie groups, and Symmetric Spaces*. Pure and Appllied Mathematics Series. Academic Press, Inc., New York, 1978.
183. M. Henderson. Computing implicitly defined surfaces: two parameter continuation. Research Report RC 18777, IBM Research Division, T.J. Watson Research Center, New York, March 1993.
184. C. Hermite. *L'Extension du Théorème de M. Sturm à un Système d'Équations Simultanées*, volume III of *Oeuvres de Charles Hermite*. Gauthier-Villars, Paris, 1912.
185. B. Von Herzen and A. Barr. Accurate triangulations of deformed, intersecting surfaces. *Computer Graphics*, 21(4), 1987. (SIGGRAPH'87).
186. N. Higham. *Accuracy and Stability of Numerical Algorithms*. Society for Industrial and Applied Mathematics (SIAM), 1996.
187. A. Hilton, A.Stoddart, J. Illingworth, and T. Windeatt. Marching triangles: range image fusion for complex object modelling. In *Proceedings of the International Conference on Image Processing*, pages 381–384. IEEE Press, 1996.
188. H. Hironaka. Subanalytic sets. In Y. Kusunoki, S. Mizohata, M. Nagata, H. Toda, M. Yamaguti, and H. Yoshizawa, editors, *Number Theory, Algebraic Geometry and Commutative Algebra*, pages 453–493. Kinokuniya Book-Store Co., Ltd., Tokyo, Japan, 1973.
189. H. Hironaka. Triangulations of algebraic sets. In R. Hartshorne, editor, *Proceedings of Symposia in Pure Mathematics: Algebraic Geometry, Vol. 29*, pages 165–185. American Mathematical Society (AMS), 1974.
190. M. Hirsch. *Differential Topology*. Number 33 in Graduate Texts in Mathematics. Springer-Verlag, New York, 1976.
191. Ø. Hjelle and M. Dæhlen. *Triangulations and Applications*. Mathematics and Visualization. Springer-Verlag, New York, 2006.
192. H. Hoppe, T. DeRose, T. Duchamp, J. McDonald, and W. Stuetzle. Surface reconstruction from unorganized points. *ACM SIGGRAPH Computer Graphics*, 26(2):71–78, July 1992.

193. K. Hormann and M. Floater. Mean value coordinates for arbitrary planar polygons. *ACM Transactions on Graphics*, 25(4):1424–1441, 2006.
194. W. Horner. A new method of solving numerical equations of all orders, by continuous approximation. *Philosophical Transactions of the Royal Society of London*, pages 308–335, July 1819.
195. R. Hosie. *Native Trees of Canada*. Fitzhenry and Whiteside Ltd., Markham, Ontario, Canada, 1990.
196. J. Huang and C. Menq. Combinatorial manifold mesh reconstruction and optimization from unorganized points with arbitrary topology. *Computer-Aided Design*, 34(2):149–165, February 2002.
197. J. Hubbard and B. Hubbard. *Vector Calculus, Linear Algebra, and Differential Forms: A Unified Approach*. Prentice-Hall, Englewood Cliffs, 1999.
198. R. Hughes. Minimum-cardinality triangulations of the d-cube for $d = 5$ and $d = 6$. *Discrete Mathematics*, 118(1–3):75–118, August 1993.
199. K. Hui and Z. Jiang. Tetrahedra based adaptive polygonization of implicit surface patches. *Computer Graphics Forum*, 18(1):57–68, March 1999.
200. I. Itenberg and M.-F. Roy. Multivariate Descartes' rule. *Contributions to Algebra and Geometry*, 36(2):337–346, 1996.
201. X. Jin, C. Tai, J. Feng, and Q. Peng. Convolution surfaces for line skeletons with polynomial weight distributions. *Journal of Graphics Tools*, 6(3):17–28, 2001.
202. J. Johnson and W. Krandick. Polynomial real root isolation using approximate arithmetic. In *Proceedings of the 1997 ACM Symposium on Symbolic and Algebraic Computations, Maui, Hawaii, USA*, pages 225–232. ACM Press, 1997.
203. K. Joy, J. Legakis, and R. MacCracken. Data structures for multiresolution representation of unstructured meshes. In G. Farin, H. Hagen, and B. Hamann, editors, *Hierarchical Approximation and Geometric Methods for Scientific Visualization*, pages 143–170. Springer-Verlag, Heidelberg, 2002.
204. Z. Kacic-Alesic and B. Wyvill. Controlled blending of procedural implicit surfaces. In *Graphics Interface 91*, pages 236–245, 1991.
205. M. Kallmann and D. Thalmann. Star-vertices: A compact representation for planar meshes with adjacency information. *Journal of Graphics Tools*, 6(1):7–18, 2001.
206. D. Kalra and A. Barr. Guaranteed ray intersections with implicit functions. *Computer Graphics (Proc. of SIGGRAPH '89)*, 23(3):297–306, 1989.
207. D. Kalra and A. Barr. Guaranteed ray intersections with implicit surfaces. *Computer Graphics*, 23(4):297–306, 1989. (SIGGRAPH'89).
208. T. Karkanis and A. Stewart. Curvature-dependent triangulation of implicit surfaces. *IEEE Computer Graphics and Applications*, 21(2):60–69, March 2001.
209. M. Kass, A. Witkin, and D. Terzopoulos. Snakes: active contour models. *International Journal of Computer Vision*, 23(4):321–332, January 1988.
210. R. Kearfott. Empirical evaluation of innovations in interval branch and bound algorithms for nonlinear systems. *SIAM Journal on Scientific Computing*, 18(2):574–594, 1997.
211. S.-J. Kim and C.-G. Song. Rendering of unorganized points with octagonal splats. In V. Alexandrov, G. Albada, P. Sloot, and J. Dongarra, editors, *Proceedings of the 6th International Conference on Computational Science*

(ICCS'06), volume 3992, Part II of *Lecture Notes in Computer Science*, pages 326–333. Springer-Verlag, New York, 2006.
212. A. Knoll, I. Wald, S. Parker, and C. Hansen. Interactive isosurface ray tracing of large octree volumes. In *Proceedings of the IEEE Symposium on Interactive Ray Tracing (IRT'06)*, pages 115–124. IEEE Press, 2006.
213. L. Kobbelt, M. Botsch, U. Schwanecke, and H.-P. Seidel. Feature sensitive surface extraction from volume data. *ACM SIGGRAPH Computer Graphics*, 35(3):57–66, August 2001.
214. N. Kojekine, I. Hagiwara, and V. Savchenko. Software tools using CSRBFs for processing scattered data. *Computers and Graphics*, 27(2):311–319, April 2003.
215. J. Kou, Y. Li, and X. Wang. Efficient continuation Newton-like method for solving systems of nonlinear equations. *Applied Mathematics and Computation*, 174(2):846–853, March 2006.
216. J. Kou, Y. Li, and X. Wang. On modified Newton methods with cubic convergence. *Applied Mathematics and Computation*, 176(1):123–127, 2006.
217. W. Krandick and K. Mehlhorn. New bounds for the Descartes' method. *Journal of Symbolic Computation*, 41(1):49–66, 2006.
218. C. Kuo and H. Yau. Reconstruction of virtual parts from unorganized scanned data for automated dimensional inspection. *Journal of Computing and Information Science in Engineering*, 3(1):76–86, March 2003.
219. C. Kuo and H. Yau. A Delaunay-based region-growing approach to surface reconstruction from unorganized points. *Computer-Aided Design*, 37(8):825–835, July 2005.
220. C. Kuo and H. Yau. A new combinatorial approach to surface reconstruction with sharp features. *IEEE Transactions on Visualization and Computer Graphics*, 12(1):825–835, January-February 2006.
221. P. Lancaster and K. Salkauskas. Surfaces generated by moving least squares methods. *Mathematics of Computation*, 37(155):141–158, July 1981.
222. J. Lane and R. Riesenfeld. Bounds on a polynomial. *BIT Numerical Mathematics*, 21(1):112–117, 1981.
223. S. Lang. *Undergraduate analysis*. Undergraduate Texts in Mathematics. Springer-Verlag, New York, 1997.
224. S. Lay. *Convex Sets and Their Applications*. Dover Publications, Mineola, 1982.
225. C. Lee. Subdivisions and triangulations of polytopes. In J. Goodman and J. O'Rourke, editors, *Handbook of Discrete and Computational Geometry*, The CRC Press Series on Discrete Mathematics and Its Applications, pages 271–290. CRC Press, 1997.
226. A. Legendre. *Nouvelles Méthodes pour la Détermination des Orbits des Comètes*. Courcier Imprimeur, Paris, France, 1805. Reissued with a supplement, 1806. Second supplement published in 1820. A portion of the appendix was translated in 1929, pages 576–579 in *A Source Book in Mathematics*, S. Smith (ed.), translated by H. Ruger and H. Walker, McGraw-Hill, New York. Reprinted in two volumes, Dover, New York, 1959.
227. C. Lemke. Bimatrix equilibrium points and mathematical programming. *Management Science*, 11(7, Series A, Sciences):681–689, May 1965.
228. C. Lemke and J. Howson. Equilibrium points of bimatrix games. *SIAM Journal on Applied Mathematics*, 12(2):413–423, June 1964.

229. D. Levin. The approximation power of moving least squares methods. *Mathematics of Computation*, 67(224):1517–1531, October 1998.
230. D. Levin. Mesh-independent surface interpolation. In G. Brunnett, B. Hamann, H. Müller, and L. Linsen, editors, *Geometric Modeling for Scientific Visualization*, Mathematics and Visualization, pages 37–50. Springer-Verlag, Berlin, 2003.
231. S. Levin. Descartes' rule of signs: how hard can it be? *American Mathematical Monthly* (submitted for publication), 2002.
232. T. Lewiner, H. Lopes, A. Vieira, and G.Tavares. Efficient implementation of marching cubes' cases with topological guarantees. *Journal of Graphics Tools*, 8(2):1–15, 2003.
233. S. Li and W. Liu. Meshfree and particle methods and their applications. *Applied Mechanics Reviews*, 55(1):1–34, 2002.
234. S. Li and W. Liu. *Mesh-free Particle Methods*. Springer-Verlag, Berlin, 2004.
235. T. Li and X. Wang. On multivariate Descartes' rule: a counterexample. *Contributions to Algebra and Geometry*, 39(1):1–5, 1998.
236. C. Liang, B. Mourrain, and J.-P. Pavone. Subdivision methods for the topology of 2D and 3D implicit curves. In *Proceedings of the Workshop on Computational Methods for Algebraic Spline Surfaces*, Oslo, Norway, 2005.
237. D. Libes. Modeling dynamic surfaces with octrees. *Computer & Graphics*, 15(3), 1991.
238. P. Lienhardt. Topological models for boundary representation: a comparison with n-dimensional generalized maps. *Computer-Aided Design*, 23(1):59–82, January 1991.
239. C. Lim, G. Turkiyyah, M. Ganter, and D. Storti. Implicit reconstruction of solids from cloud point sets. In *Proceedings of the 3rd ACM Symposium on Solid Modeling and Applications*, pages 393–402. ACM Press, 1995.
240. C. Loader. *Local Regression and Likehood*. Statistics and Computing. Springer-Verlag, New York, 1999.
241. S. Lojasiewicz. Triangulation of semianalytic sets. *Annali Della Scuola Normale de Pisa*, 3rd Series, 18:449–474, 1964.
242. S. Lojasiewicz. Ensembles semi-analytiques. Technical Report Cours Faculté des Sciences d'Orsay, Bures-sur-Yvette, Inst. Hautes Études Sci., 1965.
243. A. Lopes. *Accuracy in Scientific Visualization*. PhD thesis, The University of Leeds, School of Computer Studies, Leeds, England, March 1999.
244. A. Lopes and K. Brodlie. Improving the robustness and accuracy of the marching cubes algorithm for isosurfacing. *IEEE Transactions on Visualization and Computer Graphics*, 9(1):16–29, 2003.
245. H. Lopes, J. Oliveira, and L. Figueiredo. Robust polygonal adaptive approximation of implicit curves. *Computers & Graphics*, 26(6):841–852, 2002.
246. H. Lopes and G. Tavares. Structural operators for modeling 3-manifolds. In *Proceedings of the 4th ACM Symposium on Solid Modeling and Applications*, pages 10–18. ACM Press, May 1997.
247. W. Lorensen and H. Cline. Marching cubes: a high resolution 3D surface construction algorithm. *Computer Graphics*, 21(4):163–169, July 1987.
248. G. Lorentz. *Bernstein Polynomials*. Chelsea Publishing Company, New York, 1986.
249. Y. Lu. *Singularity Theory and An Introduction to Catastrophe Theory*. Universitext. Springer-Verlag, New York, 1976.

250. A. Lundell and S. Weingram. *The Topology od CW Complexes*. The University Series in Higher Mathematics. Van Nostrand Reinhold Company, 1969.
251. P. MacMurchy. Subdivision surfaces for plant modeling. Master's thesis, University of Calgary, Canada, 2004.
252. V. Madan. Interval Newton method: Hansen-Greenberg approach—some procedural improvements. *Applied Mathematics and Computation*, 35(3):263–276, 1990.
253. W. Madych. Miscellaneous error bounds for multiquadric and related interpolators. *Computers & Mathematics with Applications*, 24(12):121–138, 1992.
254. W. Madych and S. Nelson. Multivariate interpolation and conditionally positive definite functions. *Approximation Theory and Its Applications*, 4(4):77–89, 1988.
255. W. Madych and S. Nelson. Multivariate interpolation and conditionally positive definite functions. ii. *Mathematics of Computation*, 54(189):211–230, 1990.
256. M. Mäntylä. *An Introduction to Solid Modeling*. Computer Science Press, 1987.
257. K. Maritaud. *Rendu réaliste d'arbres vus de près en images de synthèse*. PhD thesis, University de Limoges, France, December 2003.
258. R. Martin, H. Shou, I. Voiculescu, A. Boywer, and G. Wang. Comparison of interval methods for plotting algebraic curves. *Computer Aided Geometric Design*, 19(7):553–587, July 2002.
259. R. Martin, H. Shou, I. Voiculescu, and G. Wang. A comparison of Bernstein hull and affine arithmetic methods for algebraic curve drawing. In J. Winkler and M. Niranjan, editors, *Uncertainty in Geometric Computations*, pages 143–154. Kluwer Academic Publishers, 2001.
260. T. Marzais, Y. Grard, and R. Malgouyres. LP fitting approach for reconstructing parametric surfaces from points clouds. In J. Braz, J. Jorge, M. Dias, and A. Marcos, editors, *Proceedings of the 1st International Conference on Computer Graphics Theory and Applications (GRAPP'06)*, pages 325–330, Setúbal, Portugal, February 25–26 2006. INSTICC (Institute for Systems and Technologies of Information, Control and Communication).
261. S. Matveyev. Approximation of isosurface in the marching cube: ambiguity problem. In R. Bergeron and A. Kaufman, editors, *Proceedings of the IEEE Conference on Visualization'94*, pages 288–292. IEEE Computer Society Press, 1994.
262. N. Max, P. Hanrahan, and R. Crawfis. Area and volume coherence for efficient visualization of 3D scalar functions. *ACM SIGGRAPH Computer Graphics*, 24(5):27–33, November 1990.
263. D. McLain. Drawing contours from arbitrary data points. *Computer Journal*, 17(4):318–324, 1974.
264. D. Meagher. Geometric modeling using octree encoding. *Computer Graphics and Image Processing*, 19(2):129–147, June 1982.
265. D. Meagher. Octree generation, analysis and manipulation. Technical Report IPL-TR-027, Image Processing Laboratory, Rensselaer Polytechnic Institute, Troy, New York, April 1982.
266. H. Meinhardt. *The Algorithmic Beauty of Sea Shells*. Springer-Verlag, New York, 1995.
267. E. Mencl and H. Müller. Graph-based surface reconstruction using structures in scattered point sets. In *Proceedings of Computer Graphics International (CGI'98)*, pages 298–311. IEEE Computer Society Press, 1998.

268. M. Meyer, H. Lee, A. Barr, and M. Desbrun. Generalized barycentric coordinates on irregular polygons. *Journal of Graphical Tools*, 7(1):13–22, 2002.
269. C. Micchelli. Interpolation of scattered data: distance matrices and conditionally positive definite functions. *Constructive Approximation*, 2(1):11–12, 1986.
270. D. Michelucci. Reliable representations of strange attractors. In W. Krämer and J. W. von Gudenberg, editors, *Scientific Computing, Validated Numerics, Interval Methods*, pages 379–389. Kluwer Academic Publishers, 2001.
271. A. Middleditch, C. Reade, and A. Gomes. Set combinations of the mixed-dimension cellular objects of the Djinn API. *Computer-Aided Design*, 31(11):683–694, September 1999.
272. A. Middleditch, C. Reade, and A. Gomes. Point-sets and cell structures relevant to computer aided design. *International Journal of Shape Modeling*, 6(2):175–205, 2000.
273. P. Milne. *On the Algorithms and Implementation of a Geometric Algebra System*. PhD thesis, University of Bath, England, 1990.
274. P. Milne. The zeros of a set of multivariate polynomial equations. Technical Report 90/34, University of Bath, England, 1990.
275. S. Miyajima and M. Kashiwagi. Existence test for solution of nonlinear systems applying affine arithmetic. *Journal of Computational and Applied Mathematics*, 199(2):304–309, 2007.
276. T. Möller. A fast triangle-triangle intersection test. *ACM Journal of Graphics Tools*, 2(2):25–30, 1997.
277. T. Möller and R. Yagel. Efficient rasterization of implicit functions. Technical Report OSU-CISRC-11/95-TR50, The Ohio State University, Department of Computer and Information Science, USA, November 1995.
278. R. Moore. *Interval Analysis*. Prentice-Hall, 1966.
279. R. Moore. *Methods and Applications of Interval Analysis*. SIAM (Society for Industrial and Applied Mathematics), 1979.
280. F. Morgado and A. Gomes. A derivative-free tracking algorithm for implicit curves with singularities. In M. Bubak, D. Albada, P. Sloot, and J. Dongarra, editors, *Proceedings of the 4th International Conference on Computational Conference (ICCS'04)*, volume 3039 of *Lecture Notes in Computer Science*. Springer-Verlag, New York, 2004.
281. J. Morgado and A. Gomes. A generalized false position numerical method for finding zeros and extrema of a real function. In T. Simos and G. Maroulis, editors, *Proceedings of the International Conference in Computational Methods in Sciences and Engineering (ICCMSE'05)*, volume 4 of *Lecture Series on Computer and Computational Sciences*, pages 425–428. Brill Academic Publishers, 2004.
282. B. Morse, T. Yoo, P. Rheingans, D. Chen, and K. Subramanian. Interpolating implicit surfaces from scattered surface data using compactly supported radial basis functions. In *Proceedings of the International Conference on Shape Modeling and Applications*, pages 89–98. IEEE Computer Society Press, 2001.
283. B. Mourrain and J.-P. Pavone. Subdivision methods for solving polynomial equations. Technical Report 5658, Institut National de Recherche en Informatique et en Automatique (INRIA), August 2005.
284. B. Mourrain, F. Rouillier, and M.-F. Roy. Bernstein's basis and real root isolation. Technical Report 5149, Institut National de Recherche en Informatique et en Automatique (INRIA), March 2004.

285. B. Mourrain and J.-P. Técourt. Computing the topology of real algebraic surface. In *MEGA Electronic Proceedings*, 2005.
286. B. Mourrain, M. Vrahatis, and J. Yakoubsohn. On the complexity of isolating real roots and computing with certainty the topological degree. *Journal of Complexity*, 18(2):612–640, 2002.
287. S. Mudur and P. Koparkar. Interval methods for processing geometric objects. *IEEE Computer Graphics & Applications*, 4(2):7–17, 1984.
288. D. Muller and F. Preparata. Finding the intersection of two convex polyhedra. *Theoretical Computer Science*, 7(2):217–236, 1978.
289. H. Müller and M. Wehle. Visualization of implicit surfaces using adaptive tetrahedrizations. In H. Hagen, G. Nielson, and F. Post, editors, *Proceedings of the Conference on Scientific Visualization (Dagstuhl'97)*, pages 243–250. IEEE Computer Society, 1997.
290. L. Mündermann. *Inverse Modeling of Plants*. PhD thesis, University of Calgary, 2003.
291. J. Munkres. *Elements of Algebraic Topology*. Addison-Wesley, 1984.
292. J. Munkres. *Analysis on Manifolds*. Addison-Wesley, 1991.
293. S. Muraki. Volumetric shape description of range data using "Blobby Model". *ACM SIGGRAPH Computer Graphics*, 25(4):227–235, July 1991.
294. R. Měch and P. Prusinkiewicz. Visual models of plants interacting with their environment. In *SIGGRAPH 96 Conference Proceedings*, Annual Conference Series, pages 397–410, 1996.
295. B. Natarajan. On generating topologically consistent isosurfaces from uniform samples. *The Visual Computer*, 11(1):52–62, 1994.
296. I. Navazo. Extended octree representation of general solids with plane faces: model structure and algorithms. *Computer Graphics*, 13(1):5–16, 1989.
297. I. Navazo, D. Ayala, and P. Brunet. A geometric modeller based on the exact octtree representation of polyhedra. *Computer Graphics Forum*, 5(2):91–104, 1986.
298. B. Naylor, J. Amanatides, and W. Thibault. Merging BSP trees yields polyhedral set operations. In *Proceedings of the 17th Annual Conference on Computer Graphics and Interactive Techniques (SIGGRAPH'90)*, pages 115–124. ACM Press, 1990.
299. A. Neumaier. *Interval Methods for Systems of Equations*. Cambridge University Press, 1990.
300. G. Nielson. Scattered data modeling. *IEEE Computer Graphics & Applications*, 13(1):60–70, January/February 1993.
301. G. Nielson. On marching cubes. *IEEE Transactions on Visualization and Computer Graphics*, 9(3):283–297, July-September 2003.
302. G. Nielson and R. Franke. Computing the separating surface for segmented data. In *Proceedings of the IEEE Conference on Visualization'97*, pages 229–233. IEEE Computer Society Press, 1997.
303. G. Nielson and B. Hamann. The asymptotic decider: resolving the ambiguity in marching cubes. In *Proceedings of the IEEE Conference on Visualization'91*, pages 83–91. IEEE Computer Society Press, 1991.
304. G. Nielson and J. Sung. Interval volume tetrahedrization. In *Proceedings of the IEEE Conference on Visualization'97*, pages 221–228. IEEE Computer Society Press, 1997.

305. P. Ning and J. Bloomenthal. An evaluation of implicit surface tilers. *IEEE Computer Graphics & Applications*, 13(6):33–41, November 1993.
306. H. Nishimura, M. Hirai, and T. Kawai. Object modeling by distribution function and a method of image generation. *Transactions of IECE of Japan*, 68-D(4):227–234, April 1985.
307. T. Ochotta, C. Scheidegger, J. Schreiner, Y. Lima, R. Kirby, and C. Silva. A unified projection operator for moving least squares surfaces. Technical Report UUSCI-2007-006, Scientific Computing and Imaging Institute (SCII), University of Utah, Salt Lake City, April 2007.
308. Y. Ohtake, A. Belyaev, M. Alexa, G. Turk, and H.-P. Seidel. Multilevel partition of unity implicits. *ACM Transactions on Graphics*, 22(3):463–470, 2003.
309. Y. Ohtake, A. Belyaev, and A. Pasko. Dynamic meshes for accurate polygonization of implicit surfaces with sharp features. In A. Pasko and M. Spagnuolo, editors, *Proceedings of the International Conference on Shape Modeling and Applications*, pages 74–81. IEEE Computer Society Press, May 2001.
310. Y. Ohtake, A. Belyaev, and H.-P. Seidel. A multi-scale approach to 3D scattered data interpolation with compactly supported basis functions. In Myung-Soo Kim, editor, *Proceedings of the Shape Modeling International (SMI'03)*, pages 153–161. IEEE Computer Society Press, May 2003.
311. Y. Ohtake, A. Belyaev, and H.-P. Seidel. 3d scattered data approximation with adaptive compactly supported radial basis functions. In F. Giannini and A. Pasko, editors, *Proceedings of the 6th International Conference on Shape Modeling and Applications*, pages 31–39. IEEE Computer Society Press, June 2004.
312. Y. Ohtake, A. Belyaev, and H.-P. Seidel. Multi-scale and adaptive CS-RBFs for shape reconstruction from cloud of points. In N. Dodgson, M. Floater, and M. Sabin, editors, *Advances in Multiresolution for Geometric Modelling*, pages 143–154. Springer-Verlag, June 2005. Proceedings of the MINGLE Workshop on Multiresolution in Geometric Modelling, Cambridge, United Kingdom, September 9–11, 2003.
313. P. Olver. *Equivalence, Invariants and Symmetry*. Cambridge University Press, Cambridge, England, 1995.
314. J. Ortega and W. Rheinboldt. *Iterative Solution of Nonlinear Equations in Several Variables*. Classics in Applied Mathematics. SIAM (Society for Industrial and Applied Mathematics), Philadelphia, 2000.
315. A. Paiva, L. de Figueiredo, and J. Stolfi. Chaos and graphics: robust visualization of strange attractors using affine arithmetic. *Computers and Graphics*, 30(6):1020–1026, December 2006.
316. A. Paiva, H. Lopes, T. Lewiner, and L. Figueiredo. Robust adaptive meshes for implicit surfaces. In *Proceedings of the 19th Brazilian Symposium on Computer Graphics and Image Processing (SIBGRAPI '06)*, pages 205–212. IEEE Press, October 2006.
317. A. Pasko, V. Adzhiev, A. Sourin, and V. Savchenko. Function representation in geometric modeling: concepts, implementation and applications. *Visual Computer*, 11(8):429–446, 1995.
318. G. Pasko, A. Pasko, M. Ikeda, and T. Kunnii. Bounded blending operations. In *Proceedings of the International Conference on Shape Modeling and Applications (SMI 2002)*, pages 95–103. IEEE Computer Society, May 2002.

319. M. Pauly, R. Keiser, L. Kobbelt, and M. Gross. Shape modeling with point-sampled geometry. *ACM Transactions on Graphics*, 22(3):641–650, July 2003.
320. B. Payne and A. Toga. Surface mapping brain function on 3D models. *IEEE Computer Graphics & Applications*, 10(5):33–41, September 1990.
321. P. Pedersen. *Counting Real Zeros*. PhD thesis, New York University, Department of Computer Science, Courant Institute of Mathematical Sciences, New York, 1991.
322. P. Pedersen. Multivariate Sturm theory. In *Proceedings of the 9th International Symposium on Applied Algebra, Algebraic Algorithms and Error-Correcting Codes*, volume 539 of *Lecture Notes in Computer Science*, pages 318–332. Springer-Verlag, London, 1991.
323. S. Petitjean and E. Boyer. Regular and non-regular point sets: properties and reconstruction. *Computational Geometry: Theory and Applications*, 19(2):101–126, July 2001.
324. U. Pinkall and K. Polthier. Computing discrete minimal surfaces and their conjugates. *Experimental Mathematics*, 2(1):15–36, 1993.
325. S. Plantinga and G. Vegter. Isotopic meshing of implicit surfaces. *The Visual Computer*, 23(1):45–58, January 2007.
326. H. Pottmann and S. Leopoldseder. A concept for parametric surface fitting which avoids the parametrization problem. *Computer-Aided Geometric Design*, 20(6):343–362, September 2003.
327. V. Pratt. Direct least-squares fitting of algebraic surfaces. *ACM SIGGRAPH Computer Graphics*, 21(4):145–152, July 1987.
328. F. Preparata and M. Shamos. *Computational Geometry: An Introduction*. Texts and Monographs in Computer Science. Springer-Verlag, New York, 1985.
329. W. Press, B. Flannery, S. Teukolsky, and W. Vetterling. *Numerical Recipes in C*. Cambridge University Press, second edition, 1992.
330. P. Prusinkiewicz and A. Lindenmayer. *The Algorithmic Beauty of Plants*. Springer-Verlag, New York, 1990.
331. P. Prusinkiewicz, L. Mündermann, R. Karwowski, and B. Lane. The use of positonal information in the modeling of plants. In Eugene Fiume, editor, *SIGGRAPH 2001 Conference Proceedings*, Annual Conference Series, pages 289–300. ACM SIGGRAPH, 2001.
332. V. Rajan, S. Klinkner, and R. Farouki. Root isolation and root approximation for polynomials in Bernstein form. Technical report, IBM Research Report RC14224, IBM Research Division, T.J. Watson Research Center, Yorktown Heights, New York, November 1988.
333. S. Rana. *Topological Data Structures for Surfaces: An Introduction to Geographical Information Science*. John Wiley & Sons, Chichester, 2004.
334. A. Raposo and A. Gomes. Polygonization of multi-component non-manifold implicit surfaces through a symbolic-numerical continuation algorithm. In Y.T. Lee and S. M.Shamsuddin, editors, *Proceedings of the 4th International Conference on Computer Graphics and Interactive Techniques in Australasia and South-East Asia (GRAPHITE'06)*, pages 399–406. ACM Press, 2006.
335. H. Ratschek and J. Rokne. *Computer Methods for the Range of Functions*. Mathematics and Its Applications. Ellis Horwood, Chicester, 1984.
336. Rehder. *The Audobon Society Field Guide to North American Seashells*. Alfre A. Knopf, Inc., New York, 1981.

337. X. Renbo, L. Weijun, and W. Yuechao. A robust and topological correct marching cube algorithm without look-up table. In *Proceedings of the Fifth International Conference on Computer and Information Technology (CIT'05)*, pages 565–569, Shanghai, China, September 21-23 2005. IEEE Computer Society Press.
338. A. Requicha and R. Tilove. Mathematical foundations of constructive solid geometry: general topology of closed regular sets. Production Automation Project Tech. Memo 27a, University of Rochester, 1978.
339. N. Revol. Interval Newton iteration in multiple precision for the univariate case. Research Report 4334, Institut National de Recherche en Informatique et en Automatique (INRIA), December 2001.
340. W. Rheinboldt. On a moving frame algorithm and the triangulation of equilibirum manifolds. In T. Kuper, R. Seydel, and H. Troger, editors, *Bifurcation: Analysis, Algorithms, Applications*, volume 79 of *International Series of Numerical Mathematics*, pages 256–267. Birkhauser, Boston, 1987.
341. A. Ricci. Constructive geometry for computer graphics. *Computer Journal*, 16(2):157–160, May 1973.
342. J. Rossignac, A. Safonova, and A. Szymczak. Edgebreaker on a corner table: a simple technique for representing and compressing triangulated surfaces. In G. Farin, H. Hagen, and B. Hamann, editors, *Hierarchical Approximation and Geometric Methods for Scientific Visualization*, pages 41–50. Springer-Verlag, Heidelberg, 2002.
343. F. Rouillier and P. Zimmermann. Efficient isolation of polynomial's real roots. *Journal of Computational and Applied Mathematics*, 162(1):33–50, 2004.
344. M. Sabin. Contouring: the state of the art. In R. Earnshaw, editor, *Fundamental Algorithms for Computer Graphics*, volume 17 of *NATO ASI Series F: Computer and Systems Sciences*, pages 411–482. Springer-Verlag, Berlin, 1985.
345. R. Saleh, K. Gallivan, M. Chang, I. Hajj, D. Smart, and T. Trick. Parallel circuit simulation on supercomputers. *Proceedings of the IEEE*, 77(12):1915–1931, December 1989.
346. H. Samet. The quadtree and elated hierarchical data structures. *ACM Computing Surveys*, 16(2), 1984.
347. H. Samet. *Foundations of Multidimensional and Metric Data Structures*. The Morgan Kaufmann Series in Data Management Systems. Morgan Kaufmann Publishers, 2006.
348. R. Sarraga. Algebraic methods for intersections of quadric surfaces in GM-SOLID. *Computer Vision, Graphics, and Image Processing*, 22:222–238, 1983.
349. V. Savchenko, A. Pasko, O. Okunev, and T. Kuni. Function representation of solids reconstructed from scattered surface points and contours. *Computer Graphics Forum*, 14(4):181–188, 1995.
350. T. Scavo and J. Thoo. On the geometry of Halley's method. *American Mathematical Monthly*, 102(5):417–426, 1995.
351. S. Schaefer and J. Warren. Dual marching cubes: primal contouring of dual grids. In *Proceedings of the 12th Pacific Conference on Computer Graphics and Applications (PG'04)*, pages 70–76, Seoul, Korea, October 2004. IEEE Computer Society Press.
352. M. Schmidt. Cutting cubes: visualizing implicit surfaces by adaptive polygonization. *The Visual Computer*, 10(2):101–115, 1993.

353. R. Schmidt, C. Grimm, and B. Wyvill. Interactive decal compositing with discrete exponential maps. *ACM Transactions on Graphics*, 25(3):605–613, July 2006.
354. P. Schneider. A Bézier curve-based root-finder. In A. Glassner, editor, *Graphics Gems*, pages 408–415. Academic Press, San Diego, 1990.
355. R. Seidel and N. Wolpert. On the exact computation of the topology of real algebraic curves. In *Proceedings of the 21st ACM Annual Symposium on Computational Geometry, Pisa, Italy*, pages 107–115. ACM Press, June 2005.
356. V. Shapiro. Representations of semi-algebraic sets in finite algebras generated by space decompositions. Technical Report CPA91-1 Programmable Automation, Cornell University, Ithaca, NY, February 1991.
357. V. Shapiro. Maintenance of geometric representations through space decompositions. *International Journal of Computational Geometry & Applications*, 6(4):383–418, 1997.
358. J. Sharma. A family of Newton-like methods based on an exponential model. *International Journal of Computer Mathematics*, 84(3):297–304, 2007.
359. V. Sharma. Complexity of real root isolation using continued fractions. In *Proceedings of the 2007 ACM Symposium on Symbolic and Algebraic Computation, Waterloo, Ontario, Canada*, pages 339–346, 2007.
360. R. Sharpe. *Differential Geometry*. Graduate Texts in Mathematics. Springer-Verlag, New-York, 1997.
361. A. Sheffer, E. Praun, and K. Rose. Mesh parameterization methods and their applications. *Foundations and Trends in Computer Graphics and Vision*, 2(2):105–171, 2006.
362. C. Shen, J. O'Brien, and J. Shewchuk. Interpolating and approximating implicit surfaces from polygon soup. *ACM Transactions on Graphics*, 23(3):896–904, 2004.
363. D. Shepard. A two-dimensional interpolation function for irregularly-spaced data. In *Proceedings of the 23th ACM National Conference*, pages 517–524, New York, 1968. ACM Press.
364. E. Sherbrooke and N. Patrikalakis. Computation of the solutions of nonlinear polynomial systems. *Computer-Aided Geometric Design*, 10(5):379–405, 1993.
365. M. Shiota. *Geometry of Subanalytic and Semialgebraic Sets*. Progress in Mathematics. Birkhauser, Boston, 1997.
366. P. Shirley and A. Tuchman. A polygonal approximation to direct scalar volume rendering. *ACM SIGGRAPH Computer Graphics*, 24(5):63–70, November 1990.
367. H. Shou, R. Martin, I. Voiculescu, A. Bowyer, and G. Wang. Affine arithmetic in matrix form for polynomial evaluation and algebraic curve drawing. *Progress in Natural Science*, 12(1):77–81, 2002.
368. H. Shou, R. Martin, G. Wang, A. Bowyer, and I. Voiculescu. A recursive Taylor method for algebraic curves and surfaces. In T. Dokken and B. Juettler, editors, *Computational Methods for Algebraic Spline Surfaces*, pages 135–155. Springer Verlag, New York, 2003.
369. F. Silva and A. Gomes. AIF: a data structure for polygonal meshes. In V. Kumar, M. Gavrilova, C. Tan, and P. L'Ecuyer, editors, *Computational Science and Its Applications (ICCSA'03)*, volume 2669 of *Lecture Notes in Computer Science*, pages 986–987. Springer-Verlag, New York, 2003.

370. J. Snyder. *Generative Modelling for Computer Graphics and CAD*. Academic Press, 1992.
371. J. Snyder. Interval analysis for computer graphics. *Computer Graphics*, 26(2):121–130, July 1992. (SIGGRAPH'92).
372. M. Spencer. *Polynomial Real Root Finding in Bernstein Form*. PhD thesis, Brigham Young University, 1994.
373. G. Steele. Sun Microsystems Laboratories. Personal communication, 2002.
374. J. Stoer and R. Bulirsch. *Introduction to Numerical Analysis* (third edition), volume 12 of *Texts in Applied Mathematics*. Springer-Verlag, New York, 2002.
375. J. Sturm. Mémoire sur la résolution des équations numériques. *Bulletin des Sciences de Férussac*, 11, 1829.
376. K. Suffern. An octree algorithm for displaying implicitly defined mathematical functions. *The Australian Computer Journal*, 22(1):2–10, February 1990.
377. K. Suffern. Quadtree algorithms for contouring functions of two variables. *The Computer Journal*, 33(5):402–407, October 1990.
378. K. Suffern and R. Balsys. Rendering the intersections of implicit surfaces. *IEEE Computer Graphics & Applications*, 23(5):70–77, September/October 2003.
379. K. Suffern and E. Fackerell. Interval methods in computer graphics. *Computers & Graphics*, 15(3):331–340, 1991.
380. N. Sukumar. Construction of polygonal interpolants: a maximum entropy approach. *International Journal for Numerical Methods in Engineering*, 61(12): 2159–2181, 2004.
381. N. Sukumar and E. Malsch. Recent advances in the construction of polygonal finite element interpolants. *Archives of Computational Methods in Engineering*, 13(1):129–163, 2006.
382. N. Sukumar and A. Tabarraei. Conforming polygonal finite elements. *International Journal for Numerical Methods in Engineering*, 61(12):2045–2066, 2004.
383. G. Taubin. Estimation of planar curves, surfaces, and nonplanar space curves defined by implicit equations with applications to edge and range image segmentation. *IEEE Transactions on Pattern Analysis and Machine Intelligence*, 13(11):1115–1138, November 1991.
384. W. Thibault and B. Naylor. Set operations on polyhedra using binary space partition trees. In *Proceedings of the 14th Annual Conference on Computer Graphics and Interactive Techniques (SIGGRAPH'87)*, pages 153–162. ACM Press, 1987.
385. R. Thom. La stabilité topologique des applications polynomiales. *Enseignement Math.*, 8:24–33, 1962.
386. R. Thom. Ensembles et morphisms stratifiés. *Bulletin of the American Mathematical Society*, 75:240–284, 1969.
387. M. Tigges and B. Wyvill. Texture mapping the blobtree. In *Implicit Surfaces '98*, pages 123–130, 1998.
388. M. Tigges and B. Wyvill. Python for scene and model description for computer graphics. In *IPC 2000*, Jan 2000.
389. I. Tobor, P. Reuter, and C. Schlick. Efficient reconstruction of large scattered geometric datasets using the partition of unity and radial basis functions. *Journal of WSCG*, 12(1–3):467–474, 2004. Proceedings of the 12th International Conference in Central Europe on Computer Graphics, Visualization and

Computer Vision (WSCG'04), University of West Bohemia, Campus Bory, Plzen-Bory, Czech Republic, February 2–6, 2004.
390. M. Todd. On triangulations for computing fixed points. *Mathematical Programming*, 10(1):322–346, 1976.
391. E. Tsigaridas and I. Emiris. Univariate polynomial real root isolation: continued fractions revisited. In Y. Azar and T. Erlebach, editors, *Proceedings of the 13th European Symposium on Algorithms*, volume 4168 of *Lecture Notes in Computer Science*, pages 817–828. Springer-Verlag, New York, 2006.
392. G. Turk and J. O'Brien. Shape transformations using variational implicit surfaces. *ACM SIGGRAPH Computer Graphics*, 33(3):335–342, August 1999.
393. G. Turk and J. O'Brien. Modeling with implicit surfaces that interpolate. *ACM Transactions on Graphics*, 21(4):855–873, October 2002.
394. J. Uspensky. *Theory of Equations*. McGraw-Hill, New York, 1948.
395. A. van Gelder and J. Wilhelms. Topological considerations in isosurface generation. *ACM Transactions on Graphics*, 13(4):337–375, 1994.
396. K. van Overveld and B. Wyvill. Shrinkwrap: An efficient adaptive algorithm for triangulating an iso-surface . *The Visual Computer*, 20(6):362–369, 2004.
397. L. Velho. Adaptive polygonization of implicit objects. In *Proceedings of 1990 Australian Graphics Conference (AUSGRAPH'90)*, pages 339–343, Melbourne, Australia, September 1990.
398. L. Velho. Adaptive polygonization of implicit surfaces using simplicial decomposition and boundary constraints. In C. Vandoni and D. Duce, editors, *Proceedings of 11th Annual Conference of the European Association for Computer Graphics (EUROGRAPHICS'90)*, pages 125–136, Montreux, Switzerland, September 1990. Elsevier Sciense Publishers B.V. (North-Holland).
399. L. Velho, L. Figueiredo, and J. Gomes. A unified approach for hierarchical adaptive tesselation of surfaces. *ACM Transactions on Graphics*, 18(4):329–360, 1990.
400. L. Velho, L. Figueiredo, and J. Gomes. A methodology for piecewise linear approximation of surfaces. *Journal of the Brazilian Computer Society*, 3(3):329–360, April 1997.
401. I. Voiculescu. *Implicit Function Algebra in Set-theoretic Geometric Modelling*. PhD thesis, University of Bath, England, 2001.
402. I. Voiculescu, J. Berchtold, A. Bowyer, R. Martin, and Q. Zhang. Interval and affine arithmetic for surface location of power- and Bernstein-form polynomials. In R. Cipolla and R. Martin, editors, *The Mathematics of Surfaces IX, Proceedings of the 9th IMA Conference on the Mathematics of Surfaces*, pages 410–423. Springer-Verlag, New York, 2000.
403. E. Wachspress. *Rational Finite Element Basis*. Mathematics in Science & Engineering. Academic Press, New York, 1975.
404. D. Watson. Computing the n-dimensional Delaunay tessellation with application to Voronoi polytopes,. *The Computer Journal*, 24(2):167–172, 1981.
405. D. Watson. *Contouring: A Guide to the Analysis and Display of Spatial Data*. Computer Methods in the Geosciences. Pergamon Press, Oxford, 1992.
406. S. Weerakoom and T. Fernando. A variant of Newton's method with accelerated third-order convergence. *Applied Mathematics Letters*, 13(8):87–93, November 2000.

407. C. Weigle and D. Banks. Complex-valued contour meshing. In R. Yagel and G. Nielson, editors, *Proceedings of the IEEE Conference on Visualization'96*, pages 173–181. IEEE Computer Society Press, 1996.
408. H. Wendland. Piecewise polynomial, positive definite and compactly supported radial basis functions of minimal degree. *Advances in Computational Mathematics*, 4(1):389–396, December 1995.
409. W. Wessner. *Mesh Refinement Techniques for CAD Tools*. PhD thesis, Technische Universität Wien, Austria, November 2006.
410. J. Whitehead. Combinatorial homotopy. I. *Bulletin American Mathematical Society*, 55:213–245, 1949.
411. J. Whitehead. Combinatorial homotopy. II. *Bulletin American Mathematical Society*, 55:453–496, 1949.
412. H. Whitney. Complexes of manifolds. *Proceedings of the National Academy of Sciences (U.S.A.)*, 33:10–11, 1947.
413. H. Whitney. Elementary structure of real algebraic varieties. *Annals of Mthematics*, 66(3):545–556, November 1957.
414. B. Wilson. *The Growing Tree*. The University of Massachusetts Press, Amherst, 1984.
415. W. Woolhouse. Explanation of a new method of adjusting mortality tables, with some observations upon Mr. Makeham's modifications of Gompertz's theory. *Journal of the Institute of Actuaries*, 15:389–410, 1870.
416. X. Wu. A new continuation Newton-like method and its deformation. *Applied Mathematics and Computation*, 112(1):75–78, June 2000.
417. B. Wyvill, E. Galin, and A. Guy. Extending the CSG tree. warping, blending and boolean operations in an implicit surface modeling system. *Computer Graphics Forum*, 18(2):149–158, 1999.
418. B. Wyvill, E. Galin, and A. Guy. The blob tree, warping, blending and Boolean operations in an implicit surface modeling system. *Computer Graphics Forum*, 18(2):149–158, June 1999.
419. B. Wyvill and K. van Overveld. Polygonization of implicit surfaces with constructive solid geometry. *Journal of Shape Modelling*, 2(4):257–274, 1996.
420. G. Wyvill, 2001. Personal communication.
421. G. Wyvill, T. Kunii, and Y. Shirai. Space division for ray tracing in CSG. *IEEE Computer Graphics & Applications*, 6(4):227–234, April 1986.
422. G. Wyvill, C. McPheeters, and B. Wyvill. Data structure for soft objects. *The Visual Computer*, 2(4):227–234, August 1986.
423. J. Yang, C. Zhu, and H. Zhang. Surface reconstruction with least square reproducing kernel and partition of unity. In *Workshops Proceedings of the 16th International Conference on Artificial Reality and Telexistence (ICAT'06)*, pages 375–380, Hangzhou, China, November/December 2006. IEEE Computer Society Press.
424. H. Yau, C. Kuo, and C. Yeh. Extension of surface reconstruction algorithm to the global stitching and repairing of STL models. *Computer-Aided Design*, 35(5):477–486, 2002.
425. M. Zettler and J. Garloff. Robustness analysis of polynomials with polynomial parameter dependency using Bernstein polynomials. *IEEE Transactions on Automatic Control*, 43(3):425–431, March 1998.

426. Q. Zhang and R. Martin. Polynomial evaluation using affine arithmetic for curve drawing. In *Proceedings of the Eurographics UK Conference*, pages 49–56, 2000.
427. Y. Zhou, B. Chen, and A. Kaufmann. Multiresolution tetrahedral framework for visualizing regular volume data. In *Proceedings of the IEEE Conference on Visualization'97*, pages 135–142. IEEE Computer Society Press, 1997.
428. Y. Zhou, B. Chen, and Z. Tang. An elaborate ambiguity detection method for constructing isosurfaces within tetrahedral meshes. *Computer & Graphics*, 19(3):353–364, March 1995.
429. M. Zwicker, M. Pauly, O. Knoll, and M. Gross. Pointshop 3D: an interactive system for point-based surface editing. *ACM Transactions on Graphics*, 21(3):322–329, July 2002.

Index

L_p, fitting, 230
n-cubes, 188

affine arithmetic, 90, 105
   as number approximation, 105
   conversion to interval arithmetic, 106
   operations, 105
algorithm
   adaptive Hartmann's for surfaces, 179
   adaptive marching triangles, 182
   Balsys-Suffern, 222
   bisection, 132
   CS-RBF interpolation, 253, 254
   de Casteljau, 81
   dividing cubes, 200
   dual marching cubes, 213
   Hall-Warren, 216, 217
   Hartmann's for surfaces, 175, 177
   Henderson's for surfaces, 174, 175
   integer-labelling, 154, 155
   marching cubes, 194, 196
   marching squares, 189
   marching tetrahedra, 201, 206
   marching triangles, 180
   marching triangles for surfaces, 182
   Morgado-Gomes, 167–169, 172
   moving least squares (MLS), 244, 245
   MPU approximation, 260
   multivariate false position, 135
   multivariate Newton, 125
   multivariate secant, 128
   Raposo-Gomes, 185
   real root isolation, 80
   Rheinboldt for curves, 165
   Rheinboldt for surfaces, 173
   Sturm sequence, 77
   univariate interval Newton, 139
   univariate Newton, 122
   vector-labelling, 162
   vector-valued Newton, 124
ambiguity, 191, 198, 220
   face, 198
   interior, 198
angle, external, 177
aperture, 298
approximation
   global MLS, 243
   least squares, 234
   moving least squares (MLS), 239
   MPU, 258
   weighted least squares (WLS), 238

basis
   Bernstein, 69, 99
     bivariate, 70
     coefficients, 100
     generic interval, 71
     matrix form, 70, 71
     multivariate, 100
     univariate, 69
   Gröbner, 82
   power, 68
blending
   bounded, 273
   bulge-free, 275
   controlled, 273

346   Index

blending (cont.)
  external blend group, 277
  generalised bounded blending (GBB), 281
  graph, 275
  internal blend group, 276
  operation, 274
  super-elliptic, 273
blob, 230
  Gaussian, 230
  surface, 232
blobby, model, 232
blobtree, 270
bounded blending, 272
box classification, 99
branch bark ridge, 308
bud-scale scar, 308
bump, 295

cell, 45
  -tuple data structure, 48
  complex, 44, 46
  decomposition, 45
centre, 240
closure finiteness, 46
complex
  cell, 44, 46
  CW, 46
  finite cell, 48
  simplicial, 49
component
  irreducible, 183
  symbolic, 183
  topological, 183
composition
  functional composition using f_Z functions, 271
  of implicit surfaces, 272
  operator, 291
condition
  C, 46
  frontier, 45
  W, 46
configuration, topological, 191
conservativeness, 97, 98
constructive solid geometry (CSG), 277
continuation
  piecewise linear, 146
  predictor-corrector (PC), 164

simplicial, 146
space, 221
spatial, 207
contour diagram, 267
contour, active, 230
covering, 41
  subcovering, 42
criterion
  angle, 171
  curvature, 171
  neighbour-branch, 171
cube, dividing, 200
curve
  explicit, 4
  implicit, 5
  parametric, 3

data structure
  AIF, 49
  BSP, 53
  cell-tuple, see cell-tuple data structure, 51
  corner-table, 51
  DCEL, 51
  facet-edge, 51
  half-edge, 49, 51
  handle-face, 51
  incidence graph, 50
  k-d tree, 57
  lath, 51
  nG-map, 51
  octree, 61
  quad-edge, 51
  quadtree, 60
  star-vertices, 50
  TCD graph, 51
  Whitney, 44
  winged-edge, 51
de Casteljau, 80, 82
decider, asymptotic, 192
decomposition
  5-tetrahedral, 202, 205
  6-tetrahedral, 203
  Kuhn, 203, 205
  tetrahedral, 201
deformation, 284
diagram, Voronoi, 227
difference of implicits, 273
difference, minmax, 277

disk, Henderson, 174
distance, Taubin, 260
domain of influence, 240

edge, quadtree, 58
enumeration
　exhaustive, 51
　sequential, 51
　spatial exhaustive, 188
equation, normal, 236
equivalence, topological, 9
error
　absolute, 118
　least squares (LS), 234
　relative, 118
　round-off, 118, 119
　truncation, 118

field image, 267
field, distance, 267
finiteness, closure, 46
fit, local MLS, 243
formula
　generic iteration, 119
　multivariate bisection, 132
　multivariate false position, 134, 136
　multivariate interval Newton, 139
　multivariate Newton, 125
　secant iteration, 127
　univariate bisection, 131
　univariate false position, 133
　univariate interval Newton, 137
　univariate Newton, 121
　vector-valued Newton, 124
function, 8
　C^r, 9
　C^r diffeomorphism, 9
　C^r differentiable, 9
　C^r smooth, 9
　C^∞, 9
　f_C, 270
　f_z, 270
　n-point iteration, 120
　2-point iteration, 120
　bijection, 8
　Blinn's exponential density, 234
　Blinn's Gaussian, 233
　component, 8
　deforming function, 279

diffeomorphism, 9
differentiable, 9
explicitly defined, 23
Gaussian weight, 242
global density, 234
implicitly defined, 23
injection, 8
inverse quadratic weight, 242
inverse warping, 284
mapping, 8
McLain weight, 242
R, 271
radial basis, 249
roots, 67
smooth, 9
surjection, 8
thin-plate weight, 240
warping, 284
Wendland weight, 242
zeros, 67

golden ratio, 131

hardness factor, 274
helico-spiral, 288
homeomorphism, 9
honeycomb, 213
　tetrahedral, 214
Horner's scheme, 97

interpolation
　bilinear, 189
　CS-RBF, 252
　fast RBF, 252
　moving least squares (MLS), 239
　MPU, 261
　piecewise linear (PL), 203
　RBF, 249
　trilinear, 194
intersection of implicits, 273
intersection, min, 277
interval arithmetic, 89
　as number approximation, 91
　conversion to affine arithmetic, 106
　operations, 91, 107
　root isolation, 72
interval swell, 97
isosurface, 5

Jabobian, 9

348   Index

k-d
   tree, 55
   tree data structure, 57

labelling, 152
   integer, 152
   vector, 152, 156, 158
level set, 5
local
   finiteness, 44
   topological invariance, 44

machine, precision, 118
manifold, 43
mapping, 8
   C^1-invertible, 11
   C^r, 9
   C^r diffeomorphism, 9
   C^r-invertible, 11
   C^∞, 9
   derivative, 9
   diffeomorphism, 9
   differentiable, 9
   differential, 9
   embedding, 31
   graph, 20
   image, 13
   immersion, 30
   invertible, 11
   level set, 15
   locally C^r-invertible, 12
   parametrisation, 13
   rank, 24
   regular, 24
   smooth, 9
   submersion, 30
matrix
   Jacobian, 9
   labelling, 158
mesh
   generation, 229
   partitioning, 229
method, 117
   bracketing, 131
   angular false position, 168
   bisection, 131
   bracketed secant, 133
   disambiguation, 192
   false position, 133
   fast multipole (FMM), 252
   four triangles, 191
   generalised false position, 167
   global implicit fitting, 231
   implicit fitting, 231
   interpolation, 131
   interval Newton, 136
   local implicit fitting, 231
   modified false position, 136
   Newton-like, 126
   Newton-Raphson, 120
   regula falsi, 133
   secant, 127
   Shepard's blending, 255
model, blobby, 232
molecule, 230
murex cabritii, 288–290

natural phenomenae, 287
numerical stability, 78

octree, 60
   data structure, 61
operation
   blendiing, 274
   blending union, 272
operator
   $\cap_{min}$, 277
   $\cup_{max}$, 277
   $\setminus_{minmax}$, 277
   bounded blending, 273
   controlled blending, 273
   deformation, 284
   difference, 273
   implicit modelling, 273
   intersection, 273
   precise contact modelling, 273
   projection, 247
   R-difference, 271
   R-intersection, 271
   R-union, 271
   super-elliptic blending, 273
   twist, 273
   twist and taper, 273
   twist, taper and bend, 273
   union, 273
orientation
   geometric, 49
   topological, 49

parametrisation, 229
partition, 42
partition of unity, 255
partitioning
  affine arithmetic-driven, 109
  binary space, 52
  interval arithmetic–driven, 93
phenomenon
  cycling, 171
  drifting, 171
pivoting rule, 148
  Freudenthal, 148, 149
  Todd's J_1, 150, 151
point
  cut, 29
  evaluation, 240
  fixed, 240
  quadtree, 58
  regular, 37
  self-intersection, 26
  singular, 37, 166
  turning, 166
polynomial
  Bernstein form, 67, 97, 99
  bivariate, 67
  factored form, 67, 97
  Horner form, 97
  implicit, 5
  monic, 68
  multivariate, 67
  numerical stability, 78
  power form, 67, 68, 97, 99
  resultant, 82
  trivariate, 67
  univariate, 67
populus deltoides, 305
precise contact modelling (PCM), 279
primitive, skeletal, 268
principle, door-in-door-out, 154
problem, isocontouring, 190
property, honeycomb, 213
protein data bank, 232

quadrics, 101
quadtree, 58
  data structure, 60
  edge, 58
  point, 58
  region, 59

R
  difference, 271
  function, 271
  intersection, 271
  union, 271
reconstruction
  Delaunay-based surface, 227
  implicit surface, 230
  moving least squares (MLS), 245
  multilevel CS-RBF surface, 254
  parametric surface, 228
  RBF surface, 249
  region-growing simplicial surface, 227
  simplicial surface, 227
reduction, RBF centre, 252
region
  deformation region, 279
  interpenetration region, 279
regression, local, 239
residual, 234
riblet, 290
root, 67
  isolation, 67, 72
    Bernstein base, 78
    Descartes' rule of signs, 72, 73, 78, 79
    hull approximation, 78, 79
    interval arithmetic, 72
    multivariate, 81
    Sturm sequences, 72, 74
    variation diminishing, 78, 79
  multiple, 82
root finding method, 117

saddle
  body, 195
  face, 195
shell
  geometry, 288
  wall, 288
shoot, 305
simplex, 49, 147
  adjacent, 148
  completely labelled, 152, 158
  transverse, 154, 159
simplicial
  complex, 49
  complex data structure, 50
  decomposition, 50

350   Index

singularity, 82, 169, 220
  cusp, 169
  high-curvature point, 169
  self-intersection, 170
  topological, 26
skeleton, 45
space
  partitioning, 51
  topological, 41
stage
  filling, 180
  growing, 180
stencil, 246
stratification, 43–45
  Whitney, 43
stratum, 43
subdivision, 52
  12-tetrahedral subdivision, 206
  24-tetrahedral subdivision, 206
  barycentric, 206
  octree, 211, 221
  quadtree, 208
  spatial, 208
  tetrahedral, 213
submanifold, 30
  embedded, 31
  immersed, 31
  parametric, 31
  regular, 33, 35
surface
  blob, 232
  explicit, 4
  fitting, 229
  implicit, 5, 267
    nonpolynomial, 104
  isosurface, 5
  least squares implicit, 234
  level set, 5
  Levin's MLS, 245
  moving least squares (MLS), 240
  multilevel partition of unity (MPU), 255
  offset, 267
  parametric, 3, 4, 69
  projection MLS (PMLS), 246
  radial basis function (RBF), 249
  VMLS, 246
surface fitting, 227

test
  curvature, 218
  exclusion, 218
tetrahedron
  marching, 201
theorem
  implicit function, 16, 28
    multivariate, 28
  implicit function family, 36
  implicit mapping, 18–20
  intermediate value, 120
  inverse mapping, 12
  one-circle, 80
  rank, 24
    for implicitations, 27
    for parametrisations, 25
  Sturm, 75
  two-circle, 80
topological
  equivalence, 9
  orientation, 49
topology, 41
  weak, 46
traversal, BlobTree, 284
tree
  blob, 270
  BSP, 52
  k-d, 55
  oct-, 60
triangulation, 49
  Coxeter, 147
  Delaunay, 181, 227
  Freudenthal, 148
  Henderson, 174
  maximal, 201
  minimal, 201
  Todd's J_1, 150

union of implicits, 273
union, max, 277

value
  regular, 37
  singular, 37
variety
  parametrisation, 28
  regular, 36
varix, 290, 294
voxels, 188

warping
  Barr, 284
  spatial, 284
whorl, 289

main body, 291

zero set, 145
  approximate, 153
  piecewise linear (PL), 153

MIX
Papier aus verantwortungsvollen Quellen
Paper from responsible sources
FSC® C105338

If you have any concerns about our products,
you can contact us on
ProductSafety@springernature.com

In case Publisher is established outside the EU,
the EU authorized representative is:
**Springer Nature Customer Service Center GmbH
Europaplatz 3, 69115 Heidelberg, Germany**

Printed by Libri Plureos GmbH
in Hamburg, Germany